Metallosurfactants

Metallosurfactants

From Fundamentals to Catalytic and Biomedical Applications

Edited by
Surinder K. Mehta
Ravneet Kaur

WILEY-VCH

Editors

Prof. Surinder K. Mehta
Panjab University
Department of Chemistry
160014 Chandigarh
India

Dr. Ravneet Kaur
Ivy Tech Community College
3100 Ivy Tech Drive, Valparaiso
Indiana 46383
United States

Cover Image: © Tom Biegalski/Shutterstock

Library of Congress Card No.: applied for

British Library Cataloguing-in-Publication Data
A catalogue record for this book is available from the British Library.

Bibliographic information published by the Deutsche Nationalbibliothek
The Deutsche Nationalbibliothek lists this publication in the Deutsche Nationalbibliografie; detailed bibliographic data are available on the Internet at <http://dnb.d-nb.de>.

Print ISBN: 978-3-527-34868-8
ePDF ISBN: 978-3-527-83127-2
ePub ISBN: 978-3-527-83129-6
oBook ISBN: 978-3-527-83128-9

Typesetting Straive, Chennai, India

Printed on acid-free paper

C110395_040122

Contents

Preface

This book introduces its readers to a very special class of amphiphiles termed as "metallosurfactants" from here on, incorporating a metal in the usual surfactant system. A book in this area has been long outstanding. When we started working on the metallosurfactants a few years back, we always felt the need for a holy grail or a guiding text to delve deeper into the realm of metal surfactant complexes or metallosurfactants. Something which could explain the fundamentals of the metal surfactant complexes in detail while highlighting the current progress as well as the future prospective. Back then, we would have been thrilled to get our hands on something like this. So, a lot of the topics included in this book stem from the difficulties we faced while doing the literature survey and understanding this complex topic a little better. The book has, therefore, been designed in such a way to provide a gradual evolution starting from the very basics to the more advanced concepts and then finally dealing with the applications of metallosurfactants including but not limited to catalytic, sensing, and biomedical fields. The book has been divided into 13 chapters. The first three chapters are associated with the synthesis and characterization of metallosurfactants followed by an investigation of their surface, interfacial and aggregation behavior. Accordingly, building on their properties, a variety of applications for the metallosurfactants have been discussed, highlighting the current status and development in the areas including catalysts for organic and esterolytic reactions, hydrogen generation, drug-delivery vehicles, molecular machines, transfection, nanoparticle synthesis, anticancer and antimicrobial activities, CORMs, and sensors in the subsequent Chapters 4–13. This book is intended for introductory graduate-level metallosurfactant chemistry and the researchers working in this field. We believe that fellow scientists from both academia and industry would find this book to be a concise, informative yet interesting read.

14th November 2021
Chandigarh, India
Indiana, USA

Surinder K. Mehta
Ravneet Kaur

1

Metallosurfactants, a "*Novel Portmanteau*": A Holistic Insight into the Structural–Physiognomies Relationships, Synthesis Stratagems, and Characterization

Ankush Parmar[1], Shilpee Sachar[2], and Shweta Sharma[1]

[1] *Panjab University, Institute of Forensic Science and Criminology, Chandigarh 160014, India*
[2] *University of Mumbai, Department of Chemistry, Vidyanagari, Santacruz (E), Mumbai 400098, India*

1.1 Introduction

Recently, "*surface chemistry*" has gained enormous amount of attention in bringing the vivid paradigm of synthetic chemistry [1]. Owing to the splendid attributes, being offered by it, a vast majority of the leading researchers, scientists, and physical chemists across the globe are putting their exertions to envisage noble advancements in this illustrious domain [2]. The inception of this field can be dated back to the late seventeenth century. It was through the numerous experiments (to study the phenomenon responsible for spontaneous spreading of *oil-on-water*) conducted by the famous physician *Sir Benjamin Franklin* that the early breakthroughs in this domain were achieved [3]. This innovation later on proved to be a souvenir and became the underlying fundamental basis for the upcoming research and development in this field [4].

In subsequent decades, a colossal growth has been witnessed in the dominion of surface chemistry, and the corresponding systematic findings have actually transformed variegated scientific disciplines [5]. It is because of this particular realm that the eminent fields such as colloid chemistry, interface science and surface engineering came into existence.

Several other eminent and impending disciplines *viz.* nanoscience, energy conversion, and catalysis have also momentously gained from it [6]. In other words, surface chemistry has essentially laid down the underpinning stone for voluminous superlative fields [7]. Henceforth, it can be rightly said that "*surfactants, a portmanteau of surface-active agents*" as we call them have proven to be highly potent agents (Figure 1.1) [8–11]. Amidst, all the molecules investigated in the realm of chemistry, surfactants are most exclusively studied. Owing to their exceptional solution and interfacial properties, they are currently being utilized on a widespread level in several fields.

Surfactants can be defined as the "*materials which when present at low concentrations, alters the interfacial free energies of the interfaces via adsorbing themselves onto*

Metallosurfactants: From Fundamentals to Catalytic and Biomedical Applications, First Edition.
Edited by Surinder K. Mehta and Ravneet Kaur.

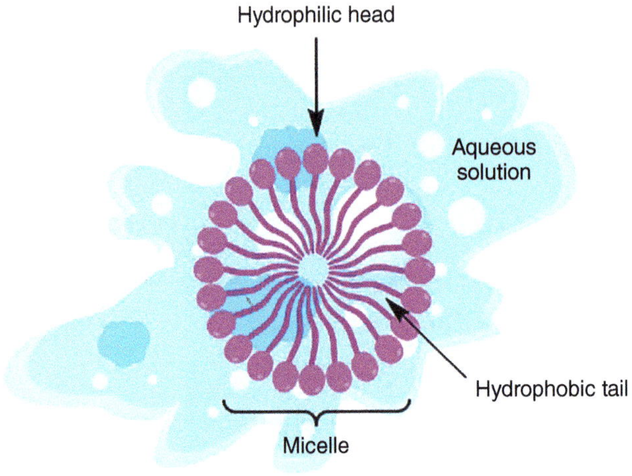

Figure 1.1 Schematic representation of surfactant.

the interface/surface of the system" [8, 12–20]. They generally fall under the category of organic compounds and are truly amphiphilic in nature [21].

When it comes to structural organization, surfactants usually comprises of two vital components the first one being the head (polar) group, which is hydrophilic in nature [22, 23], and the second part often referred to as the tail (non-polar) group is hydrophobic (lipophilic) in nature [24]. It is mostly observed that the polar head group varies greatly in structure and can be multifaceted, while the tail group demonstrates less diversity and as a rule it primarily comprises of a hydrocarbon (alkyl) chain with 8–20 carbon atoms. Broadly speaking, the tail can be branched/linear hydrocarbon, aliphatic, alkyl/aryl, short/long whereas; the head group can be ionic/non-ionic [21–23]. These chemical moieties have an innate ability to affect the air–water interface in an articulate fashion. This might be attributed to the reduction in the surface tension (interfacial tension) of water and formation of assorted assemblies (micelles) at the interface [10, 25]. Upon adsorption at the interface, the two characteristic portions of the surfactants *viz.* head and tail group align themselves according to their polarities/preferential solubilities. The polar head group align itself toward the aqueous (water) part, while the tail group orient itself away from the water (outward direction) thereby resulting in the formation of micelles [26].

When dispersed in aqueous phase, at low concentration these surfactants generally exist in monomeric/dimeric state. An increase in their concentration beyond a designated threshold (*viz. critical micellar concentration* [CMC]) leads to the spontaneous accumulation of surfactant monomers. This further results in the formation of assemblies/colloidal-sized clusters also commonly referred to as "*micelles*" [4, 7, 26, 27]. A major share of this potential breakthrough goes to James William McBain, whose discovery inevitably changed the dynamics and lead to significant advancements in the field of surface/surfactant chemistry [12, 27–29]. It was in the year 1916, while conducting his research experiments, he observed infrequent alteration in the electro-conductive physiognomies as a function of soap concentration and he coined the term "micelles" [30].

Nowadays, a dire need is felt for the upgradation and development of novel technologies, which will play an intricate role in improvising the varied aspects of human lives. This upsurge calls for a quick modification of the surfaces in order to perfectly align with these rapid advancements [31]. Herein, the innate potential to alter the interfaces as per the requisite demand plays a pivotal role in envisaging and devising innovative technological advancements, ranging from energy production to biomedical implants, which will offer promising outcomes [12, 27]. Incessant strides are being made in the field of science and technology on a regular basis to attain the aforementioned goal.

Irrespective of the fact that these chemical moieties possess such dynamic properties yet, they have not been able to secure a place for themselves among the "*catalogue of advanced materials*" [32]. In lieu of this, unique complexes with pre-selected functionalities (*viz.* proteins, carbohydrates, and metal ions) have been developed, which comprehensively aided in triumphing the lacunas associated with conventional surfactant systems. Additionally, incorporation of such modalities resulted in the fabrication of novel surfactants with engineered interfacial attributes [31]. Among all functional groups/modalities, metal ions have played a pivotal role in escalating the physicochemical attributes of the conventional surfactants exponentially. This in particular has lately led to the emergence of remarkable, and idiosyncratic complexes commonly referred to as metallosurfactants (MTS) [33].

1.2 Intrinsic Physiognomies of Metallosurfactants

Self-aggregation/association/assembly above CMC is an intrinsic property, which makes these surfactants and amphiphilic structures an invigorated tool. Length and volume of lipophilic component, size and charge of the polar head group, type of interaction with the solvent system, ionic strength, and molecular framework of the system are some of domineering factors, which tend to govern the hydrodynamic radii (particle size), and surface charge (zeta potential) of these singular entities [34–36]. The whole credit for this fundamental concept of aggregation goes to Zhulina et al. [35]. It was because of their incessant attempts that a comprehensive overview of this distinguishing phenomenon could be deciphered more than 30 years ago [37, 38].

With an advent in time, several triumphant advancements have been accomplished in the field of surfactant chemistry. These progressions have broadened up the horizons of our understanding in lieu of surfactants. Similarly, the underlying concepts providing an inclusive aftermath of the phenomenon have also upgraded in recent years. This upgradation vindicated the fact that it is the metal ion, which plays a pivotal role in regulating the self-aggregation behavior of MTS/MTSC, respectively [34]. Whilst, the positioning of the metal ion component (counter ion/integral component) in the base matrix does not seem to have an obligatory effect on the assembling phenomenon.

When dispersed in aqueous media, these MTS tends to "*diminish the characteristic qualities of metalloenzymes*" [39]. Therefore, the MTSC felicitates a self-aggregation/association of distinct metal-complex-based aggregates reciprocating the aggregation number equivalent to their counterparts *viz.* surfactants [15, 39]. This distinguishing property to self-aggregate, and formulate idiosyncratic complexes bestow these MTS with unique potential applications, which can prove to be handy in real-time scenario [39]. Additionally, the complexes so generated possess multiphasic properties, and can solubilize an array of diverse chemical moieties i.e. lipophilic/hydrophilic, and ionic components [13]. Depending upon the CMC value, distinctive structures illustrating assorted geometrical facets *viz.* micelles (spherical, oblate/prolate/ellipsoid, hybrid crystals with layered perovskite structure), and aggregates (vesicles, bilayers, helixes, tubules) are customarily formed by this special class of metallic surfactants [40–43].

As compared to their conventional counterparts, CMC of these metal-based surfactant complexes is very less. Henceforth, the presence of MTS at a miniscule concentration is potent enough to bring a substantial saturation at the interfacial point along with a significant micelle formation [44]. Apart from these facts and findings, it has also been deciphered by studies that the self-assembly of MTS is greatly affected by two key factors i.e. nature of transition metal ion and differences in coordination at atomic scale. To gain a holistic view of the aforementioned fact, let us briefly discuss the effects of ions, solvents, and structures on micellar morphology with some examples:

a) **Metal ion** – If a conventional surfactant is dispersed in water, a simple micelle is formed, while incorporation of a metal ion in the base structure results in a significant alteration in the geometrical facet, and surface charge of the MTS. In place of the micelle, formation of a vesicle takes place usually [45–47].
b) **Solvent system** – A reverse micelle having a metallic amphiphilic bilayer is formed upon the dispersion of an MTS (having multiple alkyl chains attached to a metal ion) in non-polar solvent while, such a phenomenon is absent in conventional surfactant systems [18, 48, 49]. Upon dispersion into less polar organic solvents (*viz.* chloroform, and diethyl ether), these MTS tends to aggregate, and formulation of inverted metallomicelles takes place. This phenomenon can be principally attributed to the non-covalent interactions which encourage the orientation of the reactive head group in a miniscule volume thereby ultimately resulting in the generation of a highly concentrated and catalytically active cavity of colloidal frameworks.
c) **Structural factor** – Formation of micelle takes place in case of an aqueous dispersion of single-chain MTS while vesicles are formed upon dispersion of double-chain MTS in aqueous media [50–52]. This can be further explained based upon the morphology of the metallic surfactants. Cylindrical morphology of the double-chain MTS, and interaction of metal ion with two surfactant molecules results in the formation of a pseudo-double-chain surfactant (PDCS). During the PDCS generation, the molecules orient themselves in such a manner, that the packing assumes a vesicular structure [53–55].

Apart from this, the cross-sectional area of both the vital components i.e. head and tail region is altered due to the incorporation of metallic ion in the core structure. This alteration creates a demarcating change in the packing volume, thereby inducing a series of phase transition in the metalloaggregates. The phase transitions ultimately result in the fabrication of vesicles with vivid structural forms (*viz.* unilamellar, multi-lamellar/vesicular) [56–58].

With the aid of MTS, an assortment of vivid physicochemical attributes of metal ions can be localized at air–water/oil–water interface. Henceforth, it would be apposite to state, "*the MTS are an exceptional class of surfactant system, which technically amalgamated two diverse area of science i.e. co-ordination and surface chemistry in a coherent manner*" and this leads to the emergence of a noble omnipotent scientific domain. Amidst a wide array of characteristics, redox activity is deemed to be the predominant one. The inculcated metallic ion tends to possess a variable oxidation state, and it is because of this inherent property that these MTS are able to illustrate such a noteworthy ability. The traditional surfactant systems lacks this capability as their core framework is largely composed of elements-like carbon, oxygen and sulfur [45].

On a comparative note, it has been established repeatedly, that MTS essentially bestow enhanced chemical reactivity as compared to their conventional counterparts. Charge, valency, and empty d/f orbitals entangled with metal ions are some of the pivotal factors, which escalates the chemical reactivity of these superior class of metal surfactants to many fold times [18]. In other words, it can be righteously stated that the "*base surfactant itself does not have any chemical reactivity*" [18, 45]. However, it is the embodiment of the metal ion, which brings a radical change in the chemical state of the pristine surfactant, thereby efficaciously circumventing the existing lacuna associated with the traditional surfactant systems [45].

Incorporating metal ions possessing energy levels equivalent to those of the visible transitions (taking place in the UV–visible region) imparts a chromophoric ability to the MTS. It is because of this intricate potential that these metal-based surfactants exhibit illustrious and bright colors (e.g. Cu^{2+} [blue], Fe^{2+}/Fe^{3+} [yellowish red], and Ni^{2+} [green]). This might be attributed to the d–d transition of the metal ion, which is positioned in the visible region of the electromagnetic spectrum. Additionally, it is well versed that the alteration in color is exclusively dependent upon the surrounding chemical milieu. This facilitates the user with a tranquil access where he/she can control the reaction conditions in a complacent manner [45].

In an analogous manner, inculcation of a paramagnetic/ferromagnetic metal ion within the framework of the soft matter imparts magnetic properties to the surfactants, which are fundamentally diamagnetic in nature [18]. The stimuli (magnetic field)-responsive surfactants depicts a novel class of surfactants, wherein the surface tension can be simply modulated by switching on/off the applied magnetic field. Distinct from all of this Lewis acidity, pH responsiveness, and structural diversity are some of the other striking attributes, which are exemplified by these MTS [51, 54, 57].

It seems from the aforementioned facts and findings, that the metal ion is the only component, which is responsible for allocating such eminent characteristics however, in real sense it is not so. The soft matter also adds significantly to the metal-ion chemistry. Active transfer of the above-mentioned attributes to the interface is typically accomplished due to the surface-active property of the surfactant. Henceforth, it is the reunion of the characteristic chattels of both the components, which led to the acquirement of such distinctive physicochemical attributes [48]. Owing to such exceptional attributes, this special class of building blocks have been utilized in many fields vis-à-vis (i) surface activation agents, (ii) monolayers, (iii) emulsions, (iv) contrast agents (medical resonance imaging), (v) catalysts (homogenous photo catalysis), (vi) transporting and sensing agents (fluorescent sensors), (vii) binders and metal extractors, (viii) redox agents (biological system), (ix) vectors for drug delivery (nanoparticles), (x) thin-film optoelectronic devices, (xi) solvatochromic probes, (xii) interfacial photo-physical agents, (xiii) templates for fabrication of diverse frameworks (micelles, vesicles, bilayers, liquid crystals, mesoporous materials, and metallomesogens), and (xiv) antimicrobial agents (*viz.* antibacterial, antiparasitic, anthelminthic) [50, 59, 60].

1.3 Classification of Metallosurfactants

In pertinent literature, a vast majority of systems have been outlined, which tends to intrinsically categorize these tailored/engineered hybrid surface-active moieties based upon certain distinctive characteristics. One such classification system relies upon the type of bonding between the metal ion, and the soft matter (surfactant). According to this system, surfactants comprising of inorganic moieties are generally referred to as inorganic surfactants (*viz.* I-SURFs) [18, 45]. I-SURFs can be categorically sub-divided into two major categories *viz.*

a) **I-SURFs type I** – These are the simplest type of surfactants, in which the metal ion is incorporated within the conventional surfactant. Herein, the metal ion plays a definitive role, and acts as a counter ion.
b) **I-SURFs type II** – In such kind of hybrid surfactants, the metal ion acts as an integral part of the polar head group. Type II I-SURFs can be further divided into two prominent types i.e. type II a, and type II b. In type II I-SURFS, the metal ion is localized within the polar head group component of the soft matter. In the latter one, the metal ion itself acts as the head group.

In the pertinent literature, another classification system, which shares analogous fundamental basics to the aforementioned system, has been reported. In this method, the MTS are chiefly classified into three major categories *viz.*

a) **Group I** – In this kind of metal-based surfactants, the hydrophilic head group acts as a housing cavity for the positioning of the metal ion. Further, attachment of this component with the hydrocarbon tail is accomplished via covalent binding [61].
b) **Group II** – These MTS represent a newer class of hybrid metallic surfactants. In these hybrid surfactants, the hydrophobic part of the soft matter provides accommodation to the metal ion. Herein, the metal ion acts as an integral part of the long alkyl chain, and it is with the help of co-ordination chemistry that this ionic entity is connected with the hydrophobic segment [62].

c) **Group III** – In this type of hybrid surfactants, the counter ion of the soft matter provides a housing cavity to the metallic ion. Herein, the metal ion is weakly bonded with the surfactant and is localized within the counter ion segment [63].

Based upon their dissociation in water, a third type of classification system categorizes the MTS into two chief classes *viz.*

a) **Cationic** – Upon dispersion into an aqueous phase, these MTS are primarily dissociated into two key components i.e. an amphiphilic cation, and an anion. A major share of such MTS chiefly comprises of nitrogenous compounds such as fatty amine salts, and quaternary ammoniums. These nitrogenous functional groups are further attached to elongated single or multiple alkyl tails (e.g. bis(2-phenylpyridine)(4,4-diheptadecyl-2,2-bipyridine)-iridium(III) chloride (Ir-complex)) [64].
b) **Anionic** – Unlike their predecessors, these MTS also dissociate into two major components upon dispersion into an aqueous phase. However, the demarcation lies in the fact that herein, an amphiphilic anion is generated in place of a cation followed by the generation of a cation. Anionic surfactant *viz.* sodium 1,4-bis(neopentyloxy)-3-(neopentyloxycarbonyl)-1,4-dioxobutane-2-sulfonate (TCl_4) would constitute a perfect example of such a system [65]

The fourth classification system is distinctly apart from the earlier-described systems. Herein, depending upon the localization of metal ion in the soft matter, the MTS can be classified into two broad categories as;

a) **Metal ion in hydrophilic head group** – In this class of hybrid metal-based surfactant, hydrophobic component of the MTS primarily comprises of the alkyl chain of the soft matter, while the metal ion along with its primary coordinated sphere constitutes the polar head group [15, 60].
b) **Metal ion in hydrophobic tail group** – This type of MTS depicts divergent attributes as juxtaposed to the former one. Herein, the metal ion is localized in the hydrophobic (alkyl chain) segment (tail) of the soft matter [16, 40].

The fifth classification system can be considered as the core/base structural characterization scheme. According to this classification schema, the MTS can be classified into five major categories as;

a) **Single chained** – In this class of MTS; the metal ion is incorporated with the polar head group of the soft matter. Herein, the metal ion tends to play a definitive role, and acts as a counter ion [66].
b) **Double chained** – This type of hybrid metallic surfactants possess an analogous structure to that of the single-chain MTS. However, the major demarcation lies in tail group. Alike its predecessor, the metal ion along with its primary coordination sphere acquires the functionality of the polar (hydrophilic) head group. Whereas, on the other hand, the ligand along with the alkyl (hydrophobic) tail is coordinated to the metal ion [67, 68].
c) **Pseudogemini** – With an advent in time, surfactant chemistry exemplified astonishing progressions. These momentous advancements, paved a gallant way

for the fabrication of interesting class of novel surfactants *viz.* pseudogemini surfactants. Pseudogemini MTS are single-chain analogues of gemini surfactants [69]. They possess an explanatory complexed structure, in which the amphiphilic ligands are attached to a centrally coordinated metal ion. MTS complex of an "*alkylsulfonatephosphine-functionalized alkyphenol ethoxylate*" symbolizes an interesting example of such a system [44].

d) **Mixed ligand coordination complexes** – These can be piously regarded as the exalted class of MTS. In such MTS, a transition metal complexed with an amphiphilic entity such as Schiff's base is incorporated within the core framework of a conventional surfactant [70].
e) **Organometallic surfactants** – These hybrid surfactants epitomizes the most advanced series of metallic surfactants. It can be astutely stated, "*it is because of such illustrious moieties that the dynamics of surfactant chemistry has metamorphosed colossally.*" Coming on to the schematic aspect, these MTS are analogous to the traditional surfactants, as they also tend to possess a head and tail group. Nonetheless, a major variance is generated due to the alteration in the chemical nature of the molecular entities, which constitutes both these pivotal segments. Herein, the head group usually constitutes of a transition metal embodied with one of the components of the soft matter. Whilst, in the second region, the alkyl chain is substituted with a functional group tagged π-conjugated chain, which further acquires the functional role of the hydrophobic tail [71].

1.4 Syntheses Stratagems and Characterization of Metallosurfactants

The facts and findings discussed above have provided us with a comprehensive overview of the MTS and their intricate physiognomies. Now, it has become clear that MTS are technically "*amphiphilic entities comprising of coordinated metal ion/ions.*" Coming on to the syntheses part, polar or organic solvents play a vivacious role in the syntheses of these exceptional classes of hybrid surfactants. Till date, three distinctive synthesis routes have been devised for the syntheses of MTS;

1.4.1 Metathesis Reaction

A highly sophisticated and arduous methodology becomes pre-requisite when it comes to the development of MTS possessing cationic metal species, and aliphatic tails. In such circumstances, the following stratagem i.e. metathesis reaction (MR), appears to be a unique one and offers several lucrative advantages *viz.* [72]

(i) One-pot synthesis.
(ii) Ease of extraction and cost effectiveness.
   - **Efficient recovery of catalyst** – Employment of water-soluble catalyst circumvents the need for re-extraction.

- **Minimized solvent consumption/wastage** – Tranquil extraction of catalyst can be accomplished with an aqueous phase. The aqueous phase can be reused as such for subsequent reactions (continuous batch production).
- **Ordinary reaction conditions** – Does not require any specific reaction conditions such as inert atmosphere, and degassing. The entire reaction can be carried out at room temperature.

In general terms, an MR can be defined as a "*reaction in which two distinct chemical entities react in an aqueous phase to form a novel product.*" Herein, the formation of the new product usually takes place due to the double decomposition/displacement reaction (interchanging of the radicals or ions of the reactant species). MR, in other words, can be technically referred to as "*the phenomenon of cation, and anion exchange in a polar/aqueous phase*" [73].

To understand the dynamics of the aforementioned synthesis schema, let us consider certain examples. Lipshutz et al. carried out the synthesis of a *"designer surfactant" viz.* PQS (nanomicelle-forming amphiphile) using a metathesis reaction [74]. Herein, a well-versed ring-closing ruthenium catalyst i.e. Grubbs Hoveyda 1 Ru(II) was employed as an intermediate substrate for the catalyzing the entire reaction. The present experiment exhibited promising outcomes and the results came out to be extremely significant. Proposed methodology, i.e. micellar catalysis/ring catalysis metathesis (RCM) allowed a successful covalent linking of the Ru carbene with the designated chemical moiety, thereby, leading to a successful synthesis of the novel MTS [74]. The synthesized PQS MTS offered three vivid key elements *viz.* lipophilic segment (site for solubilization of insoluble organic substrates), hydrophilic segment (solubilizes PQS in polar/aqueous solvent such as water), and free –OH residue within the hydrophobic core (site bearing the metallic part). The proposed methodology also offered a potent solution to overcome the pre-existing lacunas of the conventional methodology. Pure water acted as the solvent system, and during the entire reaction, room-temperature conditions were maintained. The catalyst used was water soluble, and no re-isolation steps were further carried out. Additionally, the proposed methodology facilitated the RCM of water-insoluble dienic substrates [74].

1.4.2 Ligand Substitution Reaction

Ligand substitution reaction (LSR) is the second most preferred type of reaction, when it comes to the synthesis of hybrid metal-based surfactants. As the name itself suggests, the method generally involves the substitution of lesser labile ligands with more labile ligands. In other words, it can be righteously said, that in these types of reactions, interchange of highly susceptible ligands takes place via the lesser ones. In lieu of this, Veeralakshmi et al. synthesized varied types of single- and double-chained MTS possessing idiosyncratic biological functionalities [75]. Facile employment of LSR led to the efficacious synthesis of hybrid ligand-anchored metallic surfactant complexes. Herein, $Co(dien)Cl_3$ acted as the precursor excipient whilst, diethylenetriamine (dien), dodecylamine (DA), and hexadecylamine (HA) were chosen as the model ligands. In the present experiment, one or two amine groups of the alkylamine ligand base played an active role and led to the substitution of one or two labile chloride ligands. This ligand-mediated substitution ultimately resulted in the synthesis of four different types of MTS *viz.* $[Co(dien)(HA)Cl_2]ClO_4$(III), $[Co(dien)(DA)Cl_2]ClO_4$(IV) $[Co(dien)(HA)_2Cl](ClO_4)_2$(V), and $[Co(dien)(DA)_2Cl](ClO_4)_2$(VI) [75].

1.4.3 Ligand Insertion Reaction

Last but not the least, ligand insertion reaction (LIR) is the third type of reaction, which is predominantly employed for the synthesis of MTS. A distinctive feature of this method is the maintenance of the integral nature of all the excipients being employed during the reaction. Unlike, its predecessor i.e. LSR, this method does not involve any substitution/elimination. The method preferred is the fabrication of hybrid metal-surfactant complexes via interjecting a chemical moiety (usually a ligand) into another molecule or a bond [19]. Following this approach, Wagay et al. synthesized a hybrid MTS named *cis*-chlorobis(ethylenediamine)dodecylamine cobalt(III) nitrate(I) (*viz.* CDCN) [67, 68, 76]. Herein, a dropwise addition of the ligand i.e. dodecylamine (ethanolic solution) to *trans*-$[Co(en)_2Cl_2]Cl$ (aqueous solution) resulted in the attainment of the desired metal surfactant complex.

In another experiment, a double-chained MTS *viz.* *cis*-bis(1,3-diaminopropane) bis(dodecylamine)cobalt(III) nitrate(II) (DDCN) was synthesized by the same group using LIR methodology [76]. Analogous to the aforementioned methodology, the preset method also involved a dropwise addition of ethanolic dodecylamine solution (ligand) to an aqueous solution of *trans*-$[Co(tmd)_2Cl_2]ClO$ (aqueous solution). This in turn led to a successful synthesis of the desired hybrid metal–surfactant complex i.e. DDCN.

Characterization plays an imperative role in deciphering the electro-physicochemical, structural, and morphological attributes of the synthesized MTS. Determination of such imperious traits can further aid in exploring the practical utility of these versatile systems in real-time scenario. Not only this, the characterization of synthesized MTS using high-tech analytical techniques gives an edge to the researchers/chemists, and allows them to develop versatile MTS with tailored properties. A wide array of analytical techniques ranging from simplistic spectrophotometer to advanced techniques such as electron/ atomic microscopy has been widely employed for the facile elucidation of intricate physiognomies. Table 1.1 illustrates the varied analytical techniques being employed for the characterization of novel MTS.

1.5 Conclusion

MTS represents a class of supra-molecular assemblies which can facilitate an effective bridging between both the explicit domains of chemistry *viz.* synthetic and inorganic. It is because of this unique bridging that these exhilarating moieties with

Table 1.1 Characterization of MTS via state-of-art analytical techniques.

Techniques	MTS explored	Characteristic attributes probed	Inference	References
Dynamic light scattering (DLS)	Tetradecylpyridinium (TP) $TP_2[MCl_4]$ (M = Mn, Co, Ni, Cu, Zn)	Hydrodynamic radii (size) and surface charge	The micelles depicted ultrafine particle size of 4–7 nm, whilst the zeta potential was found to be +45 mV, respectively.	[68]
Scanning electron microscopy (SEM)	Supramolecular MTS (SMMSs)	Surface morphology	The fabricated SMMSs illustrated a spherical (polymeric) morphology. Highly dispersed polymeric SMMSs obtained with increase in the chain length	[77]
	Biscetylpyridiniumtetrachloroplatinate (Pt-CPC) MTS	Surface morphology	Surface pattern found to be profoundly dispersed, highly homogenous, and spherical in nature	[78]
Transmission electron microscopy (TEM)	Double-tailed metallosomes (*viz.* bishexa decylpyridinium metal tetrachloride (MCPC II) [M: Fe, Co, Ni, Cu])	Morphological attributes	Fabricated MTS (50–150 nm) possessed diverse morphologies *viz.* multivesicular and spherical	[73]
	C_n–Cu–C_n ($n = 8, 12, 16$) MTS	Morphometric attributes	Primarily vesicular aggregates within a size range of 60–200 nm. Additionally microscopic studies revealed the presence of polydispersions and polyaggregates	[79]
Atomic force microscopy (AFM)	CTA–AgB MTS	Molecular geometry	The aggregates were predominantly spherical	[80]
	Copper(II)-surfactant complexes	Molecular geometry	Metallovesicles (MTVs) in the size range of 20–50 nm were formed	[81]

(Continued)

Table 1.1 (Continued)

Techniques	MTS explored	Characteristic attributes probed	Inference	References
Ultraviolet–Visible (UV–Vis) spectroscopy	Bishexadecyl pyridinium metal(II) tetrachloride (M CPC II) [M:Fe, Co, Ni]	Probabilistic mechanism responsible for binding of BSA to the corresponding MTS	Inclination in the intensity of the absorption maxima (static quenching) of BSA upon addition of MTS. Blue shift (protein stabilization/unfolding).	[82]
	Chromium (III) MTS containing various chelating ligands (ethylenediamine (en), tri-ethylenetetramine (trien), 2,2-bipyridine (bpy), and 1,10-phenanthroline (phen)), and axial amine (dodecylamine/cetylamine)	Geometrical configuration of the corresponding cations within the MTS	Escalated distortion of basal planes in case of "*phen*" complexes as compared to "*bpy*" complexes. Both the complexes (*viz.* *cis*-$[Cr(phen)_2(C_{12}H_{25}NH_2)_2]^{3+}$ and $[Cr(bpy)_2(C_{16}H_{33}NH_2)_2]^3)$ possess an identical structure	[59]
Fourier transform infra-red (FTIR) spectroscopy	Hexadecyltrimethyl ammonium chromium trichloride (CrC I), hexadecyltrimethyl ammonium chromium tetrachloride (CrC II)	Elucidation of functional group bonds generated among the parent excipients.	Interaction of metal chloride with the ammonium group of the MTS resulted in an alteration in the chemical milieu (shifting of –C–N, and NH–Cl stretching band to a lower wavenumber)	[83]
	Bistable copper(II) complex	Explication of linkage developed between the varied precursor materials	Free carbonyl groups (ν O–H) played a pivotal role in establishing linkage with the silanol (*Si*–OH) groups present in the silica walls	[84]
Nuclear magnetic resonance (NMR) spectroscopy	N-heterocyclic carbene (NHC)-MTS	Vindicating the successful fabrication (linkage of two polymer blocks) of MTS.	Covalent linkage between PS and PMMA established. Fabricated MTS comprises of core-shell-shaped particles, resulting in the synthesis of block copolymer	[33]
	FeCPC I, CoCPC I, NiCPC I MTS	—	Electronic interaction between the cationic (CP^+) and anionic (MCl_3) moieties resulted in the synthesis of corresponding MTS.	[2]

Small-angle X-ray scattering (SAXS)	Lipid mimetic MTS *viz.* $[Ru(bpy)2(dn\text{-}bpy)]^{2+}$ and $[Ir(ppy)2(dn\text{-}bpy)]^{+}$	Determination of the structural aspects of the corresponding MTS	Sharp Bragg's peaks obtained indicating the formation of crystalline, hollow, and multilamellar (composed primarily of bilayer stacks) MTS-based structures	[85]
	MTS ionogels	Elucidation of the morphological attributes of the corresponding MTS.	All the three MTS ionogels possessed lamellar morphology.	[86]
X-ray diffraction (XRD)	Cetylpyridinium chloride metallosomes (METs)	Establishment of the morphological arrangement of the as-fabricated METs	METs possessed highly crystalline geometry along with a lamellar metallosomal bilayer	[73]
Differential scanning calorimetry (DSC)	$[M\,(CH_3COO)_4]^{2-}\;[C_{12}H_{25}NH^{+}{}_{3}]_2$	Physicochemical characteristics of the MTS.	Decomposition occurs with melting in metal complexes, and metal oxides remained as final products. Activation energy order obtained was dependent on the size of transition metal ion and metal ligand bond strength	[87]
Thermogravimetry/ differential thermal analysis (TG-DTA)	Pt-CPC MTS	Thermal behavior and stability of the MTS	The fabricated MTS depicted an enhanced thermal stability. The Pt-CPC MTS illustrated an exothermic behavior prior to the transition temperature (TT) whereas; a *vice-a-versa* phenomenon was obtained on/after the TT.	[78]

tailored physiognomies can be fabricated with ease. Apart from this, the inimitable metal-phase behavior of the MTS bestows them with a unique ability to formulate an extensive array of supramolecular structures. Additionally, the existence of an interrelation between the two vivid domains *viz.* coordination chemistry and surfactant self-assembly offers tremendous opportunities for these MTS to be applied on a full-scale level in varied domains of science, and technology.

References

1 Kaur, G., Singh, P., Mehta, S. et al. (2017). A facile route for the synthesis of Co, Ni and Cu metallic nanoparticles with potential antimicrobial activity using novel metallosurfactants. *Applied Surface Science* 404: 254–262.

2 Kaur, G., Garg, P., Kaur, B. et al. (2019). Synthesis, thermal and surface activity of cationic single chain metal hybrid surfactants and their interaction with microbes and proteins. *Soft Matter* 15 (11): 2348–2358.

3 Tanford, C. (2004). *Ben Franklin Stilled the Waves: An Informal History of Pouring Oil on Water with Reflections on the Ups and Downs of Scientific Life in General.* OUP Oxford.

4 Kolasinski, K.W. (2012). *Surface Science: Foundations of Catalysis and Nanoscience.* Wiley.

5 Morra, M. (2001). *Water in Biomaterials Surface Science.* Wiley.

6 Gong, J. and Bao, X. (2017). Fundamental insights into interfacial catalysis. *Chemical Society Reviews* 46 (7): 1770–1771.

7 Somorjai, G.A. and Li, Y. (2010). *Introduction to Surface Chemistry and Catalysis.* Wiley.

8 Brown, P., Hatton, T.A., and Eastoe, J. (2015). Magnetic surfactants. *Current Opinion in Colloid and Interface Science* 20 (3): 140–150.

9 Stokes, E.C. (2017). Amphiphilic ligand architectures for s-, d- and f-block metallosurfactants towards micellar systems and microemulsions. PhD thesis. Cardiff University, School of Chemistry.

10 Rosen, M. (2004). Foaming and antifoaming by aqueous solutions of surfactants. In: *Surfactants Interfacial Phenomena*, 277–302. John Wiley & Sons.

11 Otzen, D. (2011). Protein–surfactant interactions: a tale of many states. *Biochimica et Biophysica Acta, Proteins and Proteomics* 1814 (5): 562–591.

12 Allara, D.L. (2005). A perspective on surfaces and interfaces. *Nature* 437 (7059): 638–639.

13 Bunton, C.A., Nome, F., Quina, F.H. et al. (1991). Ion binding and reactivity at charged aqueous interfaces. *Accounts of Chemical Research* 24 (12): 357–364.

14 Chang, C.H. and Franses, E. (1995). Adsorption dynamics of surfactants at the air/water interface: a critical review of mathematical models, data, and mechanisms. *Colloids and Surfaces A: Physicochemical and Engineering Aspects* 100: 1–45.

15 Griffiths, P., Fallis, I., Chuenpratoom, T. et al. (2006). Metallosurfactants: interfaces and micelles. *Advances in Colloid and Interface Science* 122 (1–3): 107–117.

16 Griffiths, P.C., Fallis, I.A., Tatchell, T. et al. (2008). Aqueous solutions of transition metal containing micelles. *Advances in Colloid and Interface Science* 144 (1, 2): 13–23.

17 Mehta, S. and Kaur, R. (2013). Self-aggregation and solution behavior of synthesized organo transition metal (Co, Fe, Zn) amphiphilic complexes. *Journal of Colloid and Interface Science* 393: 219–227.

18 Polarz, S., Odendal, J.A., Hermann, S. et al. (2015). Amphiphilic hybrids containing inorganic constituent: more than soap. *Current Opinion in Colloid and Interface Science* 20 (3): 151–160.

19 Taira, T., Yanagimoto, T., Sakai, K. et al. (2018). Self-assembling properties of an N-heterocyclic carbene-based metallosurfactant: Pd-coordination induced formation of reactive interfaces in water. *Journal of Oleo Science* 67 (9): 1107–1115.

20 Tsuchiya, K., Yajima, H., Sakai, H. et al. (2012). *Electrical Phenomena at Interfaces and Biointerfaces: Fundamentals and Applications in Nano-, Bio- and Environmental Sciences*. Hoboken, NJ: Wiley.

21 Sharma, S., Kori, S., and Parmar, A. (2015). Surfactant mediated extraction of total phenolic contents (TPC) and antioxidants from fruits juices. *Food Chemistry* 185: 284–288.

22 Kori, S., Parmar, A., Goyal, J. et al. (2018). Cloud point extraction coupled with microwave-assisted back-extraction (CPE-MABE) for determination of Eszopiclone (Z-drug) using UV–visible, HPLC and mass spectroscopic (MS) techniques: spiked and in vivo analysis. *Journal of Chromatography B* 1074: 129–138.

23 Kori, S., Parmar, A., and Sharma, S. (2019). Colloidal assemblies: effective extraction media for diazepam from impaled blood, urine and milk samples. *Australian Journal of Forensic Sciences* 51 (2): 171–181.

24 Furse, S. (2010). Bubbles, bubbles, everywhere, but not a drop to drink. *The Lipid Chronicles*.

25 Rosen, M.J. and Kunjappu, J.T. (2012). *Surfactants and Interfacial Phenomena*. Wiley.

26 Kuperkar, R.C., Kuperkar, K.C., Cohen, L. et al. (2013). *A Classical Mathematical Solution for Surfactant Induced Flow Behaviour through Porous Media*. UK: Household Products and Cleaning: Today.

27 Duke, C. (2003). The birth and evolution of surface science: child of the union of science and technology. *Proceedings of the National Academy of Sciences of the United States of America* 100 (7): 3858–3864.

28 McBain, J. (1913b). Mobility of highly-charged micelles. *Transactions of the Faraday Society* 9: 99–101.

29 McBain, J. (1913a). Micellar formation of aqueous solution. *Transactions of the Faraday Society* 9: 99–112.

30 Santhakumar, K., Kumaraguru, N., Arunachalam, S. et al. (2007). Thermodynamics and micellar properties of some surface active cobalt(III) metallosurfactants in nonaqueous medium. *International Journal of Chemical Kinetics* 39 (1): 22–31.

31 Larpent, C., Laplace, A., and Zemb, T. (2004). Macrocyclic sugar-based surfactants: block molecules combining self-aggregation and complexation properties. *Angewandte Chemie International Edition* 116 (24): 3225–3229.

32 Bhadani, A., Kafle, A., Ogura, T. et al. (2020). Current perspective of sustainable surfactants based on renewable building blocks. *Current Opinion in Colloid and Interface Science* 45: 124–135.

33 Donner, A., Trepka, B., Theiss, S. et al. (2019). NHC-metallosurfactants as active polymerization catalysts. *Langmuir* 35 (50): 16514–16520.

34 Venhuis, S.H. and Mehrvar, M. (2004). Health effects, environmental impacts, and photochemical degradation of selected surfactants in water. *International Journal of Photoenergy* 6 (3): 115–125.

35 Zhulina, E. and Borisov, O. (2005). Theory of morphological transitions in weakly dissociating diblock polyelectrolyte micelles. *Macromolecules* 38 (15): 6726–6741.

36 Koutsantonis, G.A., Nealon, G.L., Buckley, C.E. et al. (2007). Wormlike micelles from a cage amine metallosurfactant. *Langmuir* 23 (24): 11986–11990.

37 Yin, H., Lei, S., Zhu, S. et al. (2006). Micelle-to-vesicle transition induced by organic additives in catanionic surfactant systems. *Chemistry A European Journal* 12 (10): 2825–2835.

38 Owen, T., Pynn, R., Hammouda, B. et al. (2007). Metal-dependent self-assembly of a microbial surfactant. *Langmuir* 23 (18): 9393–9400.

39 Ghosh, K., Gupta, B., and Bhattacharya, S. (2016). Metallosurfactant aggregates as catalysts for the hydrolytic cleavage of carboxylate and phosphate esters. *Current Organocatalysis* 3 (1): 6–23.

40 Griffiths, P.C., Fallis, I.A., Willock, D.J. et al. (2004). The structure of metallomicelles. *Chemistry A European Journal* 10 (8): 2022–2028.

41 Zhang, J., Li, W., Wu, C. et al. (2013). Redox-controlled helical self-assembly of a polyoxometalate complex. *Chemistry A European Journal* 19 (25): 8129–8135.

42 Maran, U., Conley, H., Frank, M. et al. (2008). Giant micelles of organoplatinum(II) gemini amphiphiles. *Langmuir* 24 (10): 5400–5410.

43 Zheng, Y.Y., Wu, G., Deng, M. et al. (2006). Preparation and characterization of a layered perovskite-type organic–inorganic hybrid compound (C_8NH_6–$CH_2CH_2NH_3$) $2CuCl_4$. *Thin Solid Films* 514 (1, 2): 127–131.

44 Valls, E., Solsona, A., Suades, J. et al. (2002). Synthesis and characterization of new amphiphilic phosphines and palladium metallosurfactants. *Organometallics* 21 (12): 2473–2480.

45 Polarz, S., Landsmann, S., and Klaiber, A. (2014). Hybrid surfactant systems with inorganic constituents. *Angewandte Chemie International Edition* 53 (4): 946–954.

46 Martinez, J., Zhang, G., Holt, P. et al. (2000). Self-assembling amphiphilic siderophores from marine bacteria. *Science* 287 (5456): 1245–1247.

47 Fallis, I.A., Griffiths, P.C., Hibbs, D.E. et al. (1998). Solid state and solution behaviour of novel transition metal containing surfactants. *Chemical Communications* 6: 665–666.

48 Jaeger, D.A., Zeng, X., and Wang, Y. (2006). A surfactant Co(III) chelate containing two tridentate ligands. *Colloids and Surfaces A: Physicochemical and Engineering Aspects* 289 (1–3): 158–162.

49 Domínguez-Gutiérrez, D., Surtchev, M., Eiser, E. et al. (2006). Ru(II)-based metallosurfactant forming inverted aggregates. *Nano Letters* 6 (2): 145–147.

50 Owen, T. and Butler, A. (2011). Metallosurfactants of bioinorganic interest: coordination-induced self assembly. *Coordination Chemistry Reviews* 255 (7, 8): 678–687.

51 Brown, P., Bushmelev, A., Butts, C.P. et al. (2013). Properties of new magnetic surfactants. *Langmuir* 29 (10): 3246–3251.

52 Jaeger, D.A., Peacock, M.F., and Bohle, D.S. (2003). A surfactant transition metal chelate. *Langmuir* 19 (11): 4859–4862.

53 Jaeger, D.A., Reddy, V.B., Arulsamy, N. et al. (1998). Hydrophobic control of diastereoselectivity in the synthesis of double-chain surfactant Co(III) complexes. *Langmuir* 14 (10): 2589–2592.

54 Brown, P., Bushmelev, A., Butts, C.P. et al. (2012). Magnetic control over liquid surface properties with responsive surfactants. *Angewandte Chemie International Edition* 51 (10): 2414–2416.

55 Luo, X., Miao, W., Wu, S. et al. (2002). Spontaneous formation of vesicles from octadecylamine in dilute aqueous solution induced by Ag(I) ion. *Langmuir* 18 (24): 9611–9612.

56 Talsma, H., Jousma, H., Nicolay, K. et al. (1987). Multilamellar or multivesicular vesicles? *International Journal of Pharmaceutics* 37 (1, 2): 171–173.

57 Brown, P., Butts, C.P., Cheng, J. et al. (2012). Magnetic emulsions with responsive surfactants. *Soft Matter* 8 (29): 7545–7546.

58 Jaeger, D.A., Reddy, V.B., and Bohle, D.S. (1999). Cleavable double-chain surfactant Co(III) complexes. *Tetrahedron Letters* 40 (4): 649–652.

59 Kumaraguru, N. and Santhakumar, K. (2009). Synthesis, characterization, critical micelle concentration determination, and antimicrobial studies of some complexes of chromium(III) metallosurfactants. *Journal of Coordination Chemistry* 62 (21): 3500–3511.

60 Griffiths, P.C., Fallis, I.A., James, C. et al. (2010). Structure–property relationships in metallosurfactants. *Soft Matter* 6 (9): 1981–1989.

61 Verani, C.U.N., Driscoll, J., Keyes, P.H. et al. (2014). Cationic copper(II)-containing surfactants: molecular structures, film morphology, and influence on the alignment of nematic mesogens. *Inorganic Chemistry* 53 (11): 5647–5655.

62 Parera, E., Marín-García, M., Pons, R. et al. (2016). Supramolecular arrangement of molybdenum carbonyl metallosurfactants with CO-releasing properties. *Organometallics* 35 (4): 484–493.

63 Kaur, G., Kumar, S., Dilbaghi, N. et al. (2016). Evaluation of bishexadecyltrimethyl ammonium palladium tetrachloride based dual functional colloidal carrier as an antimicrobial and anticancer agent. *Dalton Transactions* 45 (15): 6582–6591.

64 Roldán-Carmona, C., González-Delgado, A.M., Guerrero-Martínez, A. et al. (2011). Molecular organization and effective energy transfer in iridium

metallosurfactant–porphyrin assemblies embedded in Langmuir–Schaefer films. *Physical Chemistry Chemical Physics* 13 (7): 2834–2841.
65 Trickett, K., Xing, D., Eastoe, J. et al. (2010). Hydrocarbon metallosurfactants for CO_2. *Langmuir* 26 (7): 4732–4737.
66 Garg, P., Kaur, G., and Chaudhary, G.R. (2016). Transition metal based single chained surfactants: synthesis, aggregation behavior and enhanced photoluminescence properties of fluorescein. *RSC Advances* 6 (110): 108573–108582.
67 Wagay, T.A., Charingia, A., Suting, S. et al. (2020). Aggregation and adsorption properties of benzyldimethylhexadecylammonium tetrachloromanganate(II) metallosurfactant in water–ethylene glycol medium. *Journal of Dispersion Science and Technology* 42 (12): 1843–1852.
68 Wagay, T.A., Ismail, K., and Askari, H. (2020). Assessment of aggregation and adsorption behavior of newly synthesized tetradecylpyridinium-based metallosurfactants and their interaction with bovine serum albumin. *New Journal of Chemistry* 44 (35): 15018–15030.
69 Bessie, N.A.M. and Jason, S.K. (2009). Production of gemini surfactants. In: *Handbook of Detergents Part F: Production*, vol. 142 (ed. U. Zoller), 561–575. Boca Raton/London, New York: CRC Press/Taylor & Francis Group.
70 Chandar, S.C.N., Santhakumar, K., and Arumugham, M.N. (2009). Metallosurfactant Schiff base cobalt(III) coordination complexes. Synthesis, characterization, determination of CMC values and biological activities. *Transition Metal Chemistry* 34 (8): 841.
71 Bitter, S., Kunkel, M., Burkart, L. et al. (2018). Organometallic, nonclassical surfactant with gemini design comprising π-conjugated constituents ready for modification. *ACS Omega* 3 (8): 8854–8864.
72 Atwood, D.A. (2016). *Sustainable Inorganic Chemistry*. Wiley.
73 Kaur, G., Kaur, B., Garg, P. et al. (2020). A study of synthesis, characterization and metalloplex formation ability of cetylpyridinium chloride based metallosomes. *Journal of Molecular Liquids* 300: 112326.
74 Lipshutz, B.H. and Ghorai, S. (2009). PQS: a new platform for micellar catalysis. RCM reactions in water, with catalyst recycling. *Organic Letters* 11 (3): 705–708.
75 Veeralakshmi, S., Nehru, S., Sabapathi, G. et al. (2015). Single and double chain surfactant–cobalt(III) complexes: the impact of hydrophobicity on the interaction with calf thymus DNA, and their biological activities. *RSC Advances* 5 (40): 31746–31758.
76 Wagay, T., Dey, J., Kumar, S. et al. (2016). Aggregation and surface behavior of aqueous solutions of *cis*-bis(1,3-diaminopropane)bis(dodecylamine)cobalt(III) nitrate. A double-chained metallosurfactant. *RSC Advances* 6 (71): 66900–66910.
77 Sharma, N.K., Singh, M., and Bhattarai, A. (2016). Hydrophobic study of increasing alkyl chain length of platinum surfactant complexes: synthesis, characterization, micellization, thermodynamics, thermogravimetrics and surface morphology. *RSC Advances* 6 (93): 90607–90623.
78 Sharma, N.K. and Singh, M. (2018). New class of platinum based metallosurfactant: synthesis, micellization, surface, thermal modelling and in vitro biological properties. *Journal of Molecular Liquids* 268: 55–65.

79 Zha, Q., Xie, Q., Hu, Y. et al. (2016). Metallosurfactants C_n–Cu–C_n: vesicle formation and its drug-controlled release properties. *Colloid and Polymer Science* 294 (5): 841–849.

80 Kaur, G., Kumar, S., Kant, R. et al. (2016). One-step synthesis of silver metallosurfactant as an efficient antibacterial and anticancer material. *RSC Advances* 6 (62): 57084–57097.

81 Arroyo, I.Z., Gomez, C., Alarcon, H. et al. (2018). Alkyl length effects on the DNA transport properties of Cu(II) and Zn(II) metallovesicles: an in vitro and in vivo study. *Journal of Drug Delivery* 2018: 2090–3014.

82 Kaur, G., Garg, P., Kaur, B. et al. (2018). Cationic double chained metallosurfactants: synthesis, aggregation, cytotoxicity, antimicrobial activity and their impact on the structure of bovine serum albumin. *Soft Matter* 14 (25): 5306–5318.

83 Garg, P., Kaur, G., and Chaudhary, G.R. (2018). Chromium-based metallosurfactants: synthesis, physicochemical characterization and probing of their interactions with xanthene dyes. *New Journal of Chemistry* 42 (2): 1141–1150.

84 Walace, D., Ribeiro-Santos, T.A., Jardim, I.S. et al. (2018). Bistable copper(II) metallosurfactant as molecular machine for the preparation of hybrid silica-based porous materials. *Materials and Design* 160: 876–885.

85 Mechler, A., Stringer, B.D., Mubin, M.S. et al. (2014). Labeling phospholipid membranes with lipid mimetic luminescent metal complexes. *Biochimica et Biophysica Acta (BBA) – Biomembranes* 1838 (11): 2939–2946.

86 Wang, X., Yang, Q., Cao, Y. et al. (2016). Metallosurfactant ionogels in imidazolium and protic ionic liquids as precursors to synthesize nanoceria as catalase mimetics for the catalytic decomposition of H_2O_2. *Chemistry* 5 (22): 17857–17865.

87 Mehta, S., Kaur, R., and Chaudhary, G. (2012). Self aggregation and solution behavior of copper and nickel based surfactants. *Colloids and Surfaces A: Physicochemical and Engineering Aspects* 403: 103–109.

2

Metallosurfactants: A Surface and Interface Perspective

Ravneet Kaur[1], Neena Mehta[2], and Surinder K. Mehta[3]

[1] *Ivy Tech Community College of Indiana – Valparaiso Campus, Life and Physical Science Department, 3100 Ivy Tech Drive, Valparaiso, IN, 46383, USA*
[2] *Rayat-Bahra University, Rayat Bahra Dental College and Hospital, Department of Biochemistry, Tehsil Kharar, Mohali, 140104 Punjab, India*
[3] *Panjab University, Department of Chemistry, Centre for Advanced Studies in Chemistry, Sector 14, Chandigarh, 160014, India*

2.1 Introduction

The world of surface and colloid chemistry was first introduced to "micelle" at the 1913 Faraday meeting in the context of the association of surfactant molecules in aqueous solutions by James William McBain in his discussion contribution, entitled: "on the mobility of highly-charged micelles" [1]. In 1920, McBain gave the first indication of the concept of critical micelle concentration (CMC) mentioning that these "micelles" formed only above a particular concentration, depicted by a minimum in the conductivity–concentration curve, which opened up a new arena to the researchers – the micellization and aggregation behavior of surfactants, which has been very well explored since then. As is the case with the conventional surfactants, these designer "metallosurfactants" also tend to demonstrate the classical surfactant as well as non-classical behavior. Exploring and establishing the surface-active behavior of these surfactants is vital for their perusal in several applications. Because of the extensive research in the area of surfactant chemistry, it has been concluded that the surfactants can even play a critical role in putting an end to the devastating coronavirus disease of 2019 (COVID-19) pandemic, which engulfed the whole world bringing it to a deadly standstill [2]. It has been established that the addition of anionic surfactant generates a synergistic effect with alcohols to inactivate the severe acute respiratory syndrome (SARS)-CoV-2 virus and enhance the virucidal efficiency of commonly used sanitizers, which, in fact, forms the first line of defense to combat the spread of virus. Therefore, investigation into the surface-active behavior is both pertinent and necessary. The interfacial and surface-active properties of some metallosurfactants have been brilliantly reviewed by Griffiths et al. [3, 4]. The group has done tremendous research in this area highlighting the fundamental traditional and non-traditional behavior

Metallosurfactants: From Fundamentals to Catalytic and Biomedical Applications, First Edition.
Edited by Surinder K. Mehta and Ravneet Kaur.

of metallosurfactants along with discussing their physico-chemical properties. The chapter presented here typically deals with the effect of metal incorporation on the surfactant behavior in both aqueous and non-aqueous solvents, considering the location of metal in the surfactant, i.e. head group, hydrophobic tail, or the counter ion area. To get comprehensive information on the surface activity, micellization and adsorption parameters of metallosurfactants and, hence, develop a better understanding of their surface and interfacial characteristics, metallosurfactants have been discussed here by subdividing them into three main categories (i) Category I – Metal as part of the head group, (ii) Category II – Metal as the counter ion, (iii) Category III – Metal as part of the hydrophobic chain.

2.2 Micellization and Surface Parameters

Surfactant molecules exist as monomers in aqueous solutions at low concentration, but as the concentration increases the molecules tend to self-aggregate to a variety of interesting shapes and structures such as micelles, vesicles, and lamellae (Figure 2.1). At a specific concentration range, "micelles" are formed, which is known as the CMC. The CMC value of a surfactant can be experimentally determined from the plots of physical property against the surfactant concentration. The abrupt change in a physical property marks the CMC value such as – (i) surface tension: surfactants demonstrate classical surface activity with the surface tension showing an initial decrease with increasing surfactant concentration, and then eventually tapering to a constant value. The intercept of these two different slopes defines the CMC (See Figure 2.2b); (ii) conductivity: the specific conductivity versus surfactant concentration plots again show two straight lines with different slopes. The first one corresponds to the monomers with the concentration range below the CMC showing a rapid increase in conductivity and, thereafter, micelles start forming. This phenomenon is marked by a change in slope of the conductivity–concentration graph because the conductivity still increases but at a relatively slow rate. The intersection point of these two straight lines with different slopes forms the CMC value of the surfactant. The micellization behavior for metallosurfactants becomes more interesting as in addition to the usual hydrophobic part, head group and the counter ion, metal ion also comes into play, which can be present at any of the above three mentioned sites.

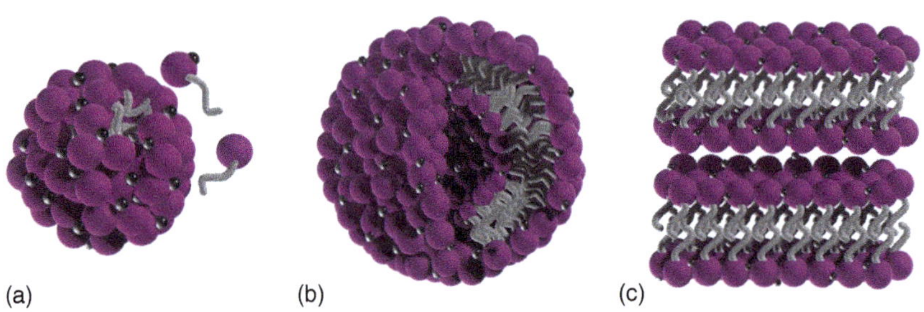

Figure 2.1 Potential self-assembly structures possible for metallosurfactants- (a) micelle, (b) vesicle, (c) lamellae. (Purple sphere: head group, shiny gray small spheres: metal.)

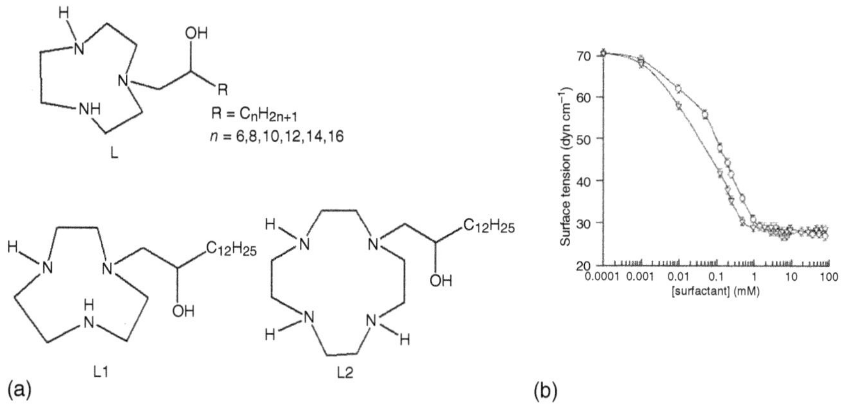

Figure 2.2 (a) Structure of ligand L where R = $C_{10}H_{21,}$ L1 and L2 (b) Surface tension versus concentration for $[M1(L1R)(H_2O)_2]Cl_2(aq)$ (R = $C_{10}H_{21}$), where M = Ni^{II} (∇) or Cu^{II} ($\bigcirc$). Source: Adapted from Fallis et al. [5], Griffiths et al. [6] with permission from The Royal Society of Chemistry.

Largely, the metallosurfactants are formed with the metal inhabiting the head group or counter ion, and a relatively small number of examples have metal in the hydrophobic alkyl chain, which we will examine individually. In general, the addition of metal leads to a considerable decrease in the CMC values of metallosurfactants when compared to the traditional/conventional surfactant or ligand systems. This has been attributed to various factors contingent with the morphology of the metallosurfactants synthesized, which may include (i) increased hydrophobicity ascribed to the presence of two alkyl chains, i.e. more non-polar chains in case of double-chained surfactants, (ii) structural similarity to gemini surfactant, leading to lower CMC values [7], (iii) decrease in overall hydrophilicity and hydration of head group, (iv) decreased charge repulsion between head groups, etc. which have been discussed in the Sections 2.2.1 and 2.2.2 in detail. The conductivity and surface tension data obtained can be further utilized to evaluate the thermodynamics of micellization and calculate various adsorption parameters.

2.2.1 Thermodynamics of Micellization

Cmc variation with temperature is analyzed in terms of phase separation or equilibrium for micellization and delivers important information on the interactions between hydrophobic part and head group, specifically leading to the computation of several thermodynamic parameters of micellization such as β, counter ion binding, $\Delta G^\circ{}_{mic}$, Gibbs energy of micellization, $\Delta H^\circ{}_{mic,}$ enthalpy of micellization, and $\Delta S^\circ{}_{mic}$, entropy of micellization. The degree of micelle ionization or counter ion binding (β) for the complexes has been estimated from the ratio of slopes above and below CMC in the conductivity versus concentration plots. $\Delta G^\circ{}_{mic}$, $\Delta H^\circ{}_{mic}$, and $\Delta S^\circ{}_{mic}$ values are calculated with a view to explore the interactions during micellization, particularly, the influence of ligand coordination on surfactant properties. The standard Gibbs free energy of the micelle formation per mole of the monomer, $\Delta G^\circ{}_{mic}$, is given by Eq. (2.1)

$$\Delta G^\circ{}_{mic} = -RT(2-\beta)\ln X_{cmc} \tag{2.1}$$

where R, T, β, and X_{CMC} represent gas constant, absolute temperature, degree of micellar ionization, and CMC, in terms of mole fraction, respectively. Knowledge of temperature-dependent CMC and β values leads to the calculation of enthalpy of micellization, ΔH°_{mic} using the Gibbs–Helmholtz relation.

$$\Delta H^{\circ}_{mic} = -RT^2 \left(\frac{\partial \ln X_{cmc}}{\partial T} \right)_P (2 - \beta) \tag{2.2}$$

The entropy of the micellization process, ΔS°_{mic}, can be calculated using Eq. (2.3) as:

$$\Delta S^{\circ}_{mic} = \frac{\Delta H^{\circ}_{mic} - \Delta G^{\circ}_{mic}}{T} \tag{2.3}$$

2.2.2 Adsorption Parameters

While conductivity studies are usually undertaken to evaluate the thermodynamics of micellization, surface tension can be used to calculate some more adsorption parameters such as the area occupied per metallosurfactant molecule at the water/air interface and surface excess concentration. When the surfactant starts dissolving in water, the surfactant concentration at the interface or surface is much higher than the volume phase due to adsorption known as the surface excess concentration or maximum surface excess (Γ_{max}). This is also reflected in the surface tension values and calculated using the Gibbs adsorption isotherm equation as given below in the units of mol cm^{-2}:

$$\Gamma_{max} = -\frac{1}{2.303nRT} \left(\frac{\delta \gamma}{\delta \log C} \right)_T \tag{2.4}$$

where n = number of species in solution, T is absolute temperature, and R is universal gas constant. Γ_{max} is obtained by calculating the slope of the linear decrease of surface tension below the CMC in the surface tension versus log C (concentration) graph. Γ_{max} leads to the calculation of another important parameter, which is A or minimum area occupied by each surfactant molecule at the air–solvent interface. A provides useful information on the degree of packing and the orientation of the adsorbed surfactant molecule and is computed using Eq. (2.5) in the units of square angstrom:

$$A = \frac{10^{16}}{N \Gamma_{max}} \tag{2.5}$$

where N is Avogadro's constant and Γ_{max} is maximum surface excess in mol cm^{-2}. A detailed investigation of the metallosurfactants and their evaluation based on these numerical factors would be made in the subsequent section categorywise.

Different Types of Metal Surfactant Complexes/Metallosurfactants and Micellization

Category I Complexes – Metal Ion as Part of the Head Group The most common type of metallosurfactants are the ones where metal exists as an integral part of the head group. Griffiths' group has presented some interesting work over the years with metallosurfactants. They have pioneered the synthesis of metallosurfactants showing a strong coordination of metal ion to the head group. The complexes of the type $[M^{II}(L)(H_2O)_2]Cl_2$ prepared with different metal (M = Ni^{II}, Cu^{II}) keeping the ligand L based upon 1,4,7-triazacyclononane ([9]aneN_3) same (Figure 2.2a) have been utilized to check for the effect of metal ion change on micellization [5]. Using the surface tension, CMC of the Ni surfactant is found to be 1.5 ± 0.2 mM while for Cu the value is 0.8 ± 0.1 mM, lower than expected for a doubly charged amphiphile. The decrease in CMC is attributed to a decrease in overall hydrophilicity due to the presence of eight carbon centers in head group itself. A lower CMC value registered for Cu surfactant as compared to Ni was attributed to the tendency for Cu to adopt five-coordinate geometries resulting in a slightly smaller head-group area, lowering the CMC. However, the adsorbed surfactant amount and area per molecule at the surface remain same for both surfactants, 50 ± 5 Å^2, undeterred by a change in the metal ion. In a subsequent report, the head groups were changed slightly with a little variation in ligands keeping the metal ion, i.e. Cu constant resulting in the formation of $[Cu^{II}(L1)(H_2O)_2]C_{12}$ and $[Cu^{II}(L2)]Cl_2$ cationic surfactants. The ligands L1 and L2 have been depicted in Figure 2.2a.

$[Cu^{II}(L2)]Cl_2$ showed a higher CMC value of 5 mM compared to the 0.4 mM for $[Cu^{II}(L1)(H_2O)_2]C_{12}$, which stems from its greater ionic character as compared to the latter [6]. Scattering studies suggest that metal ions are distributed throughout the diffuse head-group region rather than being present as a thin shell, and the ionic strength of the surfactants is determined to be dependent on the concentration of unimeric surfactant and free chloride counter ions. The group also carried out an interesting study highlighting the effect of nature of counter ion and metal ion on the micellization and CMC values by synthesizing a series of complexes [8], i.e. $[M(L1)(H_2O)_n]X_2$ type with M = Cu, X = Cl having a CMC of 0.50 ± 0.06 mM; M = Cu, X = NO_3–0.80 ± 0.05 mM; M = Cu, X = acetate – 0.65 ± 0.03 mM; M = Cu, X = SO_4 – 0.45 ± 0.07 mM; M = Ni, X = Cl – 0.45 ± 0.06 mM, respectively. As reported, all the derivatives showed almost negligible variation in CMC with a change in metal ion or counter ion. This was supportive of the fact that the hydrophilic character and charge on the head group did not vary with counter ion that was further confirmed by the electronic spectra. But the CMC values registered quite a significant decrease when compared to the conventional dodecyl sulfate (8 mM) or dodecylammonium bromide (2 mM) surfactants, indicating a strong counter ion binding to the micelle. The addition of each CH_2 group in the hydrophobic tail also resulted in a decrease of approximately ×2 in the CMC values.

Valls et al. also reported a decrease in the CMC value of the synthesized palladium complexes in comparison to the corresponding amphiphilic phosphine ligands R-$(C_6H_4)(OCH_2CH_2)_nP(Ph)CH_2CH_2SO_3$-Na (R = *tert*-octyl, n = 1.4, 5.1, 11.2; R = n-nonyl, n = 1.6, 5.6, 11.4) and $RP(Ph)CH_2CH_2SO_3Na$ (R = n-octyl, $CH_3(OCH_2CH_2)_3$) by at least one order of magnitude (Table 2.1) [7]. Structural comparison between the synthesized Pd(II) complexes and gemini surfactants suggests that the palladium metal tends to act as a spacer between the two ligand molecules similar to the structure of conventional gemini surfactants, where two surfactants are linked by a spacer group and possess lower CMC values than the conventional surfactants. The values of area occupied per molecule in the interface were found to vary with the hydrophobic character, i.e. the palladium surfactants

Table 2.1 Critical micelle concentration (CMC) and area occupied per molecule adsorbed in the water/air interface (*A*) for ligands and their Pd(II) complexes.

Structure		Ligand (L)		Palladium complex ($PdCl_2L_2$)	
		CMC (M)	*A* (Å²)	CMC (M)	*A* (Å²)
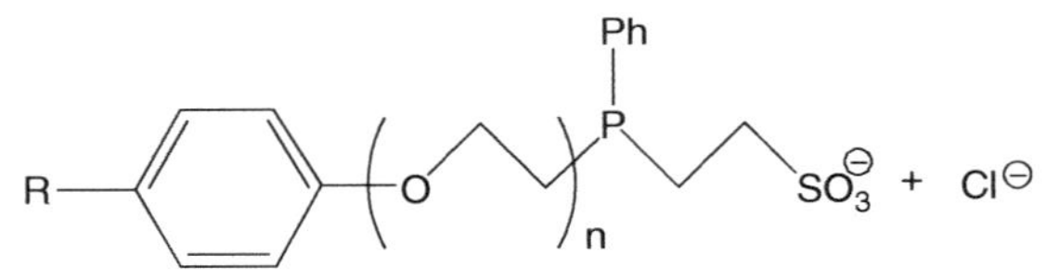 **7** (R = C_8H_{17} ; *n* = 1.4) **8** (R = C_8H_{17} ; *n* = 5.1) **9** (R = C_8H_{17} ; *n* = 11.2) **10** (R = C_9H_{19} ; *n* = 1.6) **11** (R = C_9H_{19} ; *n* = 5.6) **12** (R = C_9H_{19} ; *n* = 11.4)	7	1.0×10^{-4}	106		
	7b	9.5×10^{-4}	130		
	8	4.9×10^{-5}	122	5.4×10^{-6}	86
	9	2.4×10^{-5}	93	8.7×10^{-6}	171
	10	2.0×10^{-5}	87		
	11	7.0×10^{-5}	158	5.7×10^{-6}	54
	12	6.7×10^{-5}	197	2.2×10^{-5}	197
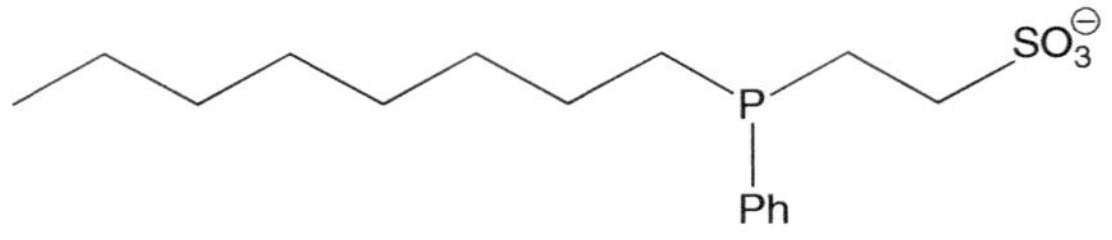	14	5.1×10^{-3}	147	1.7×10^{-4}	360
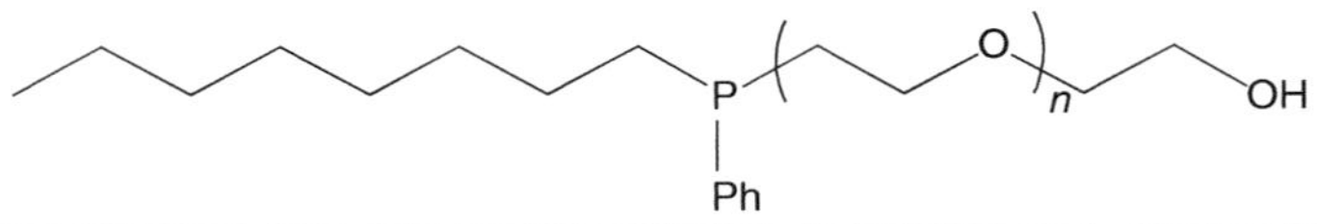	15	7.6×10^{-5}	30		

Source: Adapted with permission from Valls et al. [7]. Copyright (2002) American Chemical Society.

had higher adsorbed area per molecule in general owing to the more hydrophobic character. Comparing the structure of all the complexes, it was determined that the highly ethoxylated chains lead to larger separation between the hydrophobic chains, preventing the possible hydrophobic interactions required for compactness of the adsorbed complexes. Palladium complexes with ligands **8** and **11** as shown in Table 2.1 having moderate ethoxylation were more tightly packed at the water/air interface than those with ligands **9** and **12**, which are more ethoxylated, as indicated by the smaller Å^2 values, resulting in a higher concentration of the metal at the interface.

Quite interestingly, ligand **14** without a polyether chain in the structure shows a remarkably different behavior with the highest CMC value among the rest again attributed to the lesser hydrophobic character.

Kanappan Santhakumar's group has also been very actively involved in synthesizing and assessing the surface-active properties of metallosurfactants with metal serving as a part of the hydrophilic head group having the general structure as shown in Scheme 2.1. The complexes generated were of the type *cis*-α-[Cr(trien)AX]$(ClO_4)_2$, (X = F, Cl, Br; A = dodecyl/cetylamine), and these chromium(III)–surfactant complexes formed a coordinate bond between the head group and the hydrophobic tail resulting in a higher charge on the head group unlike conventional surfactant systems [9]. Their CMC values are lower in general compared to the conventional organic surfactants from which they are synthesized, i.e. dodecylammonium chloride (CMC = 1.5×10^{-2} mol dm^{-3}) when compared to the values given in Table 2.2.

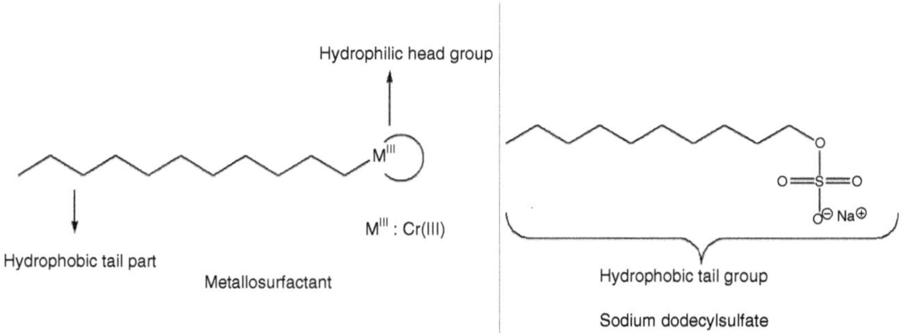

Scheme 2.1 Structure of the metal–surfactant complex or metallosurfactant compared to a classic surfactant. Source: Reprinted with permission from Kumaraguru et al. [9].

The effect of changing temperature and counter ions on the CMC was also evaluated. CMC increased with increasing temperature pertaining to two competitive effects: firstly, a decrease in hydration of the hydrophilic group, favoring micellization and, secondly, disruption of the water surrounding the hydrophobic group, which retards micellization. The change in counter ions from F^- to Br^- decreased the CMC owing to an increase in the size of ion present in the coordination sphere, making it more weakly hydrated and hence, readily adsorbed on the micellar surface. This decreases the charge repulsion between the polar groups and facilitates

Table 2.2 The CMC values of the various Cr(III) coordination complexes at 303 K.

Metallosurfactant	CMC (10^{-4} mol dm^{-3})	Metallosurfactant	CMC (10^{-4} mol dm^{-3})
cis-[Cr(en)$_2$(A)X](ClO$_4$)$_2$	45.82	*cis*-α-[Cr(trien)(A)X](ClO$_4$)$_2$	5.62
A = $C_{12}H_{25}NH_2$, X = F		A = $C_{12}H_{25}NH_2$, X = F	
A = $C_{12}H_{25}NH_2$, X = Cl	43.33	A = $C_{12}H_{25}NH_2$, X = Cl	5.37
A = $C_{12}H_{25}NH_2$, X = Br	42.11	A = $C_{12}H_{25}NH_2$, X = Br	5.09
A = $C_{16}H_{33}NH_2$, X = F	8.50	[a]A = $C_{16}H_{33}NH_2$, X = F	0.19
A = $C_{16}H_{33}NH_2$, X = Cl	7.43	[a]A = $C_{16}H_{33}NH_2$, X = Cl	0.18
A = $C_{16}H_{33}NH_2$, X = Br	6.31	[a]A = $C_{16}H_{33}NH_2$, X = Br	0.17

a) T is 298 K.

micellization. With the increase in the alkyl chain from dodecylamine to cetylamine on the polar head group, the CMC again shows a decrease due to an increase in hydrophobic character of the molecule in the coordination sphere. Thermodynamics of micellization further delivers valuable information on the head group and hydrophobic chain interactions taking place. The more negative $\Delta G°_{mic}$ values with increasing head group polarity indicate more favored micellization, while the negative values of $\Delta H°_{mic}$ depict exothermic nature of micellization suggesting the presence of London-dispersion forces in micelle formation. The positive values of $\Delta S°_{mic}$ reveal that the micellization is governed mainly by hydrophobic interactions between the surfactant cations resulting in the breakdown of the structured water surrounding the hydrophobic groups indicating that the micellization process is entropy driven confirming that increasing head group polarity favors micellization. Of all the surfactants synthesized, the type *cis*-α[Cr(trien)($C_{12}H_{25}NH_2$)X]$^{2+}$ (where trien = triethylenetetramine; X = F^-, Cl^-, Br^-) was also studied in various solvents such as *n*-alcohol and formamide at different temperatures [10]. Addition of alcohol leads to an increased penetration of formamide into the micellar interface and, consequently, a decrease in the CMC was registered for long-chain alcohols, i.e. *n*-propanol, *n*-butanol, *n*-pentanol, and *n*-hexanol due to this. The observations are considered an outcome of increased hydrophobic effect/solvophobic interactions, decreasing dielectric constant of the medium affecting interaction between polar group and medium, and increasing chain length of the alcohols. Exactly similar results were obtained when investigating Cr(III) complexes of the type *cis*-α-[Cr(trien)($C_{16}H_{33}NH_2$)X]$^{2+}$ (where trien = triethylenetetramine; X = F^-, Cl^-, Br^-) in *n*-alcohol and formamide at different temperatures. [11] Studies also revealed that the metallosurfactant systems could be further expanded by using other chelating agents such as 2,2-bipyridine (bpy) and 1,10-phenanthroline (phen) [12], and it works wonderfully by just changing the metal ion to cobalt as well [13, 14].

Wagay et al. investigated similar metallosurfactants or the ones with quite similar design, i.e. [Co(en)$_2$($C_{12}H_{25}NH_2$)Cl](NO$_3$)$_2$ *cis*-chlorobis(ethylenediamine) dodecylaminecobalt(III) nitrate (CDCN) with a CMC of 1.4 mmol kg^{-1}, which is quite low compared to the conventional surfactants owing to a reduction in the hydrophilicity of the head group ascribed to the presence of additional carbon atoms. The adsorption studies confirm A value to be 0.64 nm^2 and surface area of the bare CDCN head group as 0.14 nm^2, which suggests hydrogen bonding of the head group to water and electrostatic repulsion between the headgroups. Interestingly, a decrease in CMC was also registered upon substitution of nitro by chloride group in the coordination sphere of Co(III) [15]. In another report, the surfactant chlorobis(ethylenediamine)dodecylaminecobalt(III) perchlorate (*cis*-$[Co(en)_2(C_{12}H_{25}NH_2)Cl](ClO_4)_2$, (CDCP) prepared by Santhakumar et al. [16] was evaluated for adsorption and counter ion binding properties [17]. CDCP and CDCN have the same head group and γ_{CMC} values (27.0 mN m^{-1}), but still the CMC of CDCP (3.99 mM) is higher than that of CDCN (1.40 mM), which confirms that the CMCs of metallosurfactants with perchlorate and nitrate counter ions do not follow the trend as predicted by the positions of counter ions in the Hofmeister series. Therefore, binding of counter ions to the head groups of metallosurfactants at the micelle/solution interface is not fully controlled by the chaotropy/kosmotropy of the counter ions. This is also corroborated by the surface excess values of CDCP and CDCN, as the Γ_{max} of CDCP (2.2 μmol m^{-2}) is lower than that of CDCN (2.4 μmol m^{-2}); hence, perchlorate counter ion pushes less monomers of the metallosurfactant to the air/solution interface than nitrate counter ion. Thus, the presence of metal ion, higher electrical charge, and ethylene groups (hydrophobicity) in the head group also play a significant role in micellization process. The value of β determined is 0.8, which reveals that the number of counter ions binding to a micelle. Since the head groups have $a + 2$ charge and the counter ions have $a - 1$ charge, only 40% of micellar charge is neutralized by counter ion binding and the remaining 60% of the charge remains present on the metallomicelle. Hence, it can be inferred that the metal ion can lead to a significant change in the micellization process when present in the head group area by decreasing hydrophilicity and increasing hydrophobic interactions.

Type II Complexes: Metal Ion as Part of the Counter Ion Alkyl ammonium halides come across as excellent candidates to evaluate the effect of coordination of metals as counter ions in a metallosurfactant system. The quaternary ammonium group (C_nQAB) usually acts as the cationic surfactant with a metal entity as the counter ion. Keeping this in mind, Sharma et al. synthesized a series of supramolecular metallosurfactants with the formula: $[PtCl_4]^{2-}[C_nTA]_2{}^+$, varying the alkyl chain with different values of $n = 8, 10, 12, 14, 16$ and analyzed their effect on hydrophobicity, CMC, and various thermodynamic parameters [18]. The interplay between the ability of the metal ion to affect electrons and the length of the hydrophobic alkyl chains is responsible for a change in the CMC values, which are found to be lower for the metal complexes as compared to the conventional surfactants. Increasing alkyl chain length for the series results in a decrease in the CMC values as- ($C_{22}H_{52}C_{14}N_2Pt$ – 0.400 mM) < ($C_{26}H_{60}C_{14}N_2Pt$ – 0.372 mM) < ($C_{30}H_{68}Cl_4N_2Pt$ – 0.343 mM) < ($C_{34}H_{76}C_{14}N_2Pt$ – 0.332 mM) < ($C_{38}H_{84}Cl_4N_2Pt$ – 0.315 mM), attributed to an increased hydrophobic character. Weaker hydrophilic interactions and higher hydrophobicity induce a structured water effect and rearrange micellization at a lower concentration. The thermodynamic parameters calculated $\Delta G°_{mic}$, $\Delta H°_{mic}$, and $\Delta S°_{mic}$ help understand the micellization behavior with negative $\Delta G°_{mic}$ values indicating the spontaneity of micelle formation for the metal complexes and a decrease in the $\Delta G°_{mic}$ compared to C_nTAB attributed to the linkage between $[PtCl_4]^{2-}$and $[C_nTA]^+$. The negative values of $\Delta H°_{mic}$ predict an exothermic reaction due to the energy released by the disruption of the metal complex aggregate structures being held together by binding forces such as Van der Waals and London-dispersion forces upon micellization. Quite remarkably, $\Delta S°_{mic}$ showed negative values and unfavorable entropic changes, which could possibly be due to several factors, including the formation of solid-phase aggregates, less pronounced disordering of water molecules, or more hydrated head group compared to hydrophobic tail. In another study, biscetylpyridiniumtetrachloroplatinate (Pt-CPC) metallosurfactant was synthesized, keeping the counter ion same as $[PtCl_4]^{2-}$ but changing the cationic surfactant to cetylpyridinium chloride (CPC) [19]. The ionic bond between the two oppositely charged units coordinated into a quasi-1-D structure. As summarized in Table 2.3, the CMC is lower and $\Delta G°_{mic}$ values more negative for Pt-CPC than CPC depicting the spontaneity of micellization, whereas the negative $\Delta S°_{mic}$ for Pt-CPC illustrates unfavorable entropic changes for micellization. Therefore, the micellization process ensues with an interaction of both $\Delta H°_{mic}$ and $\Delta S°_{mic}$.

Wagay et al. also synthesized metallosurfactants with metal in the counter ion moiety, i.e. tetradecylpyridinium (TP)-based metallosurfactants, $TP_2[MCl_4]$ (M = Mn, Co, Ni, Cu, Zn) [20]. The adsorption and aggregation properties of these surfactants revealed a decrease in CMC value (1.116–1.137 mmol kg^{-1}) by coordinating metal to the counter ion when compared to a conventional surfactant. CMC also exhibits a decrease with increasing hydrocarbon chain length, but the extent of decrease in CMC per CH_2 group increase becomes less pronounced in the presence of metal ion. Metal ion also causes a reduction in the counter ion binding constant (0.33–0.35) by almost 50%, although the surface parameters registered are more or less the same with surface excess values varying from 0.64 to 0.68 μmol m^{-2} and the A values ranging between 2.45 to 2.59 nm^2.

Similar type of metallosurfactants, i.e. $[M(CH_3COO)_4]\ [C_{12}H_{25}NH_3]_2$, have been investigated by our group keeping the hydrophobic dodecyl chain same and just varying the metal ion M as cobalt, iron, and zinc [21]. The metallosurfactants dissociate into surface-active molecules and ions upon dissolution in water, i.e. the ammonium head group along with dodecyl chain and the counter ion with metal.

Table 2.3 Comparison of CMC values and thermodynamic parameters for Pt-CPC and CPC surfactants.

Complex	Cmc	B	$\Delta G°_{mic}$	$\Delta H°_{mic}$	$\Delta S°_{mic}$
Pt-CPC	0.32	0.764	−21.854	−38.45	−0.0557
CPC	2.38	0.705	−12.828	−3.21	0.0322

The $\Delta G°_{mic}$ values for the metal surfactant complexes were negative, indicating spontaneous and thermodynamically favored micellization process. Conventionally, for amphoteric and ionic surfactants, $\Delta G°_{mic}$ values range between −23 and −42 kJ mol^{-1} at 298.15 K, and the free energy values for the metal surfactant complexes show classical surfactant behavior, with the negative $\Delta G°_{mic}$ as 34.11, 30.09, and 32.40 kJ mol^{-1} for Zn, Co, and Fe, respectively. $\Delta S°_{mic}$ values obtained were negative for these metallosurfactants as well indicating head groups were more restricted in the micelle, and the surface formed by the surfactants became more ordered with micellization. The micellization process was therefore overall enthalpy driven, and the thermodynamics of micelle formation were governed by hydrogen bonding and Van der Waals interactions. An interesting observation was that CMC values varied with the change in metal ion, following the order: Zn (0.25 mM) < Fe (0.80 mM) < Co (0.85 mM) for different metals. This was due to an interplay of metal ion size as well as ionic character as the CMC values were found to increase with the hydrated metal ion size in general. In another report, a copper surfactant complex $[Cu(BrCl_2)(C_{12}H_{25}N(CH_3)_3)]$ was synthesized, and its aggregation behavior and adsorption pattern was evaluated revealing important information about the molecular organization and counter ion distribution at the interface and in bulk of the solution. The X-ray absorption fine structure technique was applied to the copper surfactant solution under both total reflection (total-reflection X-ray absorption fine structure [TRXAFS]) and transmission conditions wherein the Br and Cu K-edge absorption demonstrated the presence of Br^- ions and the absence of Cu^{2+} ions at the interface and in the inner core of micelles (Figure 2.3). Metallosurfactant dissociated, and the surfactant ion, Cl^- and Br^- ions occupied the interface while the copper cation went into the bulk, evident by the absence of a copper ion peak in X-ray absorption fine structure (XAFS) spectrum and the negative surface density of copper at the air–water interface. Furthermore, Br^- ions existed in two solvation states, designated as "free-Br" and "bound-Br" with generally bound-Br state at the interface forming ion pairs. The study helped to develop an understanding of the ion organization at the interface and in the bulk. The presence of metal in the counter ion indeed affects the micellization process and quite remarkably resulted in negative entropy values suggesting that the process remains enthalpy driven.

Type III Complexes: Metal Ion as Part of the Hydrophobic Chain The research group of Prof. Suades has done pioneering work in this area working with a family of linear surfactant phosphines $Ph_2P(CH_2)_nSO_3Na$ (1(a–c)) and utilizing them to synthesize metallosurfactants by reacting with metal moieties such as $PtCl_4$, $[Mo(CO)_5]$, and $[Mo(CO)_4]$ (Table 2.4) [22, 23]. Quite interestingly, the metal moiety is bonded to the phosphorus atom at the end of the hydrocarbon chain generating the hydrophobic part, and the polar head group consists of the $-SO_3Na$ moiety, using the organometallic approach. The study employs three different phosphine ligands with an otherwise similar structure but just the change in the number of carbon atoms. The CMC value of the metallosurfactants gets considerably reduced after the coordination of surfactant phosphine ligand to the rigid non-polar metal group ascribed to (i) the

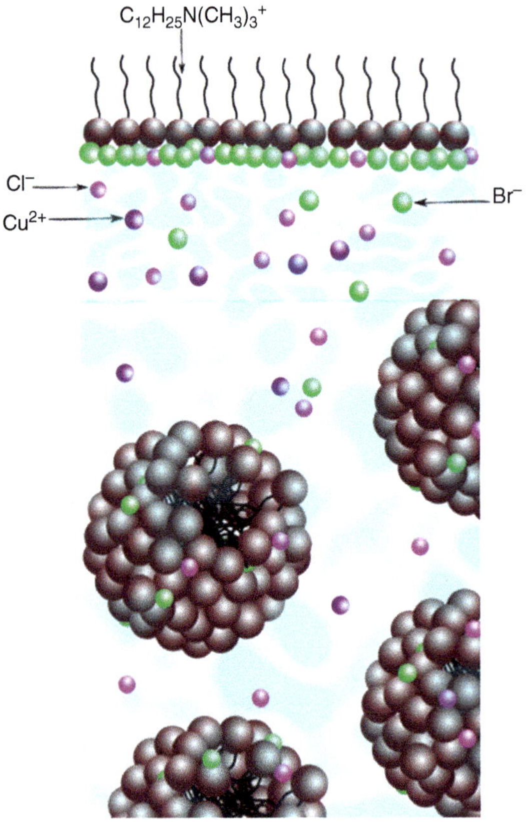

Figure 2.3 Proposed organization of copper surfactant complex (a) at the air–water interface and (b) in the bulk aqueous solution.

structural similarity between the synthesized surfactants and gemini surfactants or bolamphiphiles, which are less efficient at reducing surface tension [24]. (ii) increase in the hydrophobicity because of the addition of the lipophilic fragments such as $[Mo(CO)_5]$.

Not much difference in the values of *A* is observed for phosphines (1(a–c)) assuming a similar packing for all the aggregates with the hydrophilic sulfonate group oriented toward water and the lipophilic $(CH_2)_n$-PPh_2 group oriented toward air in an extended chain conformation. However, the values are quite different for pentacarbonyl (2(a–c)) and tetracarbonyl (3(a–c)) complexes, thanks to the existence of the voluminous hydrophobic metal carbonyl group. With an increment in the chain length of the complexes and the cross-sectional area of the hydrophobic part of the molecules, a progressive decrease in the molecular packing is also registered. Going from 3(b) to 3(c), there is a large increment in the value of *A* consistent with a double-loop conformation of the complex at the interface as depicted in Scheme 2.2 below [25].

In another study for the *cis*-$[PtCl_2L_2]$ (4(a–c)) complexes, CMC values are substantially reduced by a factor of 5 for 4(a,b) and nearly 10 for4 (c), which is comparable to the behavior shown by bolaform surfactants (Table 2.4). Hence, comparing the structure of 4(a–c) with gemini surfactants, $PtCl_2$ fragment acts essentially

Table 2.4 Calculated parameters from surface tension measurements: critical micelle concentration (CMC), surface excess concentration (*G*), estimated area occupied per molecule adsorbed in the water/air interface via Gibbs equation (*A*).

Complex	CMC(mM)	Γ_{max} (mol cm^{-2})	*A* (Å^2)
1a L1 = $Ph_2P(CH_2)_nSO_3Na$ ($n = 2$)	14	$(1.7 \pm 0.1) \times 10^{-10}$	99
1b L2 ($n = 6$)	4.0	$(1.64 \pm 0.03) \times 10^{-10}$	101
1c L3 ($n = 10$)	0.5–1.2	$(1.6 \pm 0.2) \times 10^{-10}$	100
2a	2.0	$(1.75 \pm 0.04) \times 10^{-10}$	95
2b	1.2	$(9.9 \pm 0.3) \times 10^{-11}$	167
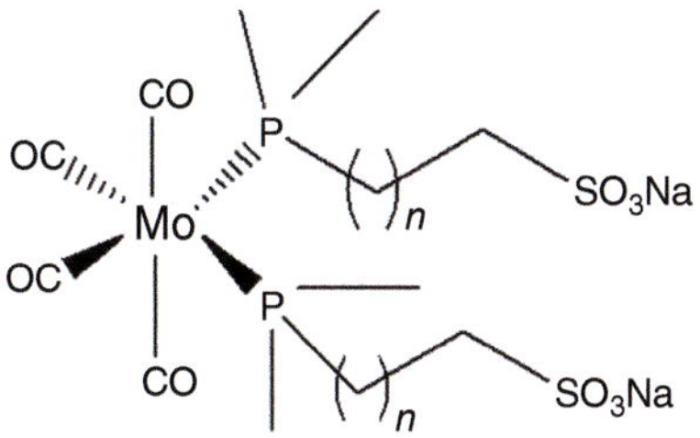			
2c	0.15	$(8.9 \pm 0.8) \times 10^{-11}$	190
3a $Mo(CO)_4L_2$	1.9	$(1.2 \pm 0.1) \times 10^{-10}$	140
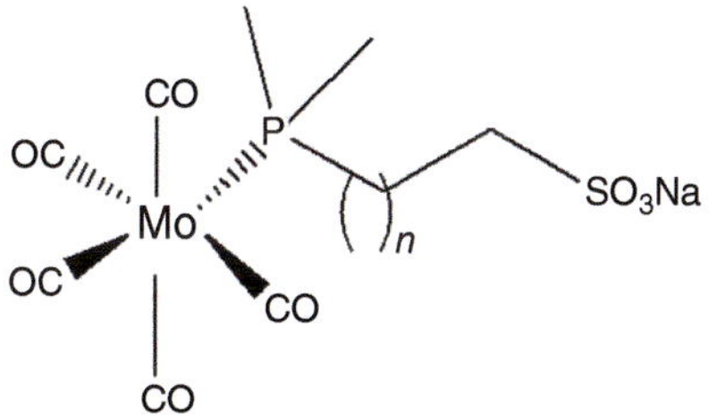			
3b	0.84	$(8.7 \pm 0.3) \times 10^{-11}$	192
3c	0.28	$(4.9 \pm 0.4) \times 10^{-11}$	340
4a *cis*-$[PtCl_2(L1)_2]$	2.9	$(6.9 \pm 0.6) \times 10^{-11}$	240
4b	0.94	$(7.7 \pm 0.9) \times 10^{-11}$	220
4c	0.062	$(4.7 \pm 0.4) \times 10^{-11}$	350

Source: Adapted from Parera et al. [22, 23] with permission from The Royal Society of Chemistry.

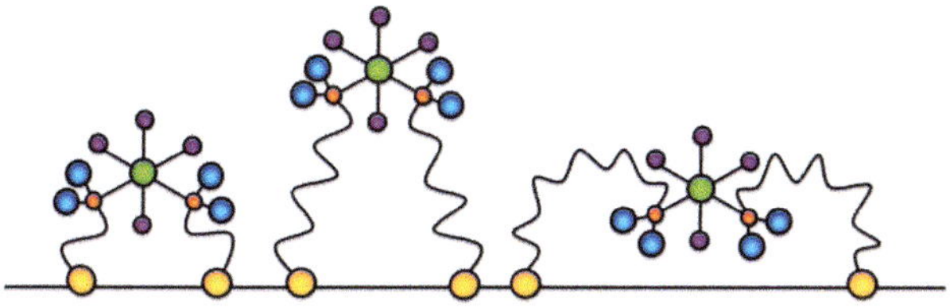

Scheme 2.2 Schematic representations of complexes in the interface showing the double-loop conformation for 4(c) (sulfonate groups are represented by yellow balls). Source: Adapted from Parera et al. [23] with permission from The Royal Society of Chemistry.

as a linker or spacer between the two hydrophobic chains, like bolaform/gemini surfactants. The results evidence that *A* value for platinum complexes is markedly higher than those of ligands L1–L3, which is in fact similar to the organic bolaform disulfonate behavior displaying larger areas than conventional monomeric sulfonates. Quite remarkably, the area per molecule occupied by 4(c) is substantially larger than 4(a) and 4(b), which points toward a different packing and structural organization specifically, a double-loop conformation. This organization is further stabilized by the presence of hydrogen bonds between the chlorine atoms bonded to the metal and water molecules. Therefore, the metal present as a part of the hydrophobic chain essentially leads to the formation of structures identical to gemini surfactants, and the thermodynamic parameters obtained follow a similar pattern as well.

2.3 Adsorption of Surfactant Monolayers

In addition to the usual micellization and aggregate formation in the bulk, surfactants tend to adsorb and form monolayers at the air water interface. Sometimes, aggregates are also observed at the air water interface, and it is worthwhile to look at the adsorption layer formed by metallosurfactants. After the surfactant adsorption, the monolayer formed at the interface can easily get transferred to a solid support yielding the Langmuir Blodgett (LB) films with a vertical dipping and Langmuir Schaefer (LS) films with a horizontal dip in the liquid. With each immersion, a uniform monolayer with a very accurate thickness is homogeneously adsorbed on the solid substrate. The monolayer formation and metallosurfactant adsorption at the interface also govern the surface tension behavior. The surface solutions of some $[Ru(bipy)_2(bipy')][Cl]_2$ complexes where (bipy′) = 5,5′-dialkyl-2,2′-bipyridine have been investigated to disclose some interesting aggregation and surface behavior by visualizing them to be structurally equivalent to gemini surfactants. The surface of aqueous solutions of $[Ru(bipy)_3][Cl]_2$-based double-chained surfactants has been studied using neutron reflectometry, and it is observed that for shorter chain lengths, i.e. $Ru_2{}^5C_{13}$ (n = 13) (CMC = 0.30 mmol dm^{-3}), studied in the range 0.01–1 mmol dm^{-3}, the surfactant adsorption takes place relatively quickly showing a maximum surface coverage of 100 Å^2 molecule^{-1}. It is comparable to the expected headgroup area of the molecule 95 Å^2 and *A* value from surface tension, i.e. 110 Å^2. However, for longer chains with n = 19, i.e. $Ru_2{}^5C_{19}$ (CMC = 0.03 mmol dm^{-3}), studied in the range 0.3–0.03 mmol dm^{-3} strong time dependence of adsorption is observed with solutions taking up to 12 hours to attain an equilibrium surface layer, arising from slow adsorption rate pertaining to alkyl chain length and diffusion limited adsorption rather than low monomer concentration. The structure obtained at the air solution interface is quite dissimilar to the interface of micelle with the solution. Also, the area per molecule values reveals that the adsorbed film possesses more than one headgroup per nominal adsorption site; and for concentrations $c >$ CMC two headgroups per nominal adsorption site exist for each monolayer adsorption suggesting a more complex and highly corrugated surface structure [26, 27].

The group has further extended their studies to check the effect of alkyl chain length and its orientation on the structure of adsorbed films [28]. The surface behavior of a range of $Ru^{p}{}_{q}C_{n}$ metallosurfactants has been reported as a function of p, substitution on bipyridine ligand (4 or 5), and q, number of substituted alkyl chains (1 or 2) for $n = 19$. The adsorption of the single-chain $Ru_1{}^4C_{19}$ and $Ru_1{}^5C_{19}$ surfactants is very slow and strongly time-dependent, taking more than 10 hours to form equilibrium films, which is probably related to the alkyl chain length rather than the low solution monomer concentration. Below the CMC, $Ru_1{}^5C_{19}$ film is denser compared to $Ru_1{}^4C_{19}$, whereas at concentrations close to and greater than CMC the inverse takes place. Close to the CMC, the average area per molecule in the adsorbed film (30 Å^2) is significantly lower than the nominal headgroup area of the surfactants (100 Å^2) by approximately a factor of 3, which implies that the surface aggregates might be the adsorbing units in this case. Although as discussed above, the adsorption of the double-chain surfactant $Ru_2{}^4C_{19}$ is weak and independent of time attributed to the orientation of the alkyl chains.

Carmona et al. formed mixed Langmuir monolayers employing the cationic metallosurfactant bis(2-phenylpyridine)(4,4′-diheptadecyl-2,2′-bipyridine)-iridium(III) chloride (Ir-complex) and the anionic tetrakis(4-sulfonatophenyl)porphyrin (TSPP) in 4 : 1 M ratio by co-spreading method at the air–water interface [29]. The homogeneous monolayer was formed using the TSPP molecules attached electrostatically underneath the amphiphilic Ir-complex matrix from low to high surface pressures. J-aggregates were formed by TSPP and Ir-complex molecules. Similarly, Langmuir monolayers containing the pure Ir-complex were also prepared for comparison. Finally, Langmuir–Schaefer (LS) films have been obtained here by successful transfer of the Ir-complex: TSPP (4 : 1) monolayers onto quartz substrates. Gonawala *et al.* assessed the amphiphilic properties of salophen-based metallosurfactants – $[Fe^{III}(LN_2O_2)Cl]$ and $[Zn^{II}(LN_2O_2)H_2O]$, using Brewster angle microscopy (BAM) and isothermal compression data (Figure 2.4) [30]. BAM images revealed a homogeneous film formation between the surface pressure of 10 and

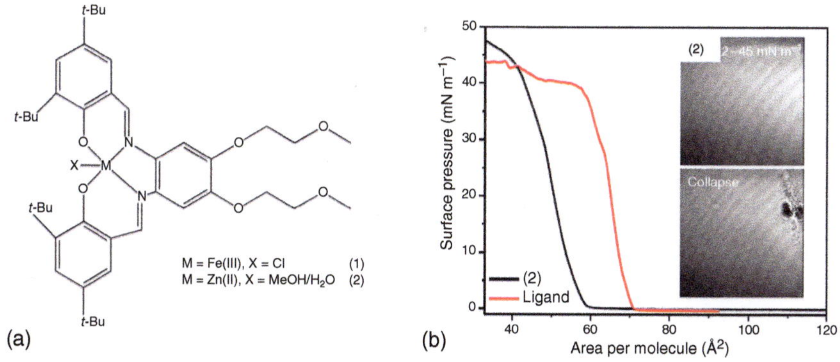

Figure 2.4 (a) The structure for redox active, amphiphilic metal complexes (b) Compression isotherm data of Zn complex and the ligand with their selected BAM micrographs. Source: Adapted from Gonawala et al. [30]/with permission from The Royal Society of Chemistry.

44 mN m^{-1} for the Zn complex. A limited film formation with a molecular area of 71–75 Å^2 and apparent collapse at 39 mN m^{-1} and was observed for the ligand. At the highest surface pressure of 45 mN m^{-1}, a critical area of 60–65 Å^2 molecule^{-1} was observed. The LB films formed were deposited onto mica substrates and successfully employed for corrosion mitigation.

2.4 Conclusions

The adsorption and micellization behavior of metallosurfactants at the interfaces and in the bulk solution is vital to the designing of various hybrid surfactant materials keeping in view their potential applications. Given the flexibility proffered by metallosurfactants in terms of molecular design, it is pertinent to investigate the micellization and surface parameters to modify the architecture of these surfactants such as the head group size, hydrophobic chain length, and orientation as desired. Micellization and thermodynamic parameters evaluated for the metallosurfactants confirm that the nature and placement of metal ion, in the hydrophobic chain, head group, or counter ion area impact the micellization characteristics as well by changing the hydrophobic character, hydrophilic nature, and polarity of the head groups. In general, all the metallosurfactants show a decreased CMC value and increase in micellization compared to conventional surfactants, which can be further fine-tuned depending upon the properties required. Also, the adsorbed surfactant films on the air water interface especially incorporated with the metal ion can be further used to create LS films and LB films on a solid substrate for numerous prospective uses, which opens a whole new world of applications. This is definitely one subject area of metallosurfactants, which needs to be investigated further and can be successfully employed to get tailor-made surfactants.

References

1 Vincent, B. (2014). Mcbain and the centenary of the micelle. *Advances in Colloid and Interface Science* 203: 51–54.

2 Edser, C. (2020). Surfactants versus COVID-19. *Focus on Surfactants* 7: 1–2.

3 Griffiths, P.C., Fallis, I.A., Tatchell, T. et al. (2008). Aqueous solutions of transition metal containing micelles. *Advances in Colloid and Interface Science* 144: 13–23.

4 Griffiths, P.C., Fallis, I.A., Chuenpratoom, T., and Watanesk, R. (2006). Metallosurfactants: interfaces and micelles. *Advances in Colloid and Interface Science* 122: 107–117.

5 Fallis, I.A., Griffiths, P.C., Griffiths, P.M. et al. (1998). Solid state and solution behaviour of novel transition metal containing surfactants. *Chemical Communications*: 665–667.

6 Griffiths, P.C., Fallis, I.A., Willock, D.J. et al. (2004). The structure of metallomicelles. *Chemistry A European Journal* 10: 2022–2028.

7 Valls, E., Solsona, A., Suades, J. et al. (2002). Synthesis and characterization of new amphiphilic phosphines and palladium metallosurfactants. *Organometallics* 21: 2473–2480.
8 Griffiths, P.C., Fallis, I.A., James, C. et al. (2010). Structure–property relationships in metallosurfactants. *Soft Matter* 6: 1981–1989.
9 Kumaraguru, N., Arunachalam, S., Arumugham, M.N., and Santhakumar, K. (2006). Metallosurfactants of chromium(III) coordination complexes. Synthesis, characterization and determination of CMC values. *Transition Metal Chemistry* 31: 250–255.
10 Kumaraguru, N., Santhakumar, K., and Kalyanasundharam, S. (2011). Studies on thermodynamics of micellization and solvophobic interactions of novel surfactant–Cr(III) complexes in non-aqueous solvents. *Journal of Solution Chemistry* 40: 1673–1686.
11 Kumaraguru, N. and Santhakumar, K. (2009). Studies on hydrophobic effect and micelle formation of some surfactant–Cr(III)-cetylamine complexes in non-aqueous solvents. *Journal of Solution Chemistry* 38: 629–640.
12 Kumaraguru, N. and Santhakumar, K. (2009). Synthesis, characterization, critical micelle concentration determination, and antimicrobial studies of some complexes of chromium(III) metallosurfactants. *Journal of Coordination Chemistry* 62 (21): 3500–3511.
13 Chandar, S.C.N., Sangeetha, D., and Arumugham, M.N. (2011). Micellization of metallosurfactant *N*-dodecyl/hexadecyl/octadecyl salicylaldimine cobalt(III) complexes in nonaqueous media. *Journal of Solution Chemistry* 40: 608–620.
14 Santhakumar, K., Kumaraguru, N., Arunachalam, S., and Arumugham, M.N. (2007). Thermodynamics and micellar properties of some surface active cobalt(III) metallosurfactants in nonaqueous medium. *International Journal of Chemical Kinetics* 39 (1): 22–31.
15 Wagay, T.A., Dey, J., Kumar, S. et al. (2016). Aggregation, adsorption, counterion binding, thermal and scattering behavior of metallosurfactant *cis*-$[Co(en)_2(C_{12}H_{25}NH_2)Cl](NO_3)_2$. *Colloids and Surfaces A: Physicochemical Engineering Aspects* 503: 61–69.
16 Santhakumar, K., Kumaraguru, N., Arumugham, M.N., and Arunachalam, S. (2006). Metallomicelles of Co(III) coordination complexes – synthesis, characterization and determination of CMC values. *Polyhedron* 25: 1507–1513.
17 Wagay, T.A. and Ismail, K. (2017). Thermal, aggregation, counterion binding, light scattering, and adsorption Behavior of *cis*-chlorobis(ethylenediamine) dodecylaminecobalt(III) perchlorate metallosurfactants in aqueous sodium perchlorate medium. *Colloid and Polymer Science* 295: 1685–1694.
18 Sharma, N.K., Singh, M., and Bhattarai, A. (2016). Hydrophobic study of increasing alkyl chain length of platinum surfactant complexes: synthesis, characterization, micellization, thermodynamics, thermogravimetrics and surface morphology. *RSC Advances* 6: 90607–90623.
19 Sharma, N.K. and Singh, M. (2018). New class of platinum based metallosurfactant: synthesis, micellization, surface, thermal modelling and in vitro biological properties. *Journal of Molecular Liquids* 268: 55–65.

20 Wagay, T.A., Ismail, K., and Askari, H. (2020). Assessment of the aggrega tion and adsorption behavior of newly synthesized tetradecylpyridinium-based metallosurfactants and their interaction with bovine serum albumin. *New Journal of Chemistry* 44: 15018–15030.

21 Mehta, S.K. and Kaur, R. (2013). Self-aggregation and solution behavior of synthesized organo transition metal (Co, Fe, Zn) amphiphilic complexes. *Journal of Colloid and Interface Science* 393: 219–227.

22 Parera, E., Comelles, F., Barnadas, R., and Suades, J. (2010). Influence of metal coordination on the critical micelle concentration and aggregation properties. New surfactant phosphine ligands and platinum(II) metallosurfactants. *Langmuir* 26 (2): 743–751.

23 Parera, E., Comelles, F., Barnadas, R., and Suades, J. (2011). Formation of vesicles with an organometallic amphiphile bilayer by supramolecular arrangement of metal carbonyl metallosurfactants. *Chemical Communications* 47: 4460–4462.

24 Eastoe, J., Rogers, S.E., Martin, L.J. et al. (2006). Fluorosurfactants at structural extremes: adsorption and aggregation. *Langmuir* 22: 2034–2038.

25 Parera, E., García, M.M., Pons, R. et al. (2016). Supramolecular arrangement of molybdenum carbonyl metallosurfactants with co-releasing properties. *Organometallics* 35 (4): 484–493.

26 Bowers, J., Danks, M.J., and Bruce, D.W. (2003). Surface and aggregation behavior of aqueous solutions of Ru(II) metallosurfactants: 1. Micellization of $[Ru(bipy)_2(bipy')][Cl]_2$ complexes. *Langmuir* 19: 292–298.

27 Bowers, J., Danks, M.J., and Bruce, D.W. (2003). Surface and aggregation behavior of aqueous solutions of Ru(II) metallosurfactants: 2. Adsorbed films of $[Ru(bipy)_2(bipy')][Cl]_2$ complexes. *Langmuir* 19: 299–305.

28 Bowers, J., Amos, K.E., and Bruce, D.W. (2005). Surface and aggregation behavior of aqueous solutions of Ru(II) metallosurfactants. 3. Effect of chain number and orientation on the structure of adsorbed films of $[Ru(bipy)_2(bipy')]Cl_2$ complexes. *Langmuir* 21: 1346–1353.

29 Carmona, C.R., Delgado, A.M.G., Martınez, A.G. et al. (2011). Molecular organization and effective energy transfer in iridium metallosurfactant–porphyrin assemblies embedded in Langmuir–Schaefer films. *Physical Chemistry Chemical Physics* 13: 2834–2841.

30 Gonawala, S., Leopoldino, V.R., Kpogo, K., and Verani, C.N. (2016). Langmuir–blodgett films of salophen-based metallosurfactants as surface pretreatment coatings for corrosion mitigation. *Chemical Communications* 52: 11155–11158.

3

Metallosurfactant Self-Assembly: Structures and Chemistry of Interfacial, Biphasic, and Phase Transfer Catalysis

Ravneet Kaur[1] and Aashima Sharma[2]

[1] *Ivy Tech Community College of Indiana – Valparaiso Campus, Life and Physical Science Department, 3100 Ivy Tech Drive, Valparaiso, Indiana, 46383, USA*
[2] *Panjab University, Department of Chemistry, Centre for Advanced Studies in Chemistry, Sector 14, Chandigarh, 160014, India*

3.1 Introduction

Metallosurfactants form the most intriguing class of surfactant family engaging the metallic and amphiphilic features in one single entity. Metallosurfactants were first discovered by Satake et al. in 1962 who synthesized bivalent metal alkyl sulfates and analyzed their micellar properties [1], while the actual reference to the term "metallosurfactant (MS)" or "metallomicelles" did not surface until the 1990s [2, 3]. Metallosurfactants are composed of metal ions, which are usually the transition metals linked, chelated, coordinated, or bonded to the ligands having primarily an alkyl chain skeleton. Supramolecular aggregates of the metallosurfactants offer great opportunity to researchers concentrating metal within the self-assemblies. As a result, they find numerous prospective applications as sensory ensembles [4], catalytic sites [5], magnetic resonance imaging (MRI) contrast agents [6], nanoreactors for nanoparticle fabrication [7, 8], drug-delivery vehicles [9], and molecular machines for the formation of more complex hybrid materials such as porous silica material and metallosomes [10]. These self-assembled structures are pretty important and they have recently been used in fighting the deadly coronavirus/COVID 2019. Nanomicelles loaded with curcumin have been administered to coronavirus disease of 2019 (COVID 19) patients, which acts as a therapeutic supplement [11]. In another study by Alavi et al., various self-assembled aggregates have been decorated with lecitin protein, which binds to the carbohydrate component of the virus and emerges as a promising strategy for targeting the coronavirus [12]. Owing to the wide application range, it becomes pertinent to look into the chemistry and morphological details of the self-assembled metallomicellar structures formed and to identify the positioning of metal and hydrophobic alkyl chain in the aggregates along with other cations and anions. The basic idea of integrating a metal ion into the surfactant molecule provides a facile means of localizing unique properties at the interface of two regions, rendering the information quite useful to further explore their potential

Metallosurfactants: From Fundamentals to Catalytic and Biomedical Applications, First Edition.
Edited by Surinder K. Mehta and Ravneet Kaur.

applications. As a result, an attempt has been made in the present chapter to discuss the range of self-assembled aggregates formed by the metallosurfactants, including but not limited to micelles, reverse micelles, worm-like micelles, vesicles, bilayers, and lamellar structures in the first section. The second part has been devoted to elaborate the applicability and role of these self-assembly structures towards different natures of catalysis.

3.2 Self-Aggregation Behavior

The metal surfactant complexes or metallosurfactants show conventional surfactant behavior as they have an inherent tendency for aggregation. These surfactants act in a similar fashion by reducing surface tension of water and self-assembling to form various micellar morphologies depending primarily on the type of ligands attached and the head group, which will be examined in this chapter. Pileni and group were among the very first researchers to work extensively on metal–sodium bis(2-ethylhexyl)sulfosuccinate (AOT) surfactants and chalk out a phase diagram for $Cu(AOT)_2$ surfactants depicting various colloidal structures. They showed the complete evolution of structures from reverse micelles to bicontinuous systems first by the generation of interconnected cylinders followed subsequently by a lamellar phase [13–16]. It becomes really important to look into the self-assembly structures as their structure greatly affects their applications. Kaur and Mehta illustrated that the metallomicelles determined the morphology of the metal oxide nanostructures, hence obtained [17]. The various self-aggregating assemblies possible for metallosurfactants have been discussed here.

3.2.1 Micelles

The simplest self-assembly structure possible is a micelle. Micelles are quite a rage owing to the emergence of cult favorite "micellar water" in beauty industry as well where micelles have been advertised as little balls of o/w, which lift dirt and grime out of the skin like magnet. Griffiths et al. have investigated these micelles in detail, actually "metallomicelles" in the present context (Figure 3.1) using scattering techniques such as small-angle X-ray scattering (SAXS) and small-angle neutron scattering (SANS) for two novel copper metallosurfactants [18]. The effect of surfactant head-group structure on the micelle morphology was determined by synthesizing two Cu(II) surfactants by reacting the ligands: 1-(2-hydroxytetradecyl)-1,4,7-triazacyclononane and 1-(2-hydroxytetradecyl)-1,4,7,10-tetraazacyclododecane with $CuCl_2{\cdot}2H_2O$ to afford $[CuII(1)(H_2O)_2]Cl_2$ and $[CuII(2)]Cl_2$ surfactants depicted as $[Cu(1)]Cl_2$ and $[Cu(2)]Cl_2$, respectively, in Figure 3.2.

The $[CuII(1)(H_2O)_2]Cl_2$, micelles were found to be prolate ellipses, rod-like while $[CuII(2)]Cl_2$ micelles were consistent with an oblate ellipse, which was attributed to a greater ionic character of $[Cu(2)]Cl_2$ as compared to $[Cu(1)(H_2O)_2]Cl_2$ resulting in an increase in the degree of ellipticity, and the micelles becoming less disc-like

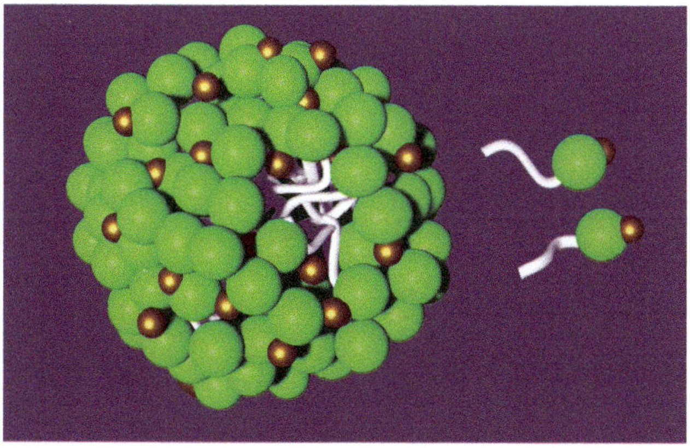

Figure 3.1 Representative structure of a metallomicelle (Green sphere: head group, golden sphere: metal; gray: tail).

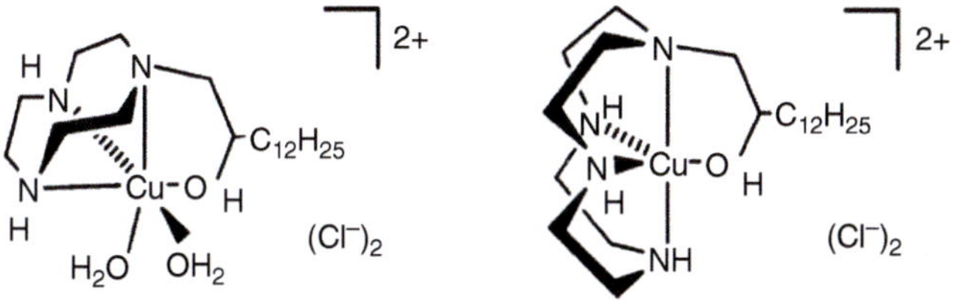

[Cu(**1**)]Cl_2 [Cu(**2**)]Cl_2

Figure 3.2 Structures of synthesized [Cu(**1**)]Cl_2 and [Cu(**2**)]Cl_2 metallosurfactants. Source: Adapted from Ref. Griffiths et al. [18] with permission from The Royal Society of Chemistry.

(oblate). Therefore, only a slight modification in the head group morphology yielded a different micellar structure. Guerrero-Martinez et al. while working on ruthenium and iridium metallosurfactants also revealed the formation of similar micelles with both the metallosurfactants containing same hydrophobic unit, i.e. dialkylbipyridine, with 17 methylene units [19]. The two surfactants bis(2,2′-bipyridine)(4,4′-diheptadecyl-2,2′-bipyridine)ruthenium(II) dichloride and bis[2-(2,4-difluorophenyl)pyridine](4,4′-diheptadecyl-2,2′-bipyridine)iridium (III) chloride made up of analogous polar transition-metal complex headgroup and hydrophobic tail resulted in no alteration toward micelle morphology. Very interestingly, the two compounds readily formed mixed-aggregate systems exhibiting excellent electronic energy transfer process from excited iridium complex to energy acceptor ruthenium metallosurfactant as illustrated in Scheme 3.1.

Concomitantly, Taira et al. presented outstanding results on Pd and Pt amphiphilic complexes obtained by a reaction of *N*-alkylbipyridinium [4,4′-bpy-*N*-$(CH_2)_{10}OC_6H_3$-3,5-$(OMe)_2$] [NO_3] with [M$(NO_3)_2$(en)] (M = Pd, Pt) [20]. The double-tailed [M{4,4′-bpy-*N*-$(CH_2)_{10}OC_6H_3$-3,5-$(OMe)_2$}$_2$(en)]-[NO_3]$_4$ Pd and Pt amphiphilic bis(bipyridinium)complexes, expressed as [2][NO_3]$_4$, M = Pd and [3][NO_3]$_4$,

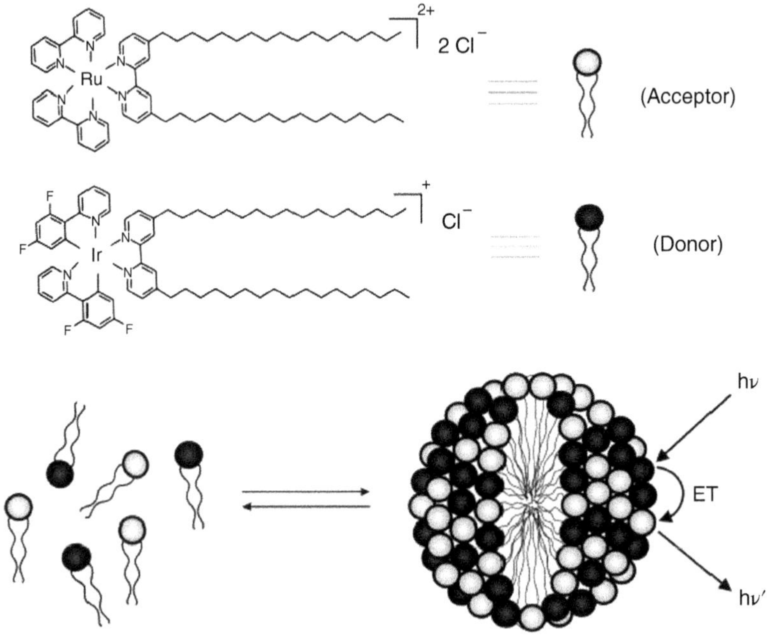

Scheme 3.1 Schematic representation of ruthenium and iridium metallosurfactants and their self-aggregation into mixed aggregates. Source: Reproduced with permission from Guerrero-Martınez et al. [19]. Copyright (2008) American Chemical Society.

M = Pt respectively (Figure 3.3), resulted in the formation of spherical micelles in water above their critical micelle concentration (CMC) values as evident from the transmission electron microscopy (TEM) images, while the hydrodynamic diameter of the micelles was determined to be 16 and 5.9 nm, respectively. Furthermore, micellar disintegration was observed upon the addition of α-circular dichroism (CD) yielding lower-order rotaxanes. This transformation or change in order with

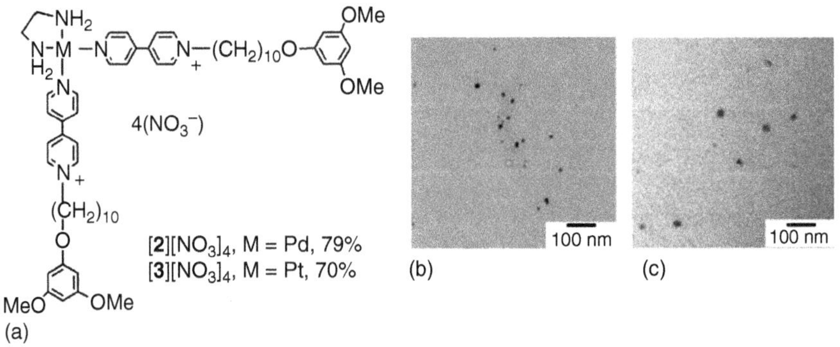

Figure 3.3 (a) Structure and (b), (c) TEM images for [**2**][NO_3]$_4$, M = Pd and [**3**][NO_3]$_4$, M = Pt. Source: Taira et al. [20]/with permission of Royal Society of Chemistry.

an external agent, cyclodextrins in this case, could be exploited in future to control micellar aggregation and functioning.

Even the diblock copolymers were found to exhibit micellization on complexation with Ru metal [21]. The association behavior of two amphiphilic metallo-supramolecular block copolymers designated with acronyms PS_{20}-[Ru]-PEO_{70} and PS_{20}-[Ru]-PEO_{375} was investigated using the dynamic light scattering (DLS) and cryo-TEM techniques.

The polymers consisted of a polystyrene (PS) block connected to a poly(ethylene oxide) (PEO) block using a bis(terpyridine)ruthenium complex as the bridge; hence, the name: PS20-[Ru]-PEO_y, with the constant length of PS block about 20 units and the two PEOs with different molecular weight. The authors reported two types of micellar populations: individual monodisperse spherical micelles and small clusters of micelles formed by the metallo block copolymer. Cryo TEM images revealed the formation of spherical micelles with a core–shell morphology for the PS_{20}-[Ru]-PEO_{375} with the external diameter of the core-shell micelles to be 35 nm, which is in good agreement with DLS results. On the other hand, covalent PS_{22}-*b*-PEO_{70} was found to form only individual micelles, but a clear image could not be obtained for the PEO_{70} counterpart, which has been previously observed for short block systems. It was also deduced that increasing the length of the PEO block had no effect on the core dimension, but it just made the visualization using cryo TEM more facile. Pileni and coworkers have done pioneering work on $Cu(AOT)_2$ surfactants and demonstrated a variation in the shape of the water droplets after the dissolution of surfactant in water, i.e. the surfactant solution with changes in water content [22–24]. The ratio of water to surfactant was represented using a parameter $\boldsymbol{w}$, which is defined as $\boldsymbol{w} = [H_2O]/[Cu(AOT)_2]$. Below $\boldsymbol{w} = 3$, spherical droplets were formed and on increasing $\boldsymbol{w}$ from 3 to 5, the droplets obtained were elongated with a cylindrical shape. This gave an evidence of metallomicellar morphology change with alteration in water content.

Koutsantonis and Nealon [25] presented an interesting report describing the formation of wormlike micelles coexisting with or interconverting to vesicles for copper and cobalt metallosurfactants derived from Cu(II) and Co(III) complexes of a macrobicyclic hexamine ("cage") (Figure 3.4). Vesicles are basically spherical self-aggregates with hollow core enclosed by a shell made up of a bilayer of the amphiphiles, which has been discussed later in the section. Cryo-TEM images reveal a series of interesting structures formed for the Co surfactant. Although the formation of cylindrical micelles is quite uncommon for triple-chain surfactants with a single headgroup, the wormlike structures obtained were highly convoluted, hundreds of nanometers in length, and showed a micellar diameter of c. (6 ± 1) nm. At this concentration, vesicles with a spherical morphology ranging from 30 to 300 nm in size were found to be coexisting with the wormlike structures. Quite intriguingly, the formation of wormlike micelles is highly unexpected because the presence of such a large charge on the headgroup (+3) should cause electrostatic repulsion and destabilization of the micellar structures. But the authors proceed with caution and attribute the wormlike micelle formation to the existence of directional attractive forces in the headgroup region as well as a lack of complete

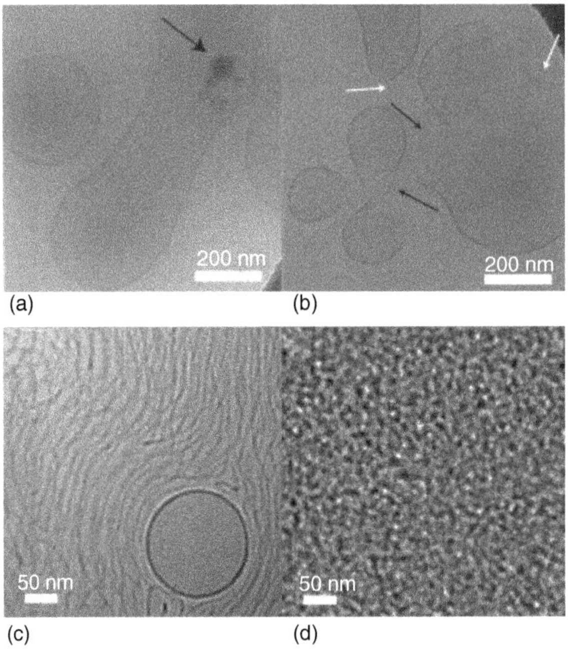

Figure 3.4 Cryo-TEM images for 1 wt% solution of Co(III) surfactant showing (a) vesicles. The arrow indicates cubic crystalline ice, (b) broken vesicles. Black arrows indicate bilayer fragments, and white arrows indicate wormlike micelles, (c) Wormlike micelles and vesicles coexisting, (d), concentrated wormlike micelles or possibly a sponge phase at 10 wt% solution (Inset: Structures of the surfactant formed). Source: Koutsantonis and Nealon [25]/with permission of American Chemical Society.

fluidity of the alkyl chains. Co(III)...Co(III) headgroups should have a large effective headgroup area expected due to heavy hydration, leading to the presence of the cylindrical micelles. Also, the counterions, i.e. the acetate anions would be readily localized within the headgroup region, owing to the extensive "H-bond chelation" by the coordinated NH groups. The model assumes a fluidlike packing of the chains in the interior of the micelle, thus leading to the formation of micelles. The authors observe a super easy interconversion between the wormlike micelles and vesicles, which might be attributed to a relatively low energy barrier between the two.

3.2.2 Inverted Micelles

Incidentally, a surfactant with small headgroup and voluminous hydrophobic substituents leads to the formation of inverted micelles. In the case of metallosurfactants, Domınguez-Gutierrez et al. became the first group to report the formation of inverted aggregates from luminescent Ru surfactants [26]. The $[Ru(bpy)_{3-n}(bpyR_x)_n]Cl_2$ complexes under study (n: 1 or 2), where $bpyR_x$ stands for a 4,4′-dialkyl 2,2′-bipyridine ligand with x alkyl chains contains large headgroups, so the formation of inverted aggregates is actually attributed to the hydrophobic moieties, which are voluminous enough to counteract the effect of the large headgroup. They extensively studied the effect of solvent polarity, rigidity, and length of the alkyl chains in ligand on the self-aggregating structures formed and drew remarkable conclusions regarding the same. The Ru surfactants aggregated in apolar solvents into inverted micelles or inverted vesicles as determined by DLS and atomic force microscopy (AFM) owing to the presence of numerous hydrophobic

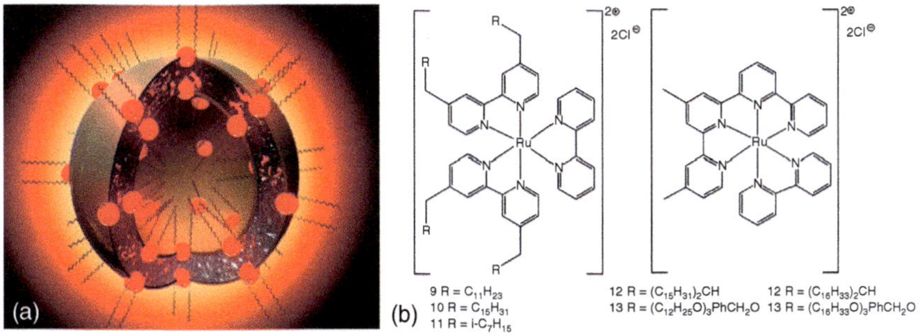

Figure 3.5 Ru metallosurfactant (a) Representation of inverted aggregates of; (b) Structure of complexes 9–11 type $[Ru(bpy)(bpyR_2)_2]Cl_2$ and complexes 12,13 type $[Ru(bpy)_2(bpyR_x)]Cl_2$. Source: Domınguez-Gutierrez et al. [26]/with permission of Royal Society of Chemistry.

tails converging into a cone geometry with the head group present in the narrow extreme of the cone.

Five different Ru metallosurfactants (complexes 9–13) have been synthesized and their self-aggregation investigated in detail by the authors as illustrated in Figure 3.5. Complex 9 did not show any self-assembly in toluene, but rather resulted in inverted vesicles (bilayers of complexes closing onto themselves, with the ruthenium groups facing each other) showing a hydrodynamic radius (R_h) of about 64 nm in hexane. Quite remarkable self-assembly process was observed for complexes 10 and 11. While complex 11 did not form any self-assemblies in hexane or toluene in spite of four large branched C_9H_{19} hydrophobic chains, suggesting, that longer chains were needed to solubilize the large $[Ru(bpy)_3]^{2+}$ head group. Solvent polarity exercised a significant effect on the self-aggregating structures of complex 10. Inverted micelles (R_h 3.7 nm) in 5×10^{-4} M hexane solutions and 100 times larger inverted vesicles ~500 nm radius in 10^{-3} M toluene solutions were obtained due to the difference in polarity of the solvents as the more polar toluene easily penetrated the interface of the surfactant, resulting in larger structures. Furthermore, the individual micelles obtained with a R_h of 3.7 nm were in agreement with the fully extended chain conformation having a length of c. 3 nm corroborating the idea that the core of the inverted micelles constituted of the divalent ruthenium headgroups in close contact with each other, while the surface of micelles was formed by the hydrophobic tails extending outwards. Also, the presence of some larger aggregates (R_h ~ 32 nm) could be a result of several small individual micelles coalescing. Comparing complexes 9 and 10, it was concluded that the shorter chains in complex 9 (about 16 methylene groups per molecule less than complex 10) could not provide efficient shielding to the headgroup, hence, the inverted vesicles formation with a larger curvature instead of inverted micelles. Complex 12 self-assembled exclusively in hexane solutions yielding spherical, monodisperse, inverted vesicles (R_h ~38 nm) primarily with a small population of micelles (~3.5 nm). These structures are not observed in case of more polar toluene-like solvents ascribed to the rigidity of the ligand. Although the chains are long, they are not flexible enough to shield the headgroup, which also hinders its arrangement into small inverted micelles even though it should be favored. With a more flexible ligand, complex 13 is able to form inverted vesicles in both hexane (~49 nm) and toluene (~57 nm).

3.2.3 Vesicles

Although there have been several examples depicting coexistence of micelles and vesicles, Owen et al. wonderfully exemplified the transition from micelles to vesicular structures with a change in coordination upon addition of different metals [27, 28]. A microbially produced amphiphile, marinobactin E (ME), was used to demonstrate the metal-induced transition from micelle to vesicle. The ferric complex of the microbially produced amphiphiles led to micelle formation with a diameter of ~3.5 nm. These micelles experience a supramolecular transformation upon the introduction of various metals such as Cd(II), Zn(II), excess Fe(III) or La(III) to monodisperse vesicles with an average diameter ranging from ~90 to 200 nm.

The marinobactins consist of an amino acid peptidic headgroup that coordinates Fe(III) and the terminal carboxylate moieties in the Fe(III)-ME headgroups bridge together on metal introduction, drawing them closer to give a vesicular arrangement (Figure 3.6a). As revealed by the SANS and cryo-TEM data, a change in morphology of Fe(III)-ME micelles to multilamellar vesicles (~100–200 nm diameter) was attributed to the ability of metals, i.e. Cd(II), Zn(II), and La(III) to form 2 : 1 ligand/metal coordination complexes. The resulting "composite surfactant" possessed a lower headgroup-area/tail-volume ratio due to the formation of a cylindrical composite with two alkyl chain tails favoring a vesicular arrangement instead of the cone-shaped, monocephalic surfactant. X-ray diffraction (XRD) indicated the interlamellar repeat distance of the Cd(II)- and Zn(II)-induced multilamellar vesicles to be ~5.3 nm, while the X-ray absorption spectroscopy (XAS) spectra of these vesicles confirmed the six-coordinate geometry of Fe(III).

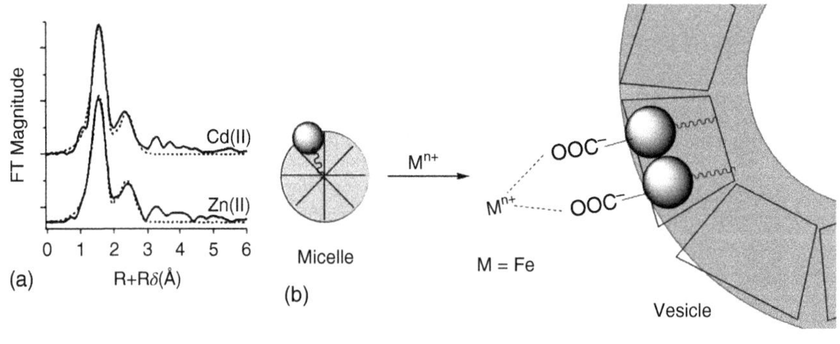

Figure 3.6 (a) Fourier transforms of Fe(III) sites in sedimented Cd(II)- and Zn(II)-induced vesicles. (b) Proposed metal cation-induced carboxylate cross-linking of Fe(III)-ME headgroups as a mechanism to promote the micelle-to-vesicle transition. Source: Reproduced with permission from Owen et al. [28]. Copyright (2008) American Chemical Society.

Consistent with the studies reported above, Suades and coworkers did some remarkable work on the impact of Pt metal coordination in the formation of metalloaggregates [29]. The authors reported the formation of medium-size vesicles in the first platinum(II) metallosurfactants of the type *cis*-$[PtCl_2L_2]$ (where L = 1–3; and L is the ligand from a new family of linear surfactant phosphines $Ph_2P(CH_2)_nSO_3Na$ {1 ($n = 2$), 2 ($n = 6$), and 3 ($n = 10$)}). The metallosurfactants synthesized by a reaction between the phosphines and $PtCl_2$ in dimethylsulfoxide resulted in the formation of spherical disperse medium-size unilamellar, bilamellar, and multilamellar vesicles. The same research group also synthesized a new set of molybdenum metallosurfactants, i.e. molybdenum carbonyl surfactants with the metal atom being located in a hydrophobic environment, as a neutral metal carbonyl (ligands remained the same as described previously) instead of the usual polar head group, while the polar head group in this case is a sulfonate linked to the phosphine ligand [30, 31]. Cryo-TEM microscopic image clearly confirmed the formation of vesicles. A wide range of aggregates with slight morphological differences as illustrated in Figure 3.7 was obtained for the metallosurfactant 5, which includes small unilamellar vesicles (diameters lower than 100 nm), large unilamellar vesicles (diameters higher than 100 nm), multilamellar vesicles (onion-like structure), and multivesicular vesicles (smaller non-concentric spheres inside a larger vesicle). The membrane structure or the wall of the vesicles obtained is actually metallic amphiphile bilayer and resembled a classical lipid bilayer conformation, which could be visualized as comprising organometallic fragments in place of hydrophobic chains. The metal is found to be located in the hydrophobic group placed in the inner part of the bilayer in the metallovesicle. Zha et al. applied this concept to rationalize the formation of vesicles by a copper metallosurfactant [32]. $Na_2\{Cu[C_nH_{2n+1}N(CH_2COO)_2]_2\}$, where: $n = 8$, 12, 16 type of copper metallosurfactants were synthesized, which spontaneously formed vesicles in water. As previously reported by Parera et al.

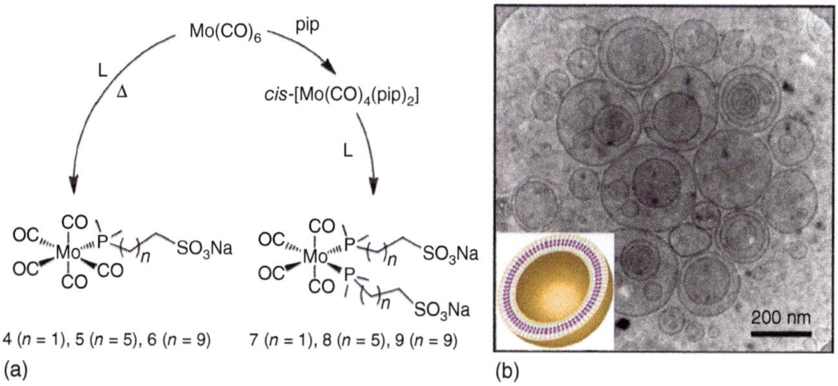

Figure 3.7 (a) Preparation of molybdenum metallosurfactants [4–9] (L = 1, 2, 3; pip = piperidine). (b) Cryo-TEM micrograph of a 7.3 mM solution of metallosurfactant 5 in water. (Inset) Schematic representations of a metallo-vesicle showing the organometallic amphiphile bilayer (yellow and purple balls represent sulfonate and $\{Ph_2PMo(CO)_5\}$ groups, respectively). Source: Parera et al. [30]/with permission of Royal Society of Chemistry.

and Owen et al., metal Cu(II) coordinated to the carboxylic acid moieties present in the surfactants' headgroups neutralizing the electrostatic repulsion between headgroups and drawing them closer together resulting in the formation of vesicles. The vesicles formed by C_{12}–Cu–C_{12} had an average size of ~60 nm displaying good stability over inorganic salt, temperature, and aging time. The data for different metallosurfactants with varying alkyl chain lengths established that the surfactant with longer chain length formed vesicles more readily. Similar behavior is observed by Sommerdijk et al. [33] while investigating the self-aggregation of a two-headed (dicephalic) surfactant. The complexation of surfactant molecules to copper(II) sulfate at pH 5.5 and 9.5 values led to the generation of metalloaggregates of the general-type Cu(imidazole)$_4{}^{2+}$. Vesicles are formed at both the pH values; however, the complexation is different due to the dicephalic nature of the surfactant. At pH 5.5, four imidazole groups from two surfactant molecules coordinate to one copper center in a bidentate fashion, whereas at pH 9.5 a two-dimensional polymeric network is formed in which each copper(II) ion is surrounded by four imidazole groups from four different surfactant molecules. Electron microscopy revealed large multilamellar vesicles with diameters of approximately 1000 nm at pH 5.5, while at pH 9.5, extended bilayer structures were noticeable with a periodicity of 63 Å. Hence, a change in pH leads to a change in coordination of the complexes, although vesicles are formed by both types of complexes.

As is the case with the formation of micelles and inverted micelles, reverse vesicles can also be obtained by dissolution of metallosurfactant in apolar media. Dong and Hao synthesized ferrum laurate [$Fe(OOCC_{11}H_{23})_3$] metallosurfactant, which could successfully self-assemble into reverse vesicles in organic media such as pure $CHCl_3$ and a mixed solvent of $CHCl_3$ and CH_3OH [34]. The metallosurfactant synthesized is composed of three long alkyl tails leading to a strong shielding of the polar headgroup in organic solvents, resulting in the formation of deformed solid reverse vesicles, including collapsed erythrocyte-like and broken hollow shells. The size of the aggregate structure was elucidated using the DLS techniques. Measurements performed at three different scattering angles of 90°, 60°, and 45° had no effect on the hydrodynamic radius (R_h) validating the formation of spherically symmetric aggregates. R_h had a fairly large size distribution, ranging from 30 to 260 nm. Solvent-dependent aggregation behavior generating normal and reverse vesicles was also studied. For this purpose, Mahato et al. employed a newly synthesized symmetrical amphiphilic Ru(II)-polypyridyl complex with three $-C_{18}H_{37}$ chains in equilateral positions [35]. The Ru(II)-polypyridyl hydrophilic head group was present as the core and each bpy-ligand further functionalized with a hydrophobic C_{18} hydrocarbon tail adopted a cone-shaped geometry and such cone-shaped complexes are usually anticipated to form inverted micelles or vesicles in nonpolar medium. Rather fascinatingly, this metallosurfactant formed both vesicles and reverse vesicles depending upon the polarity of the solvent used, i.e. normal vesicles in polar medium such as water and reverse vesicles in non-polar media like toluene and cyclohexane – which is a unique observation (Scheme 3.2a). The comprehensive DLS, static light scattering (SLS), AFM, and TEM data revealed the formation of spherical, unilamellar, hollow vesicles where the vesicle membrane is essentially a

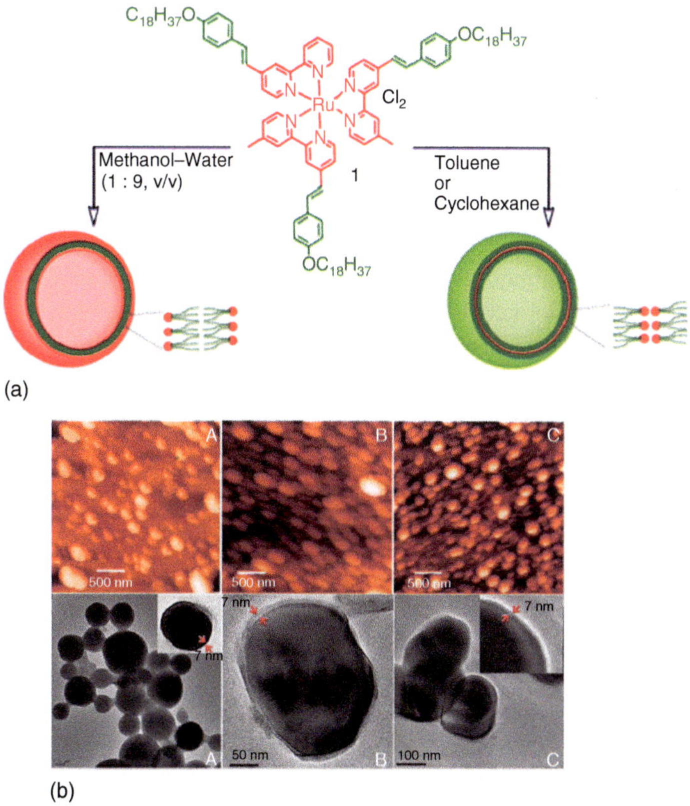

Scheme 3.2 (a) Molecular structure of Ru metallosurfactant and image representation of aggregates in different solvents. (b) upper: AFM images of cast films deposited on mica of 1 in (A) methanol: water (1 : 9; v/v), (B) toluene and in (C) cyclohexane; lower: Unstained TEM images of metallosurfactant in (A) methanol: water (1 : 9; v/v), (B) toluene and in (C) cyclohexane. (Inset (A): UO_2^{2+} stained TEM image in methanol: water (1 : 9; v/v) showing cell wall thickness; (C): magnified unstained TEM image showing cell wall thickness.) Source: Mahato et al.[35]/with permission of Royal Society of Chemistry.

bilayer. The vesicle structures are formed by exposure of the apolar aliphatic chains toward the apolar solvent medium environment, while the hydrophilic headgroup is situated at the core of the bilayers facing each other. The DLS studies for the reverse vesicles in toluene and cyclohexane indicate the mean diameter of the aggregates to be 308 and 229 nm, respectively, while in the case of normal vesicles, the mean diameter was found to be 230 nm. The solubility of the Ru metallosurfactant in toluene was found to be higher as compared to cyclohexane, credited to the higher polarity of toluene.

3.2.4 Lamellar Phases

The metallosurfactants are even capable of forming lyotropic liquid crystalline structure aggregating to lamellar phase, which potentially consists of surfactant molecules arranged in bilayers, which essentially mimics the lipid arrangement.

Each bilayer sheet is separated by layers of water. Taira et al. reported that Pd(II) induced the formation of this type of lamellar phase in N-heterocyclic carbene (NHC)-based metallosurfactant (MS), denoted as NHC-PdMS [36]. Pd metal was located between the hydrophilic and hydrophobic parts of the NHC–PdMS surfactant synthesized by stepwise ligand substitution reaction forming basically a Pd–carbene bond. NEt_3 used as a hemilabile co-ligand regulates the isomerization step to get a single isomer of NHC–PdMS. Although the MS precursor, i.e. imidazolium bromide, forms spherical micelles, the coordination with Pd(II) in case of NHC–PdMS results in self-assembly to larger aggregates, i.e. lamellar structure over a wide concentration range (>26 wt%) as corroborated by the SAXS experiment.

Hence, we can safely conclude that the metallosurfactants are capable of forming a myriad of self-assembly structures: micelles, inverse micelles, vesicles, reverse vesicles, lamellar structures, bilayers, etc. They even display a transition between various self-aggregating structures with varying metal coordination, solvent polarity, and alkyl chain length. The capacity to regulate the self-assembling properties of metallosurfactants can act as a useful tool for designing reactive interfaces.

3.3 Chemistry of Catalysis

Metallosurfactants offer innumerable advantages and applications over the conventional surfactants and are rightfully termed as “next generation surfactants” by Polarz et al. [37] owing to the presence of traditional surfactant moieties imparting amphiphilic behavior and the unique metal ion imparting redox, catalytic, or magnetic character to the molecule. The inclusion of metal into the amphiphile led to strikingly organized self-assemblies decorated with active metal centers at the interfaces – oil/water or air/water. Hence, it provided a wonderful opportunity to work with, in a constrained environment and highly advantageous to pursue their use as mini reactors for different type of reactions and catalytic processes. Incorporation of various moieties within the self-assemblies is highly beneficial also with regard to solubilization in both the aqueous and non-aqueous medium, which can further be regulated to carry out the reactions in. This addresses and resolves the persistent issue of reacting two moieties with different solvent compatibility. It is highly desirable from the green chemistry viewpoint as it helps in minimizing the use of harmful organic solvents and achieving higher yields with fewer byproducts.

The classic challenge in using metals for catalyzing organic reactions is the lack of compatibility and dissolution of reactants in a single medium to successfully carry out the reaction and at the same time extraction of the product. At this point, the self-assemblies and interfaces can be really helpful to increase cooperativity and allow diffusion of reactants. Phase transfer catalysts (PTC) having both hydrophilic and hydrophobic components easily dissolve in either of the two phases and facilitate the migration of the reactant into a favorable common phase where reaction can take place. While working with ionic reactants, PTC is an indispensable tool as the reactants are more often than not insoluble in an organic phase and solubilize only in polar media. The reaction rate can also be enhanced by forming micelles or emulsions resulting in a large interface of the immiscible phases. In the present section, we will be discussing the catalytic activity shown by metallosurfactants, which is an interplay of surfactant catalysis and transition metal catalysis wherein metallosurfactant molecules self-assemble into various organized assemblies in solution and act as efficient metal catalysts as well. Scarso and coworkers [38, 39] have excellently reviewed the micellar catalytic activity highlighting the role of metallosurfactants, while Bhattacharya and Kumari [40] focused their review on the ester hydrolysis reactions using metallomicelles. Two major approaches to metal catalysis in micellar media have been defined: (i) solubilization of metal within or at the micellar surface (ii) using metallomicellar units. For the scope of this chapter, we will be concentrating our studies in the latter area:

3.3.1 Micellar Catalysis

The reaction occurs in the metallomicelles, which have been employed as reaction vessels as well as catalytic centers. Lipshutz et al. have done some really impressive work in the field of "micellar catalysis" involving metals and surfactants [41, 42]. They very effectively carried out metal insertion into the established designer surfactant systems yielding nanomicelles for catalytic reactions with the major benefits of using water as the reaction media, product extraction with minimal organic solvent, and in-flask catalyst recycling. Initially, both Grubbs–Hoveyda (GH) 1 and 2 ruthenium carbene catalysts were covalently attached to the surfactant polyethyleneglycol ubiquinol sebacate (PQS) for carrying out olefin metathesis reactions within the spontaneous aqueous nanomicelles [43]. The catalyst-cum-reaction center was efficiently utilized in the ring-closing metathesis of both mono- and 1,1-disubstituted olefin, and N-, S-, and O-containing lipophilic dienes leading to the corresponding five-, six-, and seven-membered ring products. Taking cue, PQS was again used as the platform for micellar catalysis as it self-aggregated into nanomicelles, to which nonracemic 2,2′-bis(diphenylphosphino)-1,1′-binaphthyl (BINAP) was covalently attached and rhodium(I) was inserted resulting in "PQS-BINAP-R_h." The "PQS-BINAP-R_h" species formed is the first example of a nonracemically ligated transition metal catalyst-tethered amphiphile, utilized for Rh-catalyzed asymmetric 1, 4 conjugate addition reactions of arylboronic acids to acyclic and cyclic enones as depicted in Figure 3.8c [44]. As evident from Figure 3.8a, PQS is a truly unique molecule as it has appendages to solubilize in both organic and aqueous media and a handle to attach the catalyst as well making it an indispensable tool for micellar catalysis.

Working on similar lines, Mintogaud and coworkers prepared another type, surface active Hoveyda-type catalyst denoted as 2, 3 in Scheme 3.3, which catalyzed both ring-opening metathesis polymerization (ROMP) and ring-closing metathesis (RCM) in water with great efficiency [45]. These surfactants assembled into Langmuir monolayers at the air/water interface, which confirms the surface activity of these complexes and the fact that the ruthenium moiety is polar enough to be close to the water surface aiding catalysis. Ring-closing metathesis experiments were

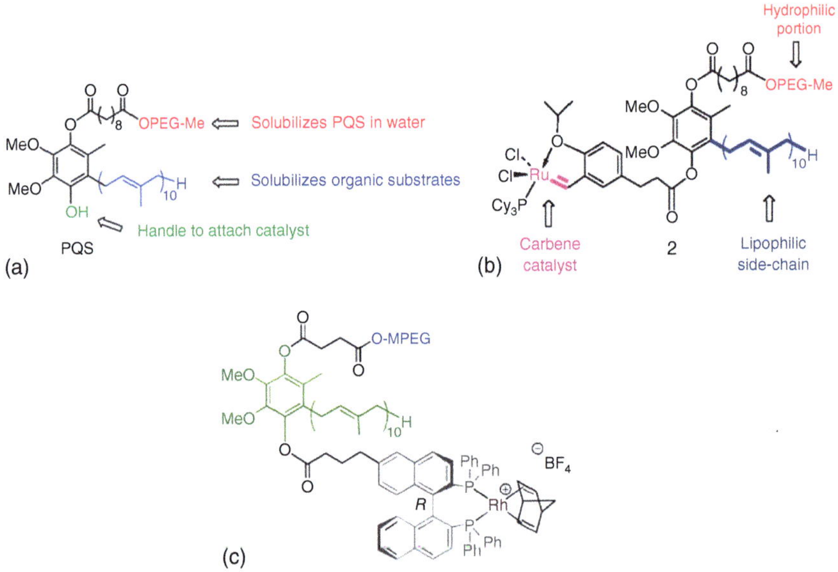

Figure 3.8 Structures of (a) PQS, (b) PQS-attached Grubbs–Hoveyda-1 metathesis catalyst, (c) PQS-BINAP-Rh. Source: (a,b) Adapted with permission from Lipshutz and Ghorai [43b]. Copyright (2009) American Chemical Society. (c) Reprinted with permission from Lipshutz et al. [44].

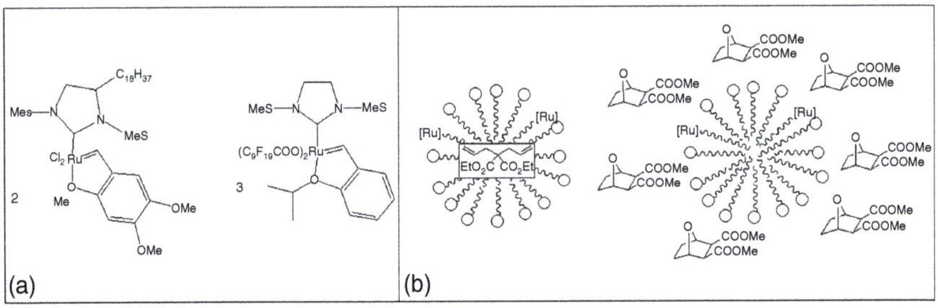

Scheme 3.3 (a) As-synthesized Hoveyda-type surface active catalysts – 2,3 (b) Schematic localization of reagents for RCM and ROMP. Source: Adapted with permission from Franc et al. [45].

conducted on diethyldiallylmalonate (DEDAM), which was localized in the core of the micelles and the active catalyst site, i.e. [Ru] is in the hydrophilic part of the micelle. Ring-opening metathesis polymerization was carried out using norbornene and diester in this case, although both water-soluble and -insoluble monomers could be envisioned. In each case, the initiator surf is present at the interface between hydrophobic and hydrophilic media. For water-soluble monomers, a dispersion polymerization could be envisaged, whereas for hydrophobic monomers, a suspension polymerization (Scheme 3.3).

Further, channelizing the catalytic activity of Rh metal, Deng and coworkers carried out reduction reactions of ketoesters and ketones in the micellar phase of

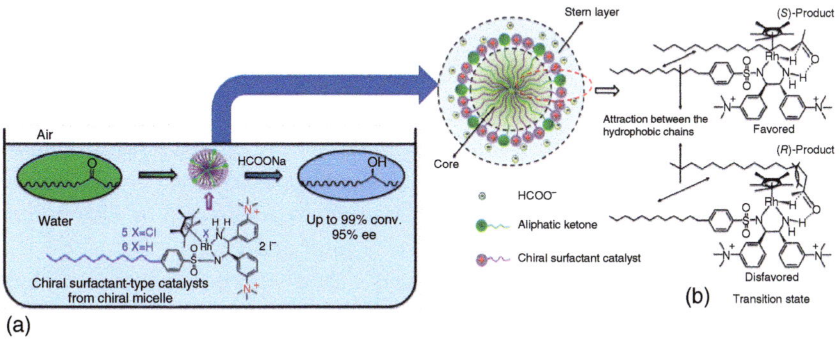

Figure 3.9 Proposed mechanism of the ATH using chiral surfactant-type catalyst 6 at the air water interface, and magnified inside the metallomicelle (a) Schematic representation of the metallomicelle. (b) Probable transition states of the reaction. Source: Adapted with permission from Li et al. [46]. Copyright (2012) American Chemical Society.

as-synthesized chiral surfactant-type Rh catalysts as depicted in Figure 3.9 with excellent conversion rate (94%) and high enantiomeric excess (84%) [46, 47]. An insight into the reaction mechanism of asymmetric transfer hydrogenation (ATH) of octan-2-one with HCO_2Na as hydrogen source in water was provided. The reaction is proposed to occur in the metallomicelles with the micelle size of the catalyst changing from 22.3 ± 5.7 nm to 66.8 ± 15.6 nm for the reaction mixture, due to the incorporation of the reactant, i.e. ketone into the metallomicelles. Alkyl length of aliphatic ketones greatly influenced the reactivity and enantioselectivity. Linear alkyl methyl ketones show increased enantioselectivity from 76% to 94% ee with an increase in alkyl chain length suggesting hydrophobic interactions between the long alkyl group of aliphatic ketones and the long chain of catalyst in the metallomicelle. Also, the positive charges on the polar head of the surfactants located at the micellar surface elevate the concentration of hydrogen source, i.e. formate ions via the electrostatic attraction, and result in the acceleration of reaction rate. It was proposed that in the transition state, stereochemistry was controlled by the steric interaction between the alkyl groups of the aliphatic ketone and the Cp* ligand of the amphiphilic catalyst. Therefore, synergistic effects between the metal-catalyzed center and the hydrophobic microenvironment of the metallomicelle core facilitate the reduction, proceeding with high stereochemical induction. ATH of esters such as methyl 2-oxodecanoate with different metal precursors and ligands was also explored to gain an insight into the mechanism.

3.3.2 Vesicular Structures in Catalysis

Vesicles are among the most common type of self-organized structures formed by the metallosurfactants and, hence, can be efficaciously exploited for performing reactions. Uozumi and coworkers have been leading the research in pincer palladium complexes and successfully utilized the self-organized structures primarily formed, i.e. vesicles for catalytically driven processes [48]. Pincer complexes are basically the coordination compounds obtained when chelating agents bind to monoionic terdentate ligand or simply to three adjacent coplanar sites of a ligand (referred to as pincer) in a meridional configuration. The researchers designed amphiphilic pincer palladium complexes by decorating the planar palladium pincer backbone with hydrophilic and hydrophobic side chains. The pincer palladium complexes as illustrated in Table 3.1 have pairs of hydrophobic dodecyl chains and hydrophilic triethylene glycol chains, located opposite to each another on the rigid planar backbone. The vesicular structures were obtained by dissolution of the pincer complex in 9 : 1 H_2O to CH_3CN ratio and concentrated at 80 °C for six hours, to afford a pale yellow aqueous suspension, which was cooled and centrifuged (4000 rpm, 15 minutes) to give nanoprecipitates. The dilution of nanoprecipitates in water resulted in vesicle formation with an average diameter of 463 nm for pincer complex A and average 550 nm for pincer complex B where metal is Pd. The comprehensive AFM, TEM, and scanning electron microscopy (SEM) analyses confirmed spherical, hollow morphologies with the thickness of the vesicle membranes to be c. 6–7 nm, consistent with their bilayer membranous structures exhibiting monomer lengths of c. 2.8 nm. The authors explored the catalytic potential of the synthesized vesicles for a number of organic reactions in water, including but not limited to (i) Miyaura–Michael reaction of 2-cyclohexene-1-one with sodium tetraphenylborate yielding arylated product [50] (ii) cyclization of alkynoic acids with catalytic amount of triethylamine to give the corresponding lactones (iii) the ring-opening reaction of vinyl epoxides, cyclization of 5-[4-(trifluoromethyl)phenyl]pent-4-ynoic acid, lactonization of 5-(het)arylpent-4-ynoic acids, cyclization of 2-[(3-phenylprop-2-yn-1-yl)oxy]phenol, [49], (iv) allylic arylation of (2*E*)-3-biphenyl-4-ylprop-2-en-1-yl acetate with sodium tetraphenylborate, allylic substitution, or the Tsuji–Trost reaction, (v) hydrosilylation of styrene and other related alkenes using dimethyl(phenyl)silane [52], (vi) arylating oxirane ring-opening reactions of the vinylepoxide with phenylboronic acid [51], (vii) the Miyaura–Michael reaction of cyclohexanone, with sodium tetraphenylborate to give the arylated product.

A detailed structural analysis of the self-assembled palladium pincer complexes points toward the formation of bilayer vesicles. Wide-angle X-ray scattering experiments and atomistic molecular dynamics calculations form a powerful tool for the structure elucidation of vesicles, which is important to map out the mechanism of reaction taking place inside vesicles. The simulated bilayer structure revealed that the membrane formed was relatively soft as compared to the normal phospholipid bilayer membrane and the degree of alignment of the complex molecules was low, which facilitates the movement of organic substrates or reactants through the bilayer assemblage. Electron density profiles also confirm the presence of space in the bilayer arrangement through which molecules can pass, hence, aiding them to act as nanoreactors. In addition, the free energy of permeation of water through the pincer complex bilayer membrane was much lower than the pure phospholipid bilayer membrane, determined to be 12 kJ mol^{-1}, which again results in facile passage of organic molecules. The catalytically active center responsible for the reaction is found to be close to the "free space," which is easily accessible to the organic molecules. Based on these results, a concept for the catalysis was

Table 3.1 Several reactions catalyzed by vesicles (vscl) and amorphous (amps) forms of different amphiphilic metal pincer complexes.

Metal Pincer complex	Reactions catalysed	amps (% yield)	vscl[ref] (% yield)
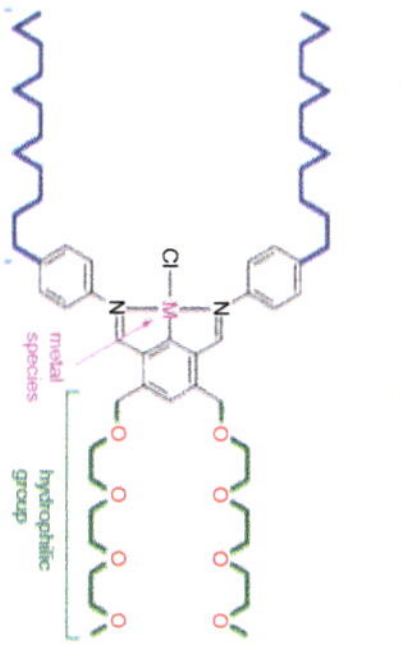	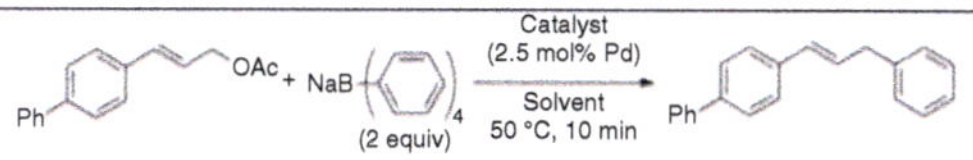	8	59[50]
	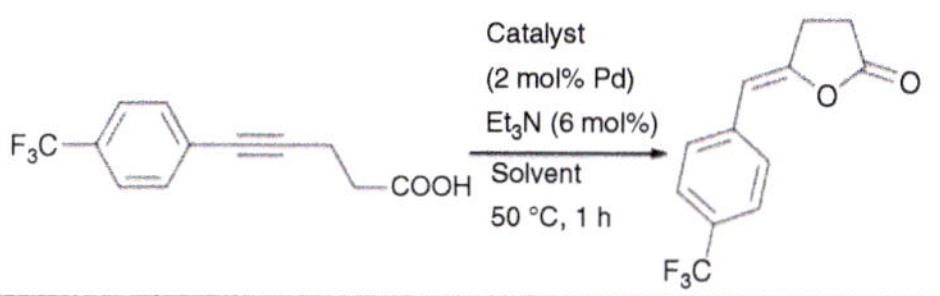	9	61[49]
	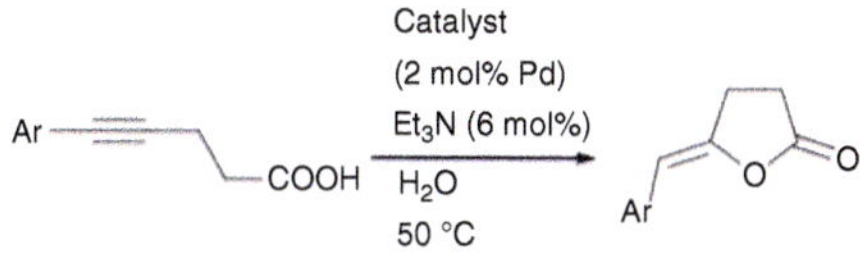		
(A)	Ar = Ph	8	66[49]
	Ar = 4-MeOC_6H_4	13	82
	Ar = l-naphthyl	10	82
	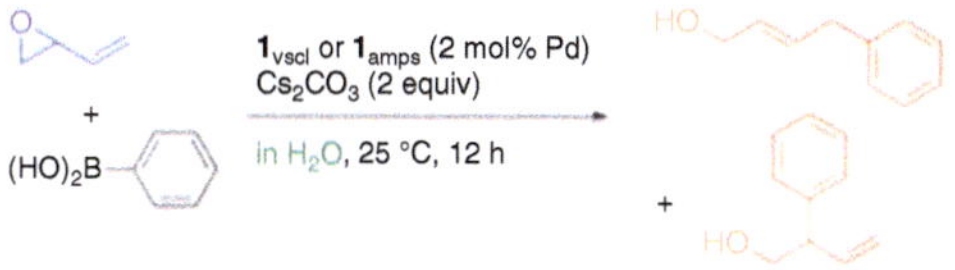	7	84[52]
	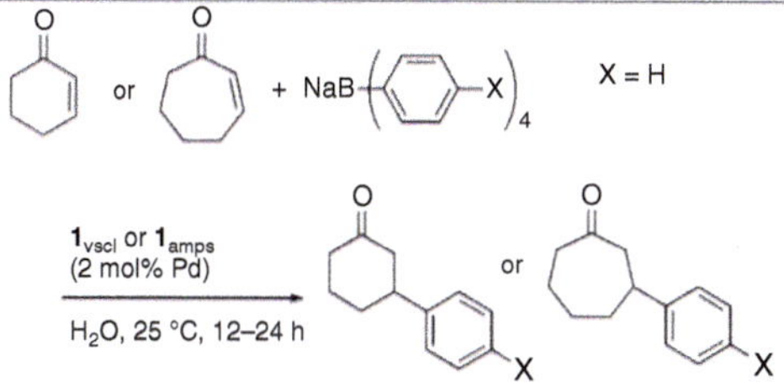	7	83[52]
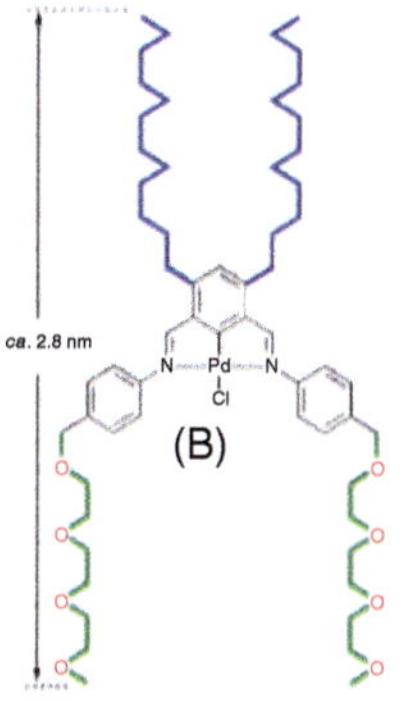	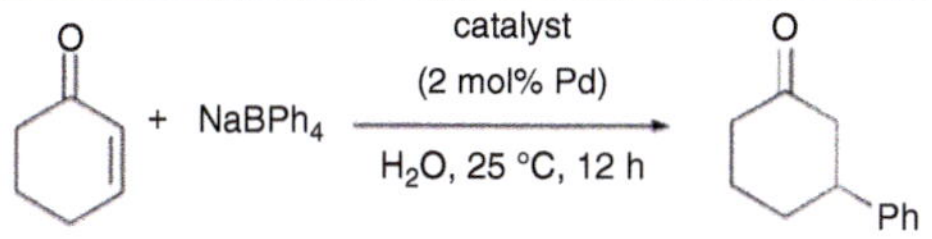	7	83[48]

Source: Adapted from Hamasaka et al. [48] and Hamasaka and Uozumi [50] with permission from The Royal Society of Chemistry. Adapted with permission from Hamasaka et al. [49]. Adapted with permission from Hamasaka et al. [51].

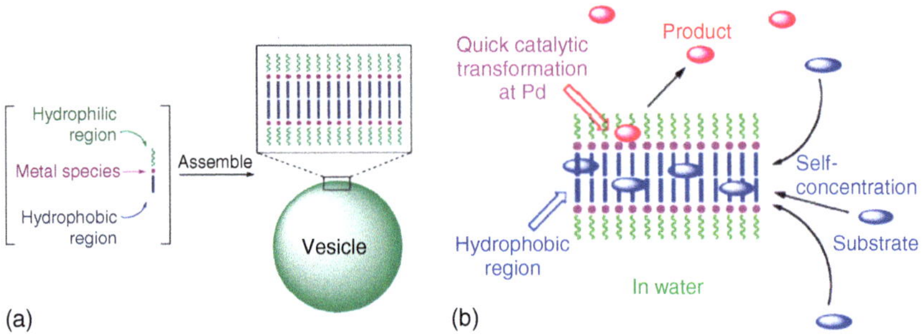

Scheme 3.4 (a) Vesicle formation via self-assembly of amphiphilic pincer complexes, (b) schematic image of the concept of catalysis within the bilayer membranes of vesicles. Source: Adapted with permission from Hamasaka et al. [51].

inferred (Scheme 3.4). The promotion of catalytic reaction inside the vesicular structure is explained by the spontaneous concentration of the organic substrate in the hydrophobic region as a result of hydrophobic interactions, where the potentially catalytic metal species is located in immediate proximity. The organic transformation proceeds rapidly as a result of the presence of high concentrations of the organic substrate near the catalytic center [53]. During the reaction, the organic substrates and the product are in a state of distribution equilibrium between the hydrophobic region of the vesicle membrane and aqueous exterior.

3.3.3 Interfacial Catalysis

In addition to carrying out reactions within the self-assembled metallomicellar units, interfacial catalysis was investigated with the possibility of regulating reactions at the interface of two phases using metallosurfactants. Recently, Toth et al. reported the interfacial catalysis in esterification of dilute acetic acid using the metallosurfactant $Ni(DBSA)_2$ obtained by anionic surfactant 4-dodecylbenzenesulfonic acid (4-DBSA) and divalent nickel ion, mimicking the two-phase Brust Schriffin method (BSM) [54, 55]. BSM has been extensively used to carry out nanoparticle synthesis employing tetraoctylammonium bromide as the phase-transfer catalyst aiding the movement of chloroaurate ions from aqueous to organic phase followed by phase separation and reduction to yield gold (Au) nanoparticles [56]. Engaging a similar methodology, the synthesized catalytic surfactant system, i.e. $Ni(DBSA)_2$, has been utilized to carry out the esterification of dilute acetic acid with solvent phase as 1-octanol, which resulted in octyl acetate in 31.3% yield, compared to 0.2% in the uncatalyzed version. While the transition-metal ions such as Ni catalyzed the esterification reactions, 4-DBSA was exploited for its ion-exchange capability to regulate the surfactant properties with transition metal ions. A biphasic (water–solvent) system is used where $Ni(DBSA)_2$ dissolved in the solvent phase gets self-assembled at the aqueous/solvent interface where the reaction takes place with acetic acid being present in the aqueous phase (Figure 3.10). The catalyst-surfactant system performs better as it lowers the interfacial tension between 1-octanol and deionized water by up to 75%, increasing the mass transfer between different

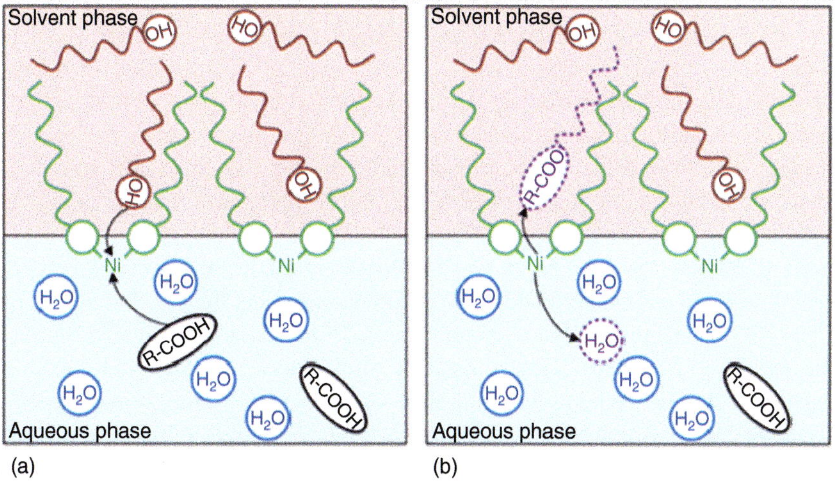

Figure 3.10 Proposed mechanism of interfacial catalysis in bifacial esterification. (a) Before the reaction and (b) after the reaction. Source: Adapted with permission from Toth et al. [54]. Copyright (2019) American Chemical Society.

phases. It was concluded that the Ni-based surfactant holds a great potential for esterification reactions, with 1-octanol serving as a reactive solvent for the esterification and extraction of carboxylic acid. As a result, biphasic esterification enhanced by interfacial catalysis was observed using this metallosurfactant.

Going a step further, Kagalwala et al. led to solar hydrogen generation employing a microemulsion-based photocatalytic water reduction metallosurfactant system [57]. Traditionally, the H_2-evolving half-reaction, or water reduction, is a light-driven process, proceeding by means of a photosensitizer (PS) and a water (or proton)-reducing catalyst (WRC), with a sacrificial reductant (SR) substituting the oxidative half. Iridium- and rhodium-based metallosurfactants, namely $[Ir(ppy)_2(dhpdbpy)]Cl$ and $[Rh(dhpdbpy)_2Cl_2]Cl$ (where ppy = 2-phenylpyridine and dhpdbpy = 4,4′-diheptadecyl-2,2′-bipyridine), were employed as photosensitizer (PS) and WRC, respectively, along with oxalic acid as a sacrificial reductant or electron donor in a toluene/water biphasic mixture. Before the onset of catalysis, the metallosurfactants and oxalic acid were present in the organic and aqueous phases, respectively. The microemulsion formation is proposed to be initiated by the addition of 1-octylamine, which forms an in situ ion pair with oxalic acid acting as an additional surfactant, bringing it in close proximity to the PS and WRC making interactions with the metallosurfactants (present in the organic layer) more favorable, assisting in the electron transfer and proton reduction. This is a great step forward in the metalloamphiphile-based photocatalytic system, which generated H_2 for ~95 hours, highlighting its stability. Donner et al. highlighted the role of Cu(I)- and Fe(II)-bearing NHC metallosurfactant in emulsion polymerizations under atom transfer radical polymerization (ATRP) conditions [58]. The synthesized systems served as all in one emulsifiers, active catalysts, and stabilization agents. The metallosurfactant stabilized the water/methyl methacrylate (MMA) emulsion forming aggregates with a size of the order of 100–110 nm per DLS and electron microscopy studies. The microemulsion droplets define the loci of polymerization and the site where polymerization takes place to yield stable poly(methyl methacrylate) (PMMA) colloids in water with the catalyst located at the surface of the colloid. The PMMA colloid chains terminated by bromine groups containing the copper surfactant at the surfaces acted as macroinitiators for the copolymerization of styrene. Surfactant is not able to migrate into the PS phase and remains bound at the PMMA surface and occupies the PMMA–PS interface.

In another study, the same group successfully catalyzed the Suzuki coupling reaction of phenylboronic acid with 4-bromoacetophenone using the micellar phases of the Pd-NHC surfactant [59]. In a separate study, Taira et al. also evaluated the catalytic activity of NHC–PdMS for aqueous Mizoroki–Heck reactions. Iodobenzene reacted with styrene in the presence of NHC–PdMS (3 mol%) catalyst in dimethylformamide (DMF), which formed an oil-in-water (O/W) emulsion. Vigorous stirring of the emulsion at 70 °C for 24 hours afforded stilbene as the quantitative coupling product [36].

3.4 Conclusions

Metallosurfactants indeed form the "next generation" of surfactants, which can be designed to include specific features and characteristics in the molecule as desired. The structural framework of the metallosurfactant can be fine tuned according to the need and desired applications by varying the transition metal ion, alkyl chains, or type of ligands attached. These surfactants give rise to various self-assembled nanostructures at the interfaces and in the aqueous/non aqueous media such as micelles, reverse micelles, vesicles, reverse vesicles, lamellae, and worm like and are also capable of showing a transition in between these structures, which can be successfully employed as nanoreactors or reaction vessels to carry out several reactions. These surfactant-cum-catalyst molecules play a significant role in directing the reactions at fluid and air/water interfaces with several advantages over conventional systems such as carrying out reactions in aqueous media, facile in-flask recyclability of catalyst, and product extraction with minimal organic solvent. Hence, the rational designing and modification of metallosurfactants finds potential applications in catalytic chemistry to conduct reactions, which otherwise would not be feasible.

Acknowledgment

Aashima Sharma is thankful to CSIR for open SRF fellowship.

References

1 Satake, I., Iwamatsu, I., Hosokawa, S., and Matuura, R. (1963). The surface activities of bivalent metal alkyl sulfates. I. On the micelles of some metal alkyl sulfates. *Bulletin of the Chemical Society of Japan* 36: 204–209.

2 Scrimin, P., Tecilla, P., Tonellato, U., and Vendrame, T. (1989). Aggregate structure and ligand location strongly influence copper(II) binding ability of cationic metallosurfactants. *Journal of Organic Chemistry* 54 (25): 5988–5991.

3 Fornasier, R., Scrimin, P., Tonellato, U., and Zanta, N. (1988). Highly enantioselective cleavage of α-amino acid p-nitrophenyl esters by chiral metallomicelles. *Journal of the Chemical Society, Chemical Communications*: 716–718.

4 Wang, Q., Li, Z., Tao, D.D. et al. (2016). Supramolecular aggregates as sensory ensembles. *Chemical Communications* 2016 (52): 12929–12939.

5 Zhang, J., Meng, X.G., Zeng, X.C., and Yu, X.Q. (2009). Metallomicellar supramolecular systems and their applications in catalytic reactions. *Coordination Chemistry Reviews* 253: 2166–2177.

6 Accardo, A., Tesauro, D., Aloj, L. et al. (2009). Supramolecular aggregates containing lipophilic Gd(III) complexes as contrast agents in MRI. *Coordination Chemistry Reviews* 253: 2193–2213.

7 Kaur, R. and Mehta, S.K. (2014). Self aggregating metal surfactant complexes: precursors for nanostructures. *Coordination Chemistry Reviews* 262: 37–54.

8 Kaur, R., Giordano, C., Gradzielski, M., and Mehta, S.K. (2014). Synthesis of highly stable, water-dispersible copper nanoparticles as catalysts for nitrobenzene reduction. *Chemistry – An Asian Journal* 9 (1): 189–198.

9 Marın-Garcıa, M., Benseny-Cases, N., Camacho, M. et al. (2017). Low-toxicity metallosomes for biomedical applications by self-assembly of organometallic metallosurfactants and phospholipids. *Chemical Communications* 53: 8455–8458.

10 Marín-García, M., Benseny-Cases, N., Camacho, M. et al. (2018). Metallosomes for biomedical applications by mixing molybdenum carbonyl metallosurfactants and phospholipids. *Dalton Transactions* 47: 14293–14303.

11 Hassaniazad, M., Inchehsablagh, B.R., Kamali, H. et al. (2020). The clinical effect of nano micelles containing curcumin as a therapeutic supplement in patients with Covid-19 and the immune responses balance changes following treatment: a structured summary of a study protocol for a randomised controlled trial. *Trials* 21 (1): 876.

12 Alavi, M., Asare-Addo, K., and Nokhodchi, A. (2020). Lectin protein as a promising component to functionalize micelles, liposomes and lipid NPs against coronavirus. *Biomedicine* 8 (12): 580–596.

13 Tanori, J. and Pileni, M.P. (1997). Control of the shape of copper metallic particles by using a colloidal system as template. *Langmuir* 13: 639–646.

14 Lisiecki, I. and Pileni, M.P. (1993). Synthesis of copper metallic clusters using reverse micelles as microreactors. *Journal of the American Chemical Society* 115 (10): 3887–3896.

15 Lisiecki, I., Billoudet, F., and Pileni, M.P. (1996). Control of the shape and the size of copper metallic particles. *Journal of Physical Chemistry A* 100 (10): 4160–4166.

16 Lisieki, I. and Pileni, M.P. (1995). Copper metallic particles synthesized "in situ" in reverse micelles: influence of various parameters on the size of the particles. *Journal of Physical Chemistry A* 99: 5077–5082.

17 Kaur, R. and Mehta, S.K. (2017). Metallomicelle templated transition metal nanostructures: synthesis, characterization, DFT study and catalytic activity. *Physical Chemistry Chemical Physics* 19: 18372–18382.

18 Griffiths, P.C., Fallis, I.A., Willock, D.J. et al. (2004). The structure of metallomicelles. *Chemistry A European Journal* 10: 2022–2028.

19 Guerrero-Martınez, A., Vida, Y., Domınguez-Gutierrez, D. et al. (2008). Tuning emission properties of iridium and ruthenium metallosurfactants in micellar systems. *Inorganic Chemistry* 47: 9131–9133.

20 Taira, T., Suzaki, Y., and Osakada, K. (2008). Pd^{II} and Pt^{II} complexes with amphiphilic ligands: formation of micelles and [5]rotaxanes with a-cyclodextrin in aqueous solution. *Chemistry – An Asian Journal* 3: 895–902.

21 Regev, O., Franc, J., Gohy, O. et al. (2004). Dynamic light scattering and cryogenic transmission electron microscopy investigations on metallo-supramolecular aqueous micelles: evidence of secondary aggregation. *Colloid and Polymer Science* 282: 407–411.

22 Petit, C., Lixon, P., and Pileni, M.P. (1991). Structural study of bimetallic bis(2-ethylhexyl) sulfosuccinate aggregates. *Langmuir* 7: 2620–2625.

23 Tumori, J. and Pileni, M.P. (1995). Change in the shape of copper nanoparticles in ordered phases. *Advanced Materials* 7 (10): 862–864.

24 Pileni, M.P. (1997). Nanosized particles made in colloidal assemblies. *Langmuir* 13: 3266–3276.

25 Koutsantonis, G.A. and Nealon, G.L. (2007). Wormlike micelles from a cage amine Metallosurfactant. *Langmuir* 23: 11986–11990.

26 Domınguez-Gutierrez, D., Paoli, G.D., Guerrero-Martınez, A. et al. (2008). Inverted aggregates of luminescent ruthenium metallosurfactants. *Journal of Materials Chemistry* 18: 2762–2768.

27 Owen, T., Pynn, R., Hammouda, B., and Butler, A. (2007). Metal-dependent self-assembly of a microbial surfactant. *Langmuir* 23: 9393–9400.

28 Owen, T., Webb, S.M., and Butler, A. (2008). XAS study of a metal-induced phase transition by a microbial surfactant. *Langmuir* 24 (9): 4999–5002.

29 Parera, E., Comelles, F., Barnadas, R., and Suades, J. (2010). New surfactant phosphine ligands and platinum(II) metallosurfactants. Influence of metal coordination on the critical micelle concentration and aggregation properties. *Langmuir* 26 (2): 743–751.

30 Parera, E., Comelles, F., Barnadas, R., and Suades, J. (2011). Formation of vesicles with an organometallic amphiphile bilayer by supramolecular arrangement of metal carbonyl metallosurfactants. *Chemical Communications* 47: 4460–4462.

31 Parera, E., Marín-García, M., Pons, R. et al. (2016). Supramolecular arrangement of molybdenum carbonyl metallosurfactants with CO-releasing properties. *Organometallics* 35 (4): 484–493.

32 Zha, Q., Xie, Q., Hu, Y. et al. (2016). Metallosurfactants C_n–Cu–C_n: vesicle formation and its drug-controlled release properties. *Colloid and Polymer Science* 294: 841–849.

33 Sommerdijk, N.A.J.M., Booy, K.J., Pistorius, A.M.A. et al. (1999). Copper(II) complexes of a dicephalic imidazole surfactant. Tunable organization of metalloaggregates. *Langmuir* 15: 7008–7013.

34 Dong, R. and Hao, J. (2012). Reverse vesicles of ferrum laurate metallosurfactant in non-aqueous solution dried to produce solid shells. *ChemPhysChem* 13 (17): 3794–3797.

35 Mahato, P., Saha, S., Choudhury, S., and Das, A. (2011). Solvent-dependent aggregation behavior of a new Ru(II)-polypyridyl based metallosurfactant. *Chemical Communications* 47: 11074–11076.

36 Taira, T., Yanagimoto, T., Sakai, K. et al. (2018). Self-assembling properties of an N-heterocyclic carbene-based metallosurfactant: Pd-coordination induced formation of reactive interfaces in water. *Journal of Oleo Science* 67 (9): 1107–1115.

37 Polarz, S., Kunkel, M., Donner, A., and Schlotte, M. (2018). Added-value surfactants. *Chemistry – An Asian Journal* 24: 18842–18856.

38 Sorella, G.L., Strukul, G., and Scarso, A. (2015). Recent advances in catalysis in micellar media. *Green Chemistry* 17: 644–683.

39 Scarso, A. (2016). Micellar nanoreactors. In: *Encyclopedia of Inorganic and Bioinorganic Chemistry* (ed. R.A. Scott), 1–16. Wiley.

40 Bhattacharya, S. and Kumari, N. (2009). Metallomicelles as potent catalysts for the ester hydrolysis reactions in water. *Coordination Chemistry Reviews* 253: 2133–2149.

41 Lipshutz, B.H. and Ghorai, S. (2012). Organocatalysis in water at room temperature with in-flask catalyst recycling. *Organic Letters* 4 (1): 422–425.

42 Lipshutz, B.H. and Ghorai, S. (2010). PQS-2: ring-closing- and cross-metathesis reactions on lipophilic substrates; in water only at room temperature, with in-flask catalyst recycling. *Tetrahedron* 66 (5): 1057–1063.

43 (a) Kingsbury, J.S., Harrity, J.P.A., Bonitatebus, P.J., and Hoveyda, A.H. (1999). A recyclable Ru-based metathesis catalyst. *Journal of the American Chemical Society* 21 (4): 791–799. (b)Lipshutz, B.H. and Ghorai, S. (2009). PQS: a new platform for micellar catalysis. RCM reactions in water, with catalyst recycling. *Organic Letters* 11: 705–708.

44 Lipshutz, B.H., Isley, N.A., Moser, R. et al. (2012). Rhodium-catalyzed asymmetric 1,4-additions, in water at room temperature, with in-flask catalyst recycling. *Advanced Synthesis and Catalysis* 354 (17): 3175–3179.

45 Franc, A., Mingotaud, O., Kramer, M., and Mingotaud, C. (2007). Catalytic surfactants for ring-opening metathesis polymerization and ring-closing metathesis in non-degassed micellar solutions. *Journal of Molecular Catalysis A: Chemical* 263: 39–47.

46 Li, J., Tang, Y., Wang, Q. et al. (2012). Chiral surfactant-type catalyst for asymmetric reduction of aliphatic ketones in water. *Journal of the American Chemical Society* 134 (45): 18522–18525.

47 Li, Z., Li, J., Huang, Q. et al. (2015). Chiral surfactant-type catalyst: enantioselective reduction of long-chain aliphatic ketoesters in water. *Journal of Organometallic Chemistry* 80 (9): 4419–4429.

48 Hamasaka, G., Mutoa, T., and Uozumi, Y. (2011). A novel amphiphilic pincer palladium complex: design, preparation and self-assembling behavior. *Dalton Transaction* 40: 8859–8868.

49 Hamasaka, G., Sakuraia, F., and Uozumi, Y. (2015). A vesicular self-assembled amphiphilic palladium NNC-pincer complex-catalyzed allylic arylation of allyl acetates with sodium tetraarylborates in water. *Tetrahedron* 71 (37): 6437–6441.

50 Hamasaka, G. and Uozumi, Y. (2014). Cyclization of Alkynoic acids in water in the presence of a vesicular self-assembled amphiphilic pincer palladium complex catalyst. *Chemical Communications* 50: 14516–14518.

51 Hamasaka, G., Muto, T., and Uozumi, Y. (2011). Molecular-architecture-based administration of catalysis in water: self-assembly of an amphiphilic palladium pincer complex. *Angewandte Chemie International Edition* 50: 4876–4878.

52 Hamasaka, G. and Uozumi, Y. (2016). The development of a vesicular self-assembled amphiphilic platinum NCN-pincer complex and its catalytic application to hydrosilylation of alkenes in water. *Chemistry Letters* 45: 1244–1246.

53 Hamasaka, G., Muto, T., Andoh, Y. et al. (2017). Detailed structural analysis of a self-assembled vesicular amphiphilic NCN-pincer palladium complex by using wide-angle X-ray scattering and molecular dynamics calculations. *Chemistry A European Journal* 23: 1291–1298.

54 Toth, A., Schnedl, S., Painer, D. et al. (2019). Interfacial catalysis in biphasic carboxylic acid esterification with a nickel-based Metallosurfactant. *ACS Sustainable Chemistry & Engineering* 7: 18547–18553.

55 Graham, T.R., Renslow, R., Govind, N., and Saunders, S.R. (2016). Precursor ion–ion aggregation in the Brust–Schiffrin synthesis of alkanethiol nanoparticles. *Journal of Physical Chemistry A* 120 (35): 19837–19847.

56 Perala, S.R.K. and Kumar, S. (2013). On the mechanism of metal nanoparticle synthesis in the Brust–Schiffrin method. *Langmuir* 29 (31): 9863–9873.

57 Kagalwala, H.N., Chirdon, D.N., Mills, I.N. et al. (2017). Light-driven hydrogen generation from microemulsions using metallosurfactant catalysts and oxalic acid. *Inorganic Chemistry* 56: 10162–10171.

58 Donner, A., Trepka, B., Theiss, S. et al. (2019). NHC-metallosurfactants as active polymerization catalysts. *Langmuir* 35: 16514–16520.

59 Donner, A., Hagedorn, K., Mattes, L. et al. (2017). Hybrid surfactants with N-heterocyclic carbene heads as a multifunctional platform for interfacial catalysis. *Chemistry A European Journal* 23: 18129–18133.

4

Hydrolytic Metallosurfactants: Nanocatalysts for Esterolytic Reactions

Bhanushree Gupta[1], Rahul Sharma[2], and Kallol K. Ghosh[1,3]

[1] *Pt. Ravishankar Shukla University, Center for Basic Sciences, GE Road, Amanaka, Raipur 492010, Chhattisgarh, India*
[2] *Chhattisgarh Council of Science & Technology, Intellectual Property Facilitation Centre, Patent Information Centre, Daldal Seoni, Mowa, Raipur 492014, Chhattisgarh, India*
[3] *Pt. Ravishankar Shukla University, School of Studies in Chemistry, GE Road, Amanaka, Raipur 492010, Chhattisgarh, India*

4.1 Introduction

Metallosurfactants, as implied from the name itself, belong to special class of surfactants in which a metal atom is present as an integral part of the surfactant architecture. They form spherical nano-assembly and provide a perfect blend of both amphiphilic character and surface activity [1]. Generally d- and f-block metal complexes are incorporated to amphiphilic moiety, which supplies extraordinary catalytic activity to it [2]. The metal atom is chelated to several structurally different lipophilic ligands. The motivation of designing metallosurfactants originated from outstanding catalytic action of metalloenzymes. Metalloenzymes are enzyme proteins and responsible for many important biological reactions [3]. The skeletal structure of metalloenzymes contains a metal ion; co-factor and enzyme/protein [4]. Carboxypeptidase A is the best example of metalloenzyme with Zn^{2+} metal ion; it hydrolyzes peptide bond. Hence, this class of surfactants mimics the functioning action of enzyme active site and exhibits outstanding catalytic efficiency.

The metal complexes have several interesting features (like magnetic properties, variable charge, and promising catalytic activity), which, after insertion into amphiphilic moiety, are exhibited at interfaces [5]. The most fascinating aspect of metallosurfactants is that it possesses similar number of complexation sites with respect to its aggregation number in the micelles, hence proves useful in various applications [6]. Figure 4.1 represents the comparative structural aspects of a common surfactant and metallosurfactant with its types. Metallosurfactants can be broadly classified into three types [7] (presented in Figure 4.1B1–B3). In first type, metal ion with its coordinated sphere is present in hydrophilic head group and hydrophobic part is represented by alkyl chain. In second class, metal atom is situated in the hydrophobic alkyl chain part. The last classification involves

Metallosurfactants: From Fundamentals to Catalytic and Biomedical Applications, First Edition.
Edited by Surinder K. Mehta and Ravneet Kaur.

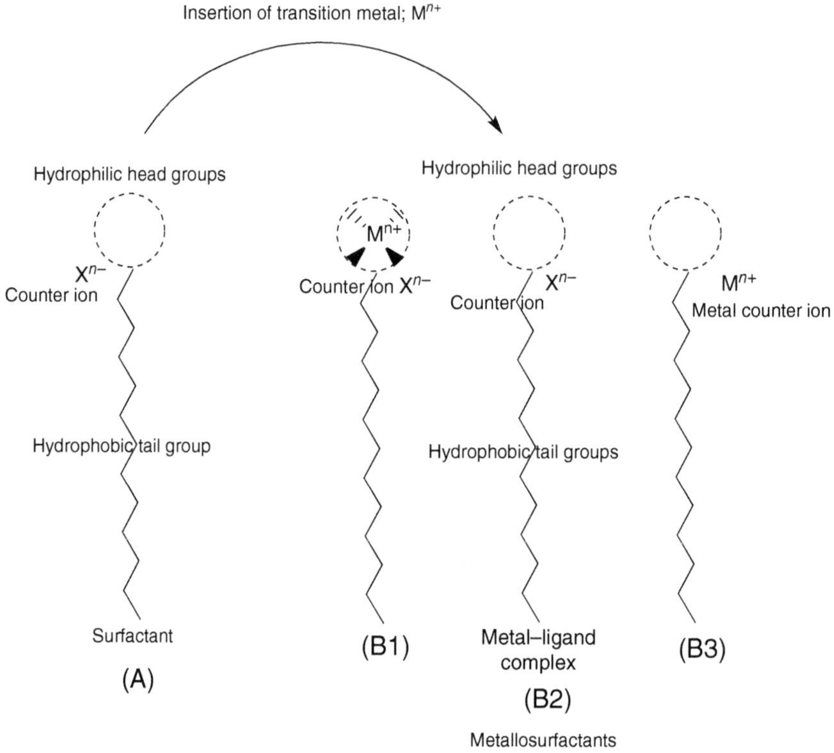

Figure 4.1 (a) Structure of common surfactant, (b) structure of metallosurfactant with types; (B1) metal ion coordinated to hydrophilic head group, (B2) metal ligand-coordinated sphere situated in hydrophobic tail group, (B3) metal present as counterion.

the presence of metal as counterion in the architecture of surfactant. The CMC (critical micelle concentration) of metallosurfactants is very low as compared to conventional surfactants, and their amphiphilic architecture forms a nanosized assembly, which act as nanoreactor for several important biochemical reactions [8]. Generally, the first class of metallosurfactants is applied as a catalyst to study various esterolytic reactions.

4.2 Metallosurfactants as Nanocatalyst for Esterolytic Reactions

Hydrolytic cleavage of various biologically important esters (carboxylate/phosphate) is essential for several biochemical processes. In metallosurfactants, metal ions (Figure 4.1B1) are endowed with ligands at polar head group to produce effective chelated architecture. These surfactants imitate certain characteristic properties of hydrolytic metalloenzymes upon aggregation in water. Generally, the metal ions are present at the active site of metalloenzyme and coordinate with H_2O molecule [5, 6]. This results in increased acidity of bound H_2O molecule by polarizing H—O

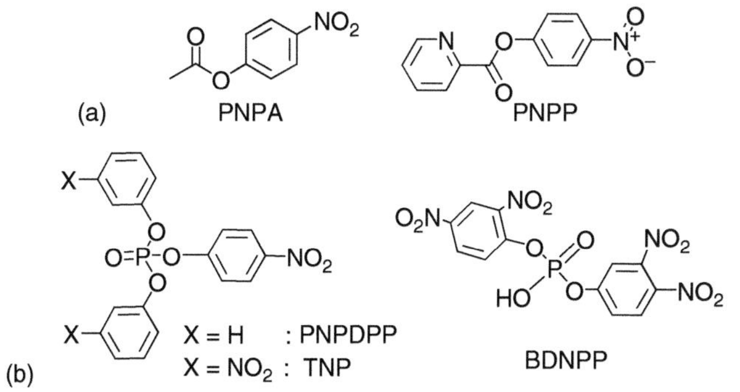

Figure 4.2 Commonly investigated carboxylate and phosphate esters: (a) PNPA; *p*-nitrophenylacetate, PNPP; *p*-nitrophenylphosphate. (b) PNPDPP; *p*-nitrophenyl diphenylphosphate, TNP; trinitro phosphate, BDNPP; bis-2,4 dinitrophenyl phosphate.

bond and thus further enhances the nucleophilicity. Thus, in metallomicelles, metal ion acts as Lewis acid, which binds and activates C=O or P=O centers of ester/amide/phosphate substrates. This in turn makes C atom of carbonyl group or P atom of phosphoryl group more prone to nucleophilic attack. During the course of hydrolysis, it further supplies the metal-bound HO– nucleophile to attack on C=O or P=O centers of the substrates [5, 6]. The major benefit of such metalloaggregates is that it can attract ionic substances and can fit in hydrophobic solutes too.

In past decades, extensive efforts have been made to investigate hydrolytic cleavage of carboxylate and phosphate esters with different catalytic systems [9]. Chemical structure of some common esters has been presented in Figure 4.2. The first hydrolytic metallosurfactant was reported by Gutsche and coworkers in 1978 [10, 11]. The group studied hydrolytic cleavage of acetyl phosphate with metal-chelated polyamine and polyamide micelles. In metallomicelles imidazole, pyridine, trizole, crown ether, phenanthroline-based ligands are generally applied with Cu^{2+}, Zn^{2+}, Ni^{2+}, and Co^{2+} metal ions. Some f-block elements (lanthanum, cerium, etc.) are also utilized in preparation of metallosurfactants.

4.3 Catalytic Hydrolysis of Carboxylate Esters

Metallosurfactants are broadly applied as catalysts to study various carboxylate esters. In various metalloenzymes, imidazole attached to a histidine moiety is commonly noticed as a ligand of Zn^{2+} ion, which promotes enzyme-catalyzed reactions [5]. Following this idea, several systems have been devised, which can exhibit efficiency of both imidazole and OH- groups in the presence of Zn^{2+} ion. Tagaki and coworkers [12] worked with the same motive and developed imidazole-based ligands (**1–7**) (Figure 4.3) and incorporated them with Zn^{2+} metal ions to investigate hydrolytic cleavage of PNPP; *p*-nitrophenyl picolinate. The group also prepared a series with Cu^{2+} metal ions. Ligands **1** and **2** exhibited outstanding activity for hydrolysis reaction. The observed result was obtained due to –OH group, which was

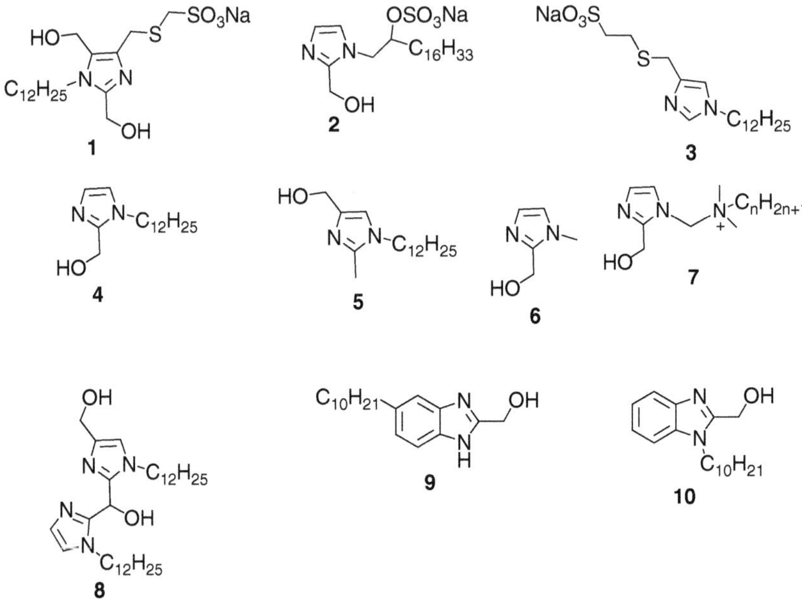

Figure 4.3 Imidazole-based metallosurfactant ligands prepared by Tagaki et al. Source: Adapted from Tagaki et al. [12].

activated by metal ions for transacylation process. They also compared the catalytic ability of **4** and **5** and observed **4** to be more potent.

The group [13] further investigated the co-micellar effect for Zn^{2+} and Cu^{2+} complexes formed with ligands **4–10** in the presence of various cationic, anionic, and non-ionic surfactants. The observed rate constants indicate that hydrolytic cleavage by Cu-**4** was hundred times greater than Cu-**6**. Figure 4.4 displays the assumed selective activation of primary and secondary –OH groups of **8** to coordinate with Zn^{2+} and Cu^{2+} metal ions to participate in the hydrolytic cleavage. Benzimidazole-based ligands (**9** and **10**) complexed with Zn^{2+} ion were also applied to study the breakdown of PNPP with co-micelles of cetyltrimethylammonium bromide (CTAB) [14].

In view of developing proficient catalytic systems, several modifications were carried out and pyridine-based surfactant ligands were prepared. The contributions

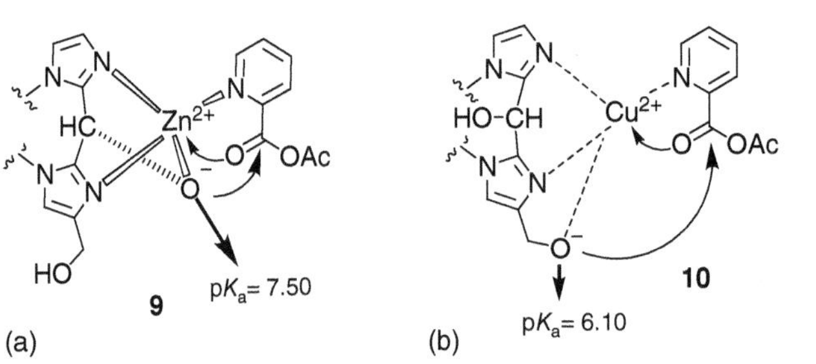

Figure 4.4 Selective activation of primary and secondary –OH groups of ligand **9** by Zn^{2+} (a) and Cu^{2+} (b).

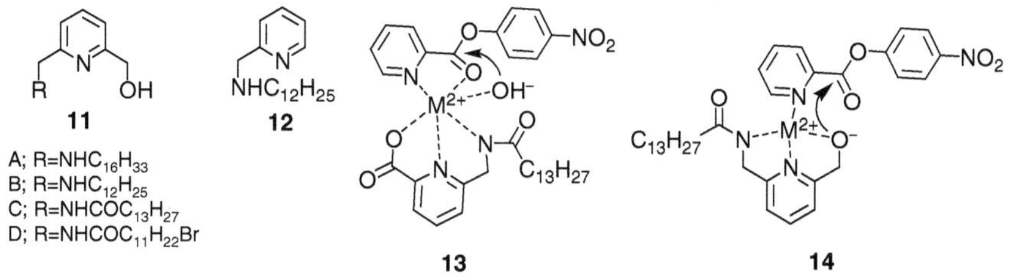

Figure 4.5 Pyridine-based ligands prepared by Tonellato and coworkers. Source: Adapted from Fornasier et al. [16].

of Tonellato and coworkers [15, 16] are considerable in this regard. They designed and synthesized a series of structurally different pyridine-based ligands (Figure 4.5) to investigate their hydrolytic strength against carboxylic esters in co-micellar conditions.

The hydrolysis rate enhanced several times with the incorporation of metal ions to pyridine surfactant ligands. Due to the presence of –OH group in ligand **11**, the generated metallomicelles were found to be more potent for hydrolysis than **12**. Investigation also indicated the necessity of optimal chain length of ligand for desired catalytic strength. The formation of ternary complexes (**13** and **14**) was also suggested based on Job plots. The complex contains metal ion, ligand, and substrate and plays the role of nucleophile. As a result, the increment in rate was recorded [17, 18].

The co-micelles formed by CTAB and Cu-**15a** and Zn-**15a** demonstrated considerable catalytic activity for the hydrolysis of PNPP. The methylated ligand **15b** possessed least catalytic power as compared to **15a** and **15c**, which has no hydrophobic chain. The authors suggested three steps for the hydrolytic mechanisms. (i) The creation of a ternary complex between the substrate, metal ion, and lipophilic ligand. (ii) The formation of transacylation intermediate and liberation of *p*-nitrophenol due to the pseudo-intramolecular nucleophilic attack of the hydroxyl ligand. (iii) The hydrolysis of intermediate by metal ion and regeneration of the catalytic moiety [16, 19].

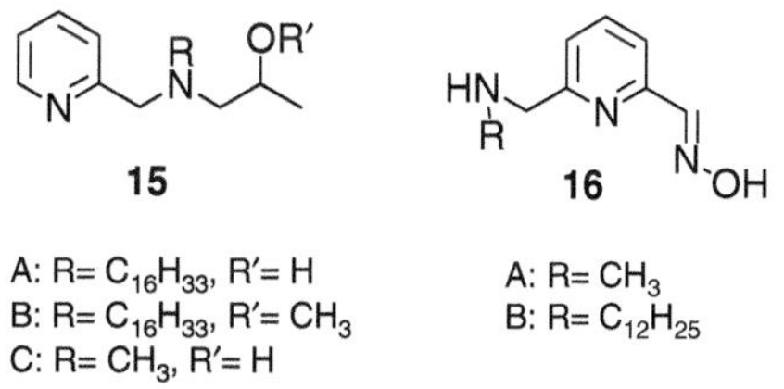

Moreover, oxime-based pyridine ligands (**16**) were reported with their Ni^{2+} and Zn^{2+} complexes to investigate the breakdown of *p*-nitrophenyl hexanoate; PNPH and acetate; PNPA in the co-miceller system of CTAB [20]. Appreciable increment in rate was observed for acidic medium than in neutral medium.

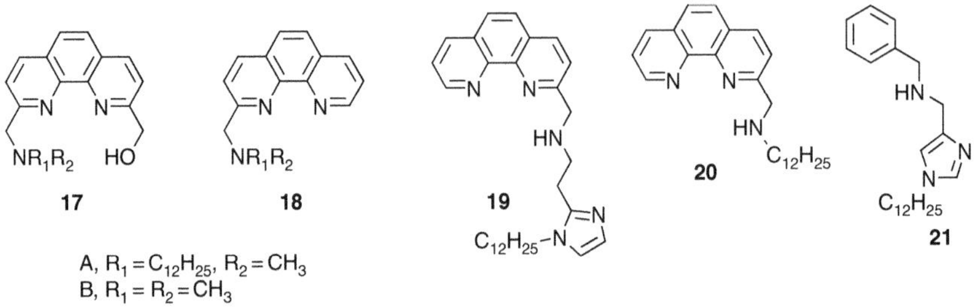

Figure 4.6 Phenanthroline-based ligands synthesized by Engbersen and coworkers. Source: Adapted from Weijnen et al. [21].

Phenanthroline-based novel ligands (Figure 4.6) were also reported by Engbersen [21, 22] and team. Several metals were used for complexation in which 1 : 1 type complex was formed with ligand **17a**. The tendency of catalytic activity followed the order – $Zn^{2+} > Co^{2+} > Cd^{2+} > Ni^{2+}$. On the other hand, hydrolytic potency of **17b** was found to be in 1 : 1 and 2 : 1 for complexation stoichiometry with Co^{2+} ion [21]. The incorporation of imidazole ring resulted in the formation of **19** and **21**, which demonstrated appropriate potential to form comicelles with different surfactants (like CTAB and Brij 35). Hence, it played a role of competent catalyst for the hydrolysis of PNPP as compared to ligands **17, 18**.

The assumed reaction mechanism for the hydrolytic breakdown of PNPP with complex of Zn with ligand **20** is shown in Figure 4.7. In model **A** of Figure 4.7, coordinated carbonyl oxygen is complexed with a transition metal ion and provided a smooth nucleophilic attack of free hydroxide anion. Model **B** displays the coordination of Zn^{2+} ion with free hydroxide anion to encourage an intramolecular nucleophilic attack.

Further, several macrocyclic metal complexes were also reported to study esterolytic reactions. Kimura et al. [23] investigated the catalytic activity of long-chain macrocyclic complex **22** toward the hydrolysis of PNPA. The ability of **22a** to form co-micelles with Triton X-100 exhibited ~50 times greater activity than that of the water-soluble Zn^{2+} complex of **22b**. The Cu^{2+} complexes (**23**, **24**) of lipophilic macrocyclic polyamines were developed by Griffiths et al. [24]. The authors studied the morphology of formation of aggregates by small-angle neutron scattering (SANS) and small-angle-X-ray scattering (SAXS). Hughes and

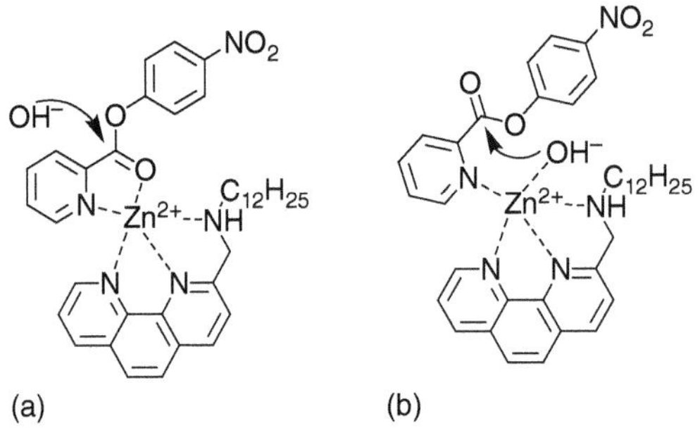

Figure 4.7 Proposed mechanisms of hydrolytic cleavage of PNPP catalyzed by Zn-**20**.

coworkers [25] also synthesized a triamine-based copper complexed metallosurfactant **25**; Cu(II)-1-tetradecyl diethylenetriamine; Cu(II) TDET and tested its esterolytic activity toward the cleavage of several esters.

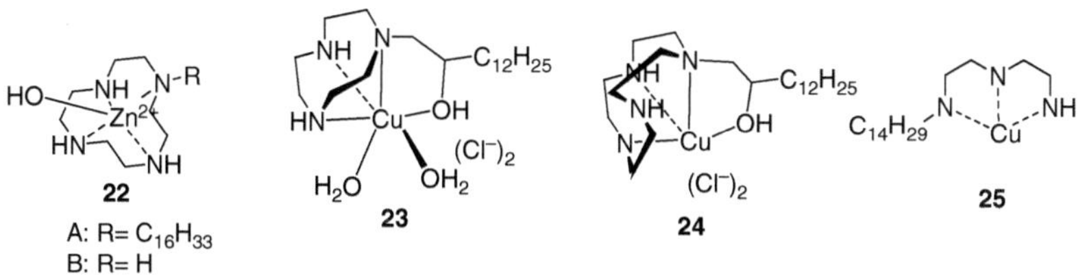

A novel class of triazole-based ligand (**26**) has been reported by Qiu et al. [26] to investigate the hydrolytic breakdown of PNPP in comiceller media of CTAB. The ligand was coordinated with two metal ions Ni^{2+} and Cu^{2+} of which Cu^{2+} complex displayed better catalytic action. Tendency of coordination and formation of ternary complexes increased with lengthening of lipophilic tail.

Moreover, the group [27] employed this ligand with gemini surfactants ($C_{12}H_{25}$-NMe_2^+-$(CH_2)_S$-NMe_2^+-$C_{12}H_{25}$, where $S = 2,3,6$), which resulted in effective catalytic system for breakdown of PNPP. In addition, the authors have also investigated the potency of ligand **27** for hydrolytic cleavage of PNPP in the presence of gemini and single-chain conventional surfactants [28].

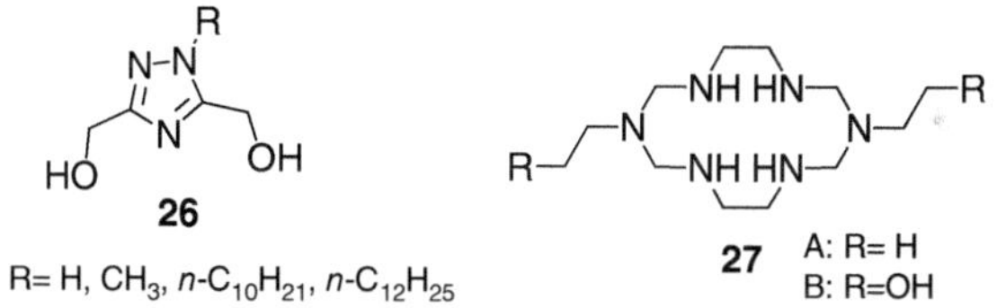

Few more amphiphilic ligands were developed (**28**) and coordinated with different metal ions (like Cu^{2+}, Ni^{2+}, and Zn^{2+}) to monitor the hydrolysis of PNPP in comicellar medium of several cationic and non-ionic surfactants [29]. The observed trend of construction of complexes with the change of medium is presented in (**29**).

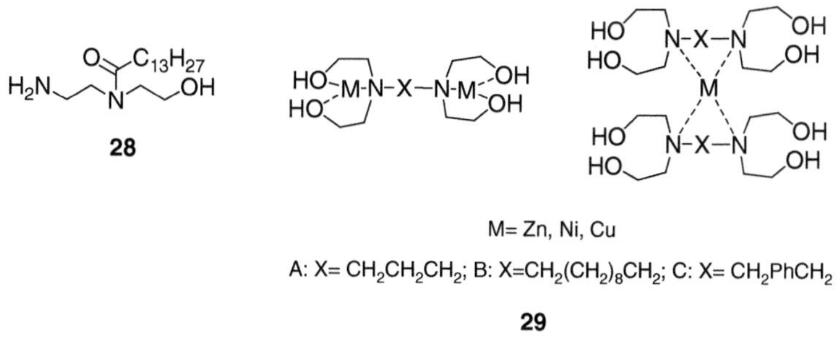

Some crown ether–based aromatic Schiff base (**30, 31**) complexes of cobalt were also prepared [30] and investigated for catalytic hydrolysis of PNPA in comicellar medium of cationic (CTAB) and gemini surfactant ($C_{16}H_{33}$-NMe_2^+-$(CH_2)_6$-NMe_2^+ -$C_{16}H_{33}$). As per the obtained results, gemini micellar medium was recognized as a more effective catalytic system than CTAB. In gemini micellar media, complex of **30** displayed two times greater activity than **31**, which indicates that steric bulk of aza-crown group creates steric hindrance and blocks the active site in micellar media for compound **31** and thus suppresses the hydrolysis rate.

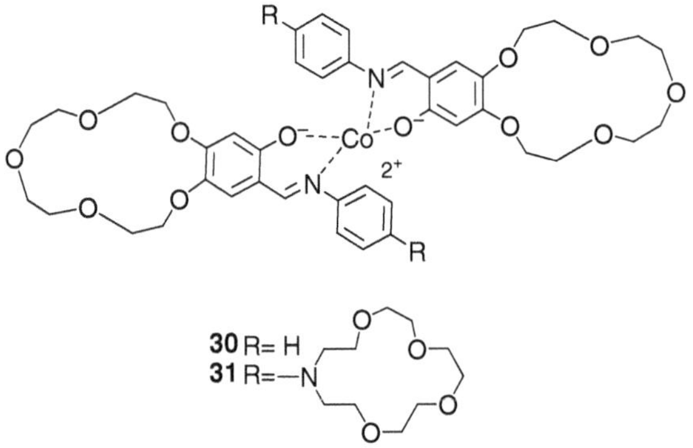

Dinuclear copper complexes bridged with alkoxo/hydroxide moieties (**32, 33**) were reported by Xu et al. [31]. The group further examined these complexes for hydrolytic breakdown of PNPP in the co-micellar system of some conventional and gemini surfactants. The activity of dinuclear copper complexes toward the hydrolysis of ester was fourfold higher than mononuclear copper complexes. The co-micelles of gemini surfactants appreciably enhanced the rate of hydrolysis and exhibited better activity than conventional CTAB. The proposed mechanism for breakdown of PNPP via metallosurfactant **33** in co-micellar medium of gemini surfactant is shown in Figure 4.8.

4.4 Catalytic Hydrolysis of Phosphate Esters

In view of biological significance of phosphate esters, a lot of studies [32] have been carried out to understand the metallosurfactant-catalyzed breakdown of phosphate esters. Phosphate esters also exist in the form of deleterious nerve agents and pesticides. Hence, their catalytic breakdown into non-toxic molecules is highly important. Gellman et al. first time reported metallosurfactant-catalyzed hydrolysis of a triester PNPDPP; *p*-nitrophenyl diphenylphosphate in 1986 [33]. The authors applied complex of Zn^{2+} ion with ligand **34** and identified it as an effective catalyst at a little alkaline pH.

Menger et al. [34] prepared copper-based metallosurfactants with ligand **35** and studied the hydrolytic breakdown of PNPDPP. The system displayed considerable rate increment with significant turnover. Kimura et al. [23] reported a complex **36** of Zn with hexadecylcyclen and compared its catalytic ability with **34** against the hydrolytic breakdown of TNP; tris (*p*-nitrophenyl) phosphate in comicelles of Triton-X.

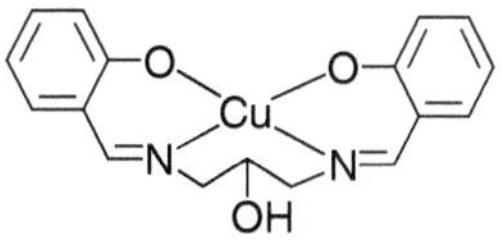

32

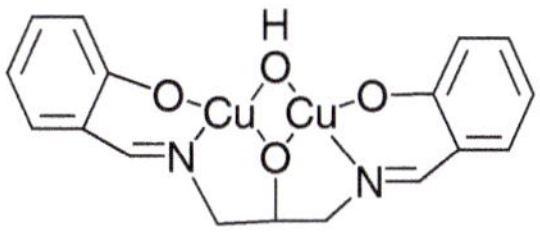

33

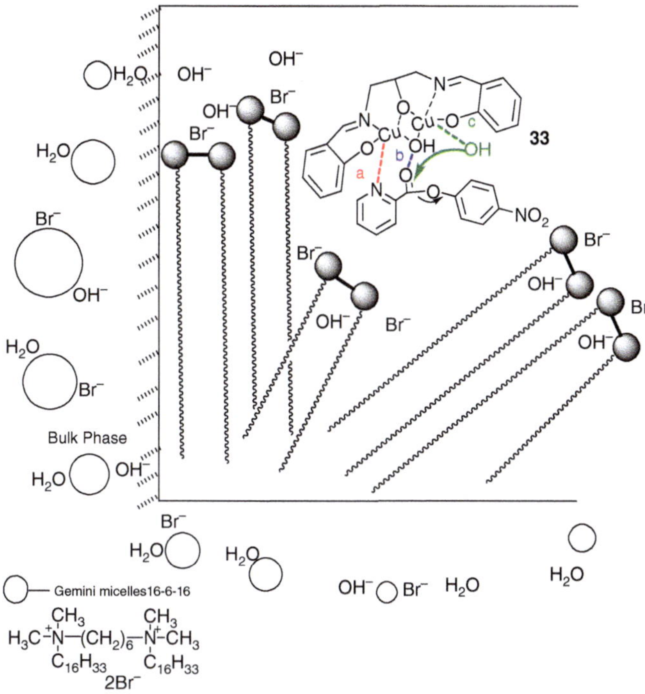

Figure 4.8 PNPP hydrolysis by Cu^{2+}-metallomicelles in the presence of gemini comicelles.

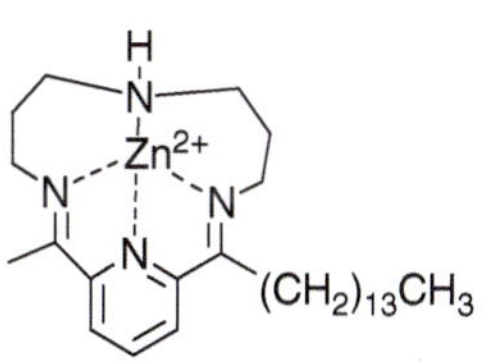

34

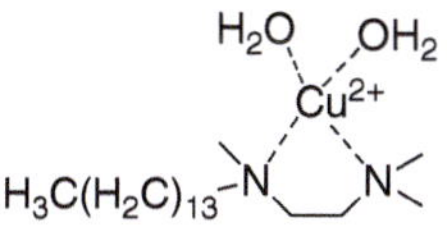

35

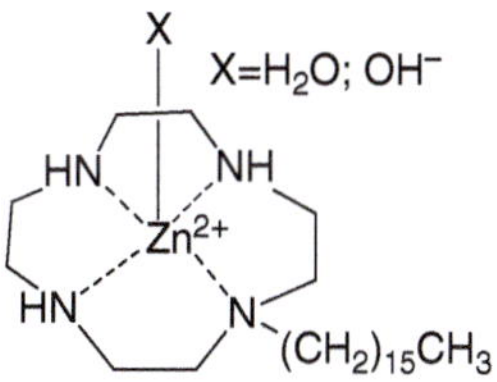

36

The rate increment in comicellar media is attributed to increased hydrophobicity of substrate TNP, which facilitated smooth entry to micelles and nucleophilic attack of **36** to supply intramolecular hydrolysis. Generally, the catalytic efficiency of metallosurfactants is dissimilar for structurally dissimilar esters, i.e. a ligand may act as a proficient catalyst for phosphate ester but may have poor catalytic activity for carboxylate or other esters. Scrimin et al. [17] verified this observation through their study in which they examined pyridine-based metallosurfactants for the hydrolytic cleavage of different esters.

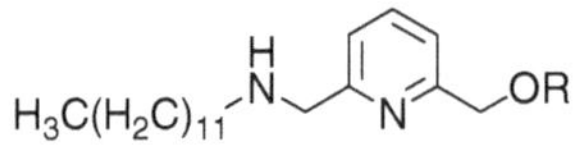

37: R= CH_3

The group has drawn a conclusion that the catalytic strength of metallosurfactants is greatly affected by the type of ester. The breakdown of carboxylate ester was supported by ligand **11d**, which contained –OH group. The cleavage of PNPH was insignificant by methylated ligands **37** and **12**; on the other hand, ligand **37** has shown superior reactivity against breakdown of PNPDPP. It was concluded from the investigation that the presence of –OH group is imperative for the breakdown of studied carboxylate esters but not for phosphate esters. Similar findings were reported by Menger et al. [34, 35] and Morrow and Trogler [36] independently. The ability of metallosurfactants for the hydrolysis of several different esters is also dependent on their individual binding constants. The binding constants of CH_3COOCH_3 with dipyridylamine complex of copper were determined by Chin and Jubian [37] and lower affinity toward C=O center of carboxylate ester and greater affinity toward center of phosphate ester was observed [38].

Scrimin et al. [39] also addressed the matter of different reactivity of metallosurfactants using two complexes of copper in which copper was coordinated to ligand **38**; tetramethyl ethylenediamine and **39**; *N*-hexadecyl-*N, N′, N′*-trimethyl ethylene diamine in co-micellar medium of $CTANO_3$. They monitored comparative breakdown of different phosphate esters (di- and -trimesters) in the above-mentioned medium.

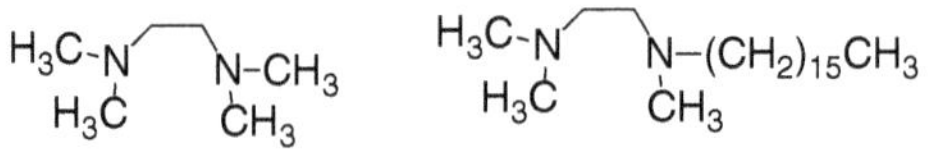

38 **39**

The observed values of first-order rate constants were higher for copper complexes with ligand **39** than **40**. In case of second-order rate constants, Cu(II) complexes of ligand **42** possessed greater values than ligand **39** and showed better reactivity [40, 41].

Tecilla and coworkers [20] have reported pyridine aldoxime ligand **16** and studied their complexes with Ni^{2+} and Zn^{2+} for the breakdown of PNPH and PNPA. The group undertook this study in comicellar media of CTAB too. The observed data indicated hydrolysis was pH dependent and found to be effective because of increased concentration of substrate.

Contribution of Bhattacharya et al. [42, 43] for the synthesis of few 4,4′-(dialkylamino) pyridine-based ligands (**40–42**) is worth mentioning. They investigated catalytic ability of prepared ligands for hydrolytic cleavage of PNPDPP and PNPH at pH 7.6. Slight rate enhancement was observed in comicellar medium of

CTAB in comparison to only CTAB possibly due to dimethylamino pyridine entity and increased Lewis character of copper(II) ion. However, ligands **40** and **41** have demonstrated burst-type kinetics with excess substrate.

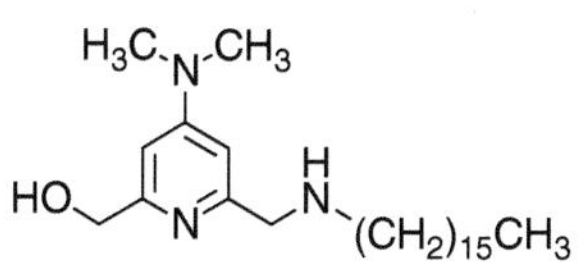

40

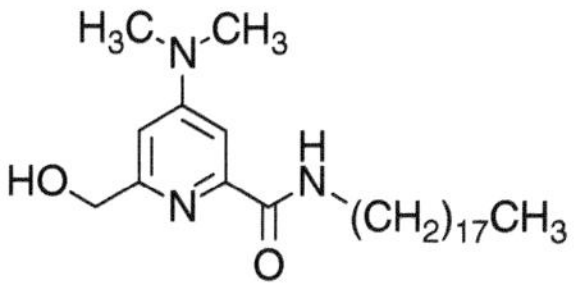

41

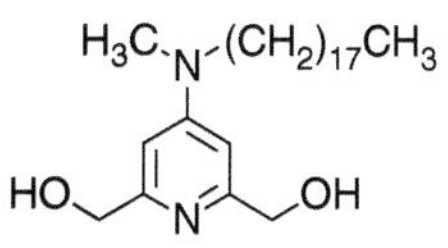

42

Moreover, the team [44] undertook a comparative analysis to check the catalytic power of ligands **43** and **44** (tetradentate) with that of ligands **48** and **49** (tridentate). The study reported modest rates with CTAB and without metal ion by ligand **44** for hydrolytic breakdown of PNPDPP and PNPH. However, remaining ligands exhibited average rate enhancement for phosphate ester in CTAB medium. This observation can be accounted on the basis of the participation of supernucleophilic 4-Dimethylaminopyridine (DMAP) entity.

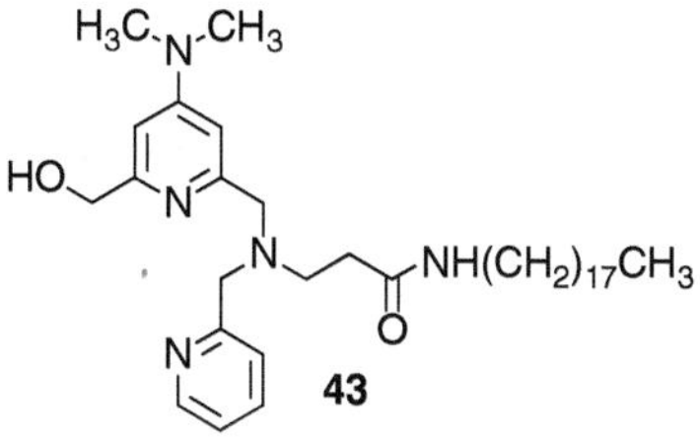

43

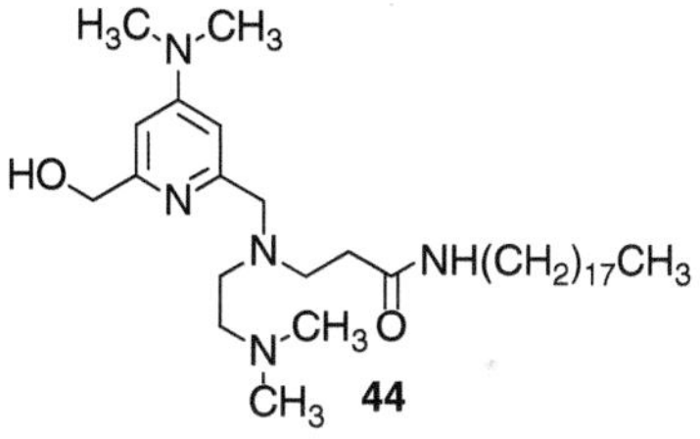

44

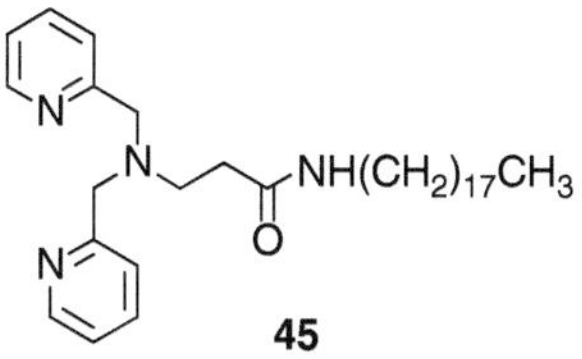

45

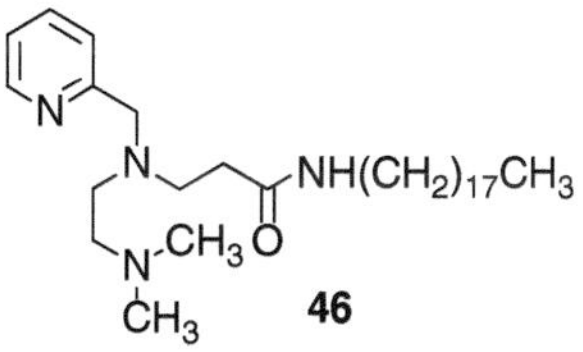

46

Overall, it was concluded that tridentate ligands **45, 46** have demonstrated inferior catalytic strength than tetradentate ligands **43, 44**. The responsible entity for the activation of hydrolytic cleavage is shown in Figure 4.9. The investigation reported 1 : 1 type complex of metal and ligand.

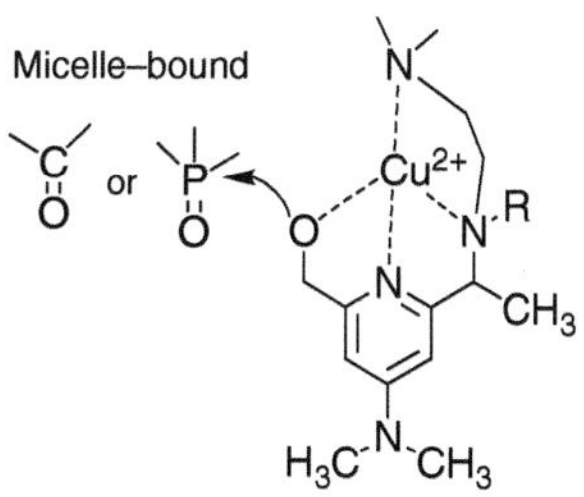

Figure 4.9 Active species formed by the tetradentate ligand.

Jiang et al. developed several different ligands having pyridine moiety [45, 46] and alkanol imidazole group [47]. Complexation of imidazole-based ligands (**47, 48**) with copper, zinc, and nickel displayed excellent catalytic activity for the hydrolysis of BNPP: bis (*p*-nitrophenyl) phosphate. The observed rate increment was due to lipophilic interaction, which raised the concentration of catalyst. The enhancement in rate of reaction was due to hydrophobic interaction, which increased the local concentration of catalyst. In between ligands **49** and **50**, **49** was recognized as most competent ligand.

47: R= n-C_4H_9
48: R= n-$C_{12}H_{25}$

49: R= n-$C_{12}H_{25}$
50: R= CH_3

The general observations obtained for both the pyridine- and imidazole-based ligands were: (i) superior lipophilic interaction was recorded with micellar assembly when longer hydrophobic chain was used and (ii) increased involvement in lowering the activation energy barrier for hydrolytic cleavage of ester. The team proposed a plausible reaction pathway through which the ternary complex formed (Figure 4.10).

The octahedral geometry of metal–ligand complex stimulates hydroxyl group with reduction in pK_a and generation of Nu^- nucleophile. Finally, the release of *p*-nitrophenol (PNPOH) followed by formation of *p*-nitrophenyl phosphate (NPP) in the subsequent step regenerated the catalyst [48]. The trend of catalytic strength was in order of $Cu^{2+} \sim Zn^{2+} > Co^{2+} \sim Ni^{2+}$. Complexes of ligands **51–53** were also studied in CTAB media for the breakdown of monoester NPP and diester BNPP [46].

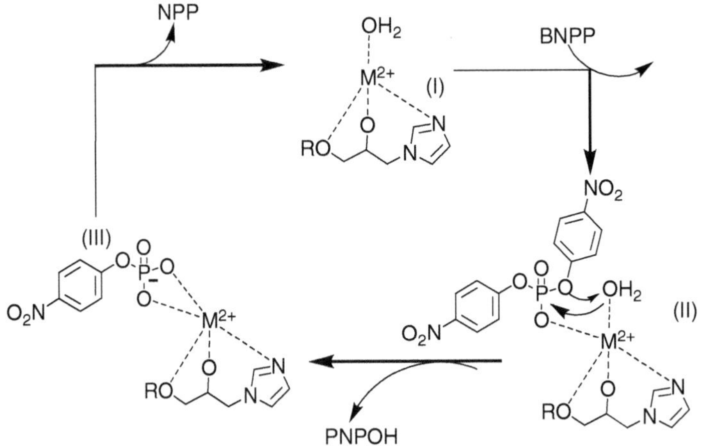

Figure 4.10 Proposed mechanism of the BNPP hydrolysis by metallomicelles of **50** and **51**.

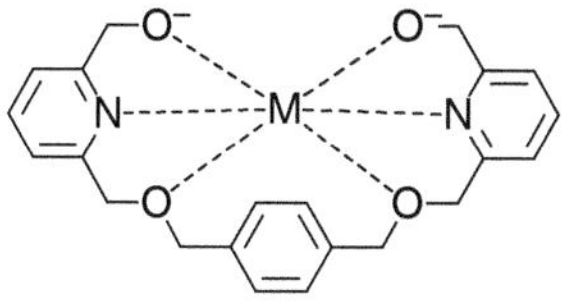

51

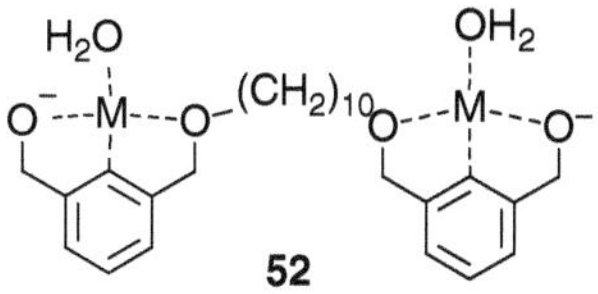

52

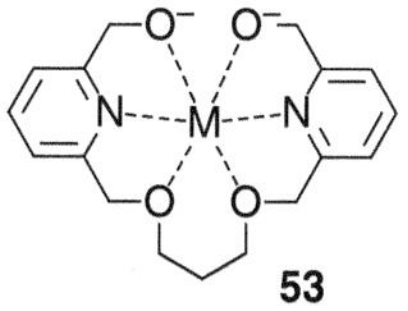

53

M= Cu^{2+}, Zn^{2+}

Complexes of crowned Schiff base ligands (**54**, **53**) with manganese were applied to examine the hydrolytic breakdown of BNPP in various comicellar media of cationic, anionic, and gemini surfactants [49]. The comparative catalytic activity of complexes was evaluated at pH 7.5. Observed rates were significantly affected by structure of metal complexes and surfactants used. Gemini surfactants proved to be best co-micellar medium for breakdown of ester, while complex of ligand **59** exhibited superior reactivity than complex of ligand **54**.

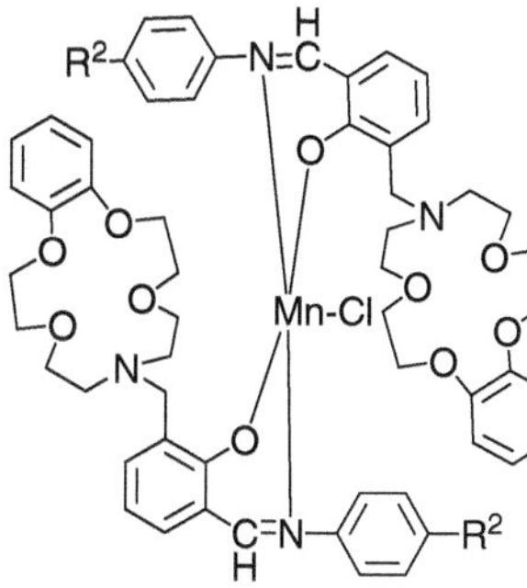

54: $MnL^1{}_2Cl$, L^1: R^1= Br, R^2= Cl
55: $MnL^2{}_2Cl$, L^2: R^1= Cl, R^2= CH_3

A new class of imidazole propanamide-based ligand **56 (L^2)** was reported by Meng and coworkers [50]. Ligand **56** was complexed with several d-block metal ions (Cu^{2+}, Zn^{2+}, and Co^{2+}) and f-block metal ions (La^{3+}). Further, a comparative study was performed with ligand histamine; L^1 to investigate their catalytic potency toward the breakdown of BNPP at 35 °C and pH 7.0. The study recognized complex ML^2 to be more proficient for hydrolytic reaction than complex ML^1.

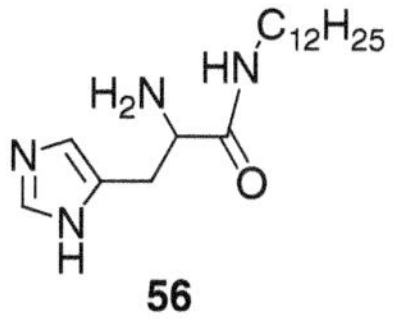

56

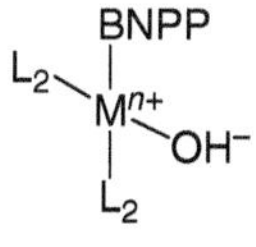

57

Interestingly, the complex of f-block metal lanthanum LaL^2 showed better ability for hydrolysis than its complex with d-block metals. It is assumed that the breakdown of ester BNPP is governed by the construction of a ternary complex (presented in **57**) in which active hydroxyl group is attached. The breakdown pathway is demonstrated in Figure 4.11.

Additionally, several lanthanide metal (Ce and La) complexes were reported with aza-macrocyclic ligand **58** by Xie and co-workers[51]. The team investigated the hydrolytic breakdown of BNPP with prepared metallosurfactants in LSS *n*-lauryl sarcosine comicellar medium. The catalytic ability of LaL assembly was less than that of CeL assembly. The finest catalytic strength for both the metallosurfactant systems (CeL and LaL) was recorded at pH 8.0 and pH 8.2, respectively.

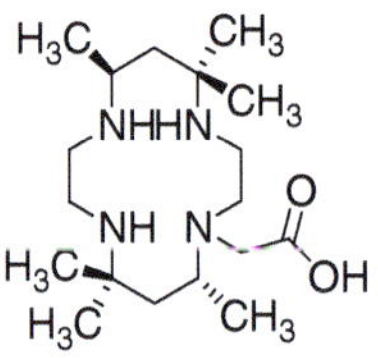

58

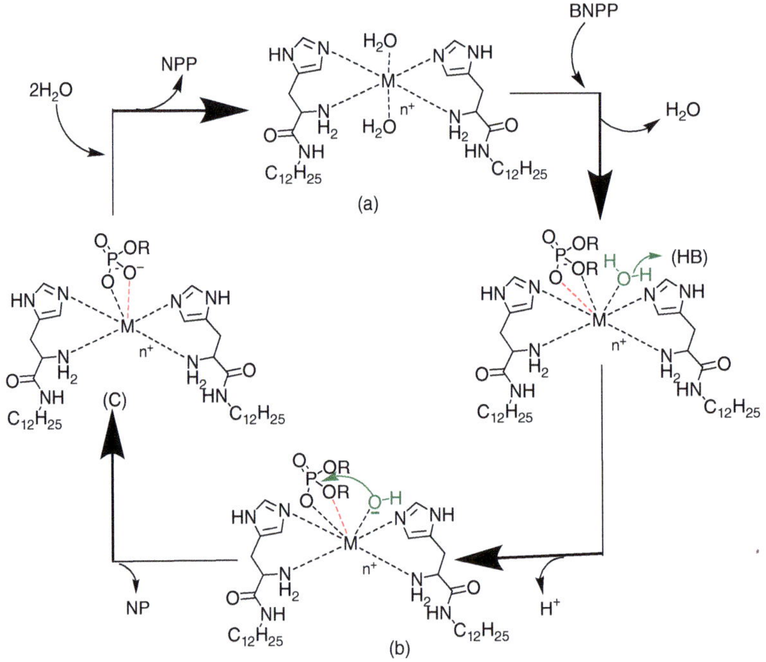

Figure 4.11 Proposed mechanism for metallosurfactant-mediated breakdown of BNPP by La complex.

4.5 Quantitative Treatment of Observed Rates: Application of Kinetic Models

The quantitative treatment of observed rates for metallosurfactant-mediated hydrolytic cleavage of esters is seldom documented but has great significance in providing a general idea of nucleophilicity of Nu^- against substrates. Tagaki et al. [13] proposed a binary complex model for quantitative rate treatment of some metallosurfactant-catalyzed reactions. The model assumes metallosurfactant-catalyzed reactions as an intermolecular reaction in which weak interaction occurs between metal ion and substrate as a ligand base in aqueous buffer medium. A ternary complex kinetic model was proposed by Zeng et al. [52], which suggests strong interaction between metal ion M^{n+}, ligand L, and substrate S in a metallosurfactant-catalyzed reaction.

Another approach named as phase separation model (Scheme 4.1a) was proposed particularly for micellar-catalyzed systems. It presumes the existence of two separate phases in a micellar solution. One phase is micellar pseudophase and second is bulk phase. However, it is proposed that the hydrolytic cleavage takes place concurrently in both the phases with formation of the product (Scheme 4.1b).

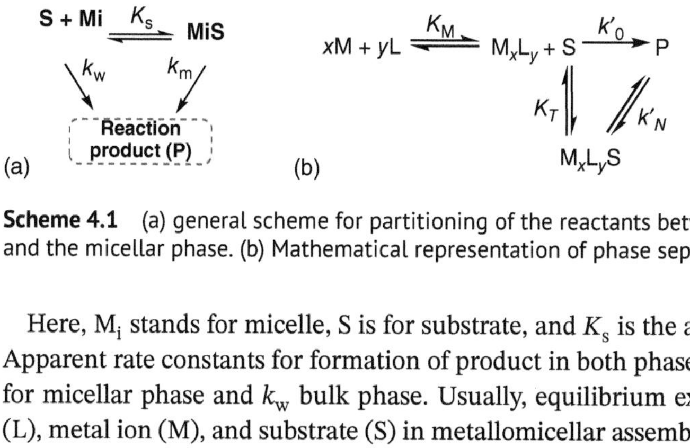

Scheme 4.1 (a) general scheme for partitioning of the reactants between the bulk phase and the micellar phase. (b) Mathematical representation of phase separation model.

Here, M_i stands for micelle, S is for substrate, and K_s is the association constant. Apparent rate constants for formation of product in both phases are denoted by k_m for micellar phase and k_w bulk phase. Usually, equilibrium exists between ligand (L), metal ion (M), and substrate (S) in metallomicellar assembly.

In Scheme 4.1b, association constant K_M shows association between metal ions x and y ligands, K_T is the association constant for interaction between binary complex $[M_xL_y]$ and substrate S. The apparent first-order rate constants k'_N and k'_0 denote the formation of the product in metallomicellar phase and bulk phase, respectively. The hydrolytic breakdown of substrates, i.e. ester in only buffer medium, is presented by rate constant k_0. Notations k_L and k_M correspond to second-order rate constants for ligand and metal ion alone. The active concentrations of metal ion, ligand, and substrate in bulk phase are symbolized by [M], [L], and [S]. $[M_xL_y]$ illustrates the concentration of x metal ion and y ligands in to bulk phase, and $[M_xL_yS]$ demonstrates the concentration of substrate in metallomicellar phase.

$$[S] = [S]_T - [M_xL_yS] \tag{4.1}$$

$$[M] = [M]_T - x[M_xL_y] \tag{4.2}$$

$$[L] = [L]_T - y[M_xL_y] \tag{4.3}$$

where $[L]_T$, $[M]_T$, and $[S]_T$ represent total concentrations of ligand, metal ion, and substrate, respectively. The rate equation for Scheme 4.1b may be derived as:

$$r = k_{obs}[S]_T = k'_N[M_xL_yS] + k'_0[S] \tag{4.4}$$

Finally, following Eq. (4.5) is derived.

$$\frac{1}{k_{obs} - k'_0} = \frac{1}{K_T(k'_N - k'_0[M_xL_y])} + \frac{1}{k'_N - k'_0} \tag{4.5}$$

On ignoring higher-order terms in Eqs. (4.5) and (4.6) is achieved.

$$\frac{1}{k_{obs} - k'_0} = \frac{1}{K_T(k'_N - k'_0)K_M[M]_T^x\,[L]_T^y} + \frac{y^2}{K_T(k'_N - k'_0)[L]_T} + \frac{x^2}{K_T(k'_N - k'_0)[M]_T} + \frac{1}{k'_N - k'_0} \tag{4.6}$$

In case where $y = 1$ and $x = 1$, above equation becomes Eq. (4.7).

$$\frac{1}{k_{obs} - k'_0} = \frac{1}{K_T(k'_N - k'_0)}\left(\frac{1}{K_M[M]_T} + 1\right) + \frac{1}{[L]_T} + \frac{1}{K_T(k'_N - k'_0)[M]_T} + \frac{1}{k'_N - k'_0} \tag{4.7}$$

The resulting three equations are commonly known as ternary complex kinetic equations. The effect of pH can also be monitored with the help of Eq. (4.8).

$$\frac{1}{k'_N} = \frac{1}{k_N} + \frac{1}{k_N K_a}[H^+] \tag{4.8}$$

Thus, the quantitative rate treatment approaches have been effectively applied in the form of binary complex model and ternary complex model [19, 25, 46, 53] to metallomicellar catalysis. The foundation of such quantitative outcomes was based on phase separation theory.

4.6 Conclusion

Hydrolytic metallosurfactants mimic the function of natural hydrolase enzymes. Owing to this, noteworthy development of metallomicelles has already been documented and described. The motive of designing such hydrolase-based assembly was to apply them effectively in catalytic cleavage of several esters, breakdown of lethal phosphate esters, and in enantioselective catalysis. Many important investigations have been conducted, which are reported in this chapter to address catalytic breakdown of C=O and P=O based esters. A number of developed surfactant ligands are presented separately for carboxylate and phosphate esters. Several models, which depict the quantitative treatment of observed rate data, are also discussed with special emphasis. In spite of this, there exists a huge scope of developing hydrolytic metallosurfactants with preferred reactivity and specificity.

References

1 Griffiths, P.C., Fallis, I.A., Chuenpratoom, T., and Watanesk, R. (2006). Metallosurfactants: interfaces and micelles. *Advances in Colloid and Interface Science* 122 (1–3): 107–117.

2 Owen, T. and Butler, A. (2011). Metallosurfactants of bioinorganic interest: coordination-induced self assembly. *Coordination Chemistry Reviews* 255 (7–8): 678–687.

3 Christianson, D.W. and Lipscomb, W.N. (1989). Carboxypeptidase A. *Accounts of Chemical Research* 22 (2): 62–69.

4 Herries, D. (1985). Enzyme structure and mechanism (second edition). *Biochemical Education* 13 (3): 146.

5 Bhattacharya, S. and Kumari, N. (2009). Metallomicelles as potent catalysts for the ester hydrolysis reactions in water. *Coordination Chemistry Reviews* 253 (17–18): 2133–2149.

6 Griffiths, P.C., Fallis, I.A., Tatchell, T. et al. (2008). Aqueous solutions of transition metal containing micelles. *Advances in Colloid and Interface Science* 144 (1–2): 13–23.

7 Polarz, S., Landsmann, S., and Klaiber, A. (2014). Hybrid surfactant systems with inorganic constituents. *Angewandte Chemie International Edition* 53 (4): 946–954.

8 Mancin, F., Prins, L.J., Pengo, P. et al. (2016). Hydrolytic metallo-nanozymes: from micelles and vesicles to gold nanoparticles. *Molecules* 21 (8).

9 Ghosh, K., Gupta, B., and Bhattacharya, S. (2015). Metallosurfactant aggregates as catalysts for the hydrolytic cleavage of carboxylate and phosphate esters. *Current Organocatalysis* 3 (1): 6–23.

10 Melhado, L.L. and Gutsche, C.D. (1978). Association phenomena. 2. Catalysis of the decomposition of acetyl phosphate by chelate micelles and by amine-ammonium micelles. *Journal of the American Chemical Society* 100 (6): 1850–1856.

11 Peng, L.H. and Gutsche, C.D. (1978). Association phenomena. 3. Polyfunctional catalysis of acetyl phosphate decomposition. *Journal of the American Chemical Society* 100 (6): 1857–1865.

12 Tagaki, W., Ogino, K., Fujita, T. et al. (1993). Hydrolytic metalloenzyme models. Catalysis in the hydrolysis of *p*-nitrophenyl 2-pyridinecarboxylate by copper and zinc ion complexes of anionic surfactants having a functional imidazole and hydroxyl moieties. *Bulletin of the Chemical Society of Japan* 66 (1): 140–147.

13 Tagaki, W., Ogino, K., Tanaka, O. et al. (1991). Hydrolytic metalloenzyme models. Micellar effects on the activation of the hydroxyl groups of *N*-alkyl-2-(hydroxymethyl)-imidazole ligands by Cu^{2+} in the transacylation of *p*-nitrophenyl picolinate. *Bulletin of the Chemical Society of Japan* 64 (1): 74–80.

14 Faivre, V., Brembilla, A., Roizard, D., and Lochon, P. (1991). Zn^{++}-complexes as models of metalloenzymes in micellar esterolysis: ligand structure-dependent stoichiometry of the complexes. *Tetrahedron Letters* 32 (2): 193–196.

15 Fornasier, R., Scrimin, P., Tecilla, P., and Tonellato, U. (1989). Bolaform and classical cationic metallomicelles as catalysts of the cleavage of *p*-nitrophenyl picolinate. *Journal of the American Chemical Society* 111 (1): 224–229.

16 Fornasier, R., Scrimin, P., Tonellato, U., and Zanta, N. (1988). Highly enantioselective cleavage of α-amino acid *p*-nitrophenyl esters by chiral metallomicelles. *Journal of the Chemical Society, Chemical Communications* (11): 716–718.

17 Scrimin, P., Tecilla, P., and Tonellato, U. (1991). Metallomicelles as catalysts of the hydrolysis of carboxylic and phosphoric acid esters. *The Journal of Organic Chemistry* 56 (1): 161–166.

18 Fornasier, R., Milani, D., Scrimin, P., and Tonellato, U. (1986). Functional micellar catalysis. Part 8. Catalysis of the hydrolysis of *p*-nitrophenyl picolinate by metal-chelating micelles containing copper(II) or zinc(II). *Journal of the Chemical Society, Perkin Transactions 2* (2): 233–237.

19 Scrimin, P., Tecilla, P., and Tonellato, U. (1994). Chiral lipophilic ligands. 1. Enantioselective cleavage of α-amino acid esters in metallomicellar aggregates. *The Journal of Organic Chemistry* 59 (15): 4194–4201.

20 Mancin, F., Tecilla, P., and Tonellato, U. (2000). Metallomicelles made of Ni(II) and Zn(II) complexes of 2-pyridinealdoxime-based ligands as catalyst of the cleavage of carboxylic acid esters. *Langmuir* 16 (1): 227–233.

21 Weijnen, J.G.J., Koudijs, A., Schellekens, G.A., and Engbersen, J.F.J. (1992). Functionalised 1,10-phenanthroline metallocatalysts as models for hydrolytic metalloenzymes. *Journal of the Chemical Society, Perkin Transactions 2* (5): 829–834.

22 Weijnen, J.G.J., Koudijs, A., and Engbersen, J.F.J. (1991). Carboxylic and phosphate ester hydrolysis catalysed by bivalent zinc and copper metallosurfactants. *Journal of the Chemical Society, Perkin Transactions 2* (8): 1121–1126.

23 Kimura, E., Hashimoto, H., and Koike, T. (1996). Hydrolysis of lipophilic esters catalyzed by a zinc(II) complex of a long alkyl-pendant macrocyclic tetraamine in micellar solution. *Journal of the American Chemical Society* 118: 10963–10970.

24 Griffiths, P.C., Fallis, I.A., Willock, D.J. et al. (2004). The structure of metallomicelles. *Chemistry A European Journal* 10 (8): 2022–2028.

25 Polyzos, A., Hughes, A.B., and Christie, J.R. (2007). Catalysis of aryl ester hydrolysis in the presence of metallomicelles containing a copper(II) diethylenetriamine derivative. *Langmuir* 23 (4): 1872–1879.

26 Qiu, L.-G., Xie, A.-J., and Shen, Y.-H. (2005). Hydrolysis of *p*-nitrophenyl picolinate catalyzed by metal complexes of *N*-alkyl-3,5-bis(hydroxymethyl)-1,2,4-triazole in CTAB micelles. *Journal of Colloid and Interface Science* 290 (2): 475–480.

27 Qiu, L.-G., Xie, A.-J., and Shen, Y.-H. (2006). Complexes of triazole-based ligands in cationic gemini surfactant micelles. *Journal of Molecular Catalysis A: Chemical* 244: 58–63.

28 Qiu, L.G., Jiang, X., Gu, L.N., and Hu, G. (2007). Gemini metallomicellar catalysis: hydrolysis of *p*-nitrophenyl picolinate catalyzed by Cu(II) and Ni(II) complexes of macrocyclic ligands in gemini surfactant micelles. *Journal of Molecular Catalysis A: Chemical* 277 (1–2): 15–20.

29 Fubin, J., Bingying, J., Xiaoqi, Y., and Xiancheng, Z. (2002). Metallomicellar catalysis. Effects of bridge-connecting ligands on the hydrolysis of PNPP catalyzed by Zn(II), Co(II), and Ni(II) complexes of ethoxyl-diamine ligands in micellar solution. *Langmuir* 18 (18): 6769–6774.

30 Jiang, W., Xu, B., Li, J. et al. (2007). Metallomicelle catalysis: hydrolysis of p-nitrophenyl picolinate induced by schiff base Co(II) complexes in a gemini surfactant micellar solution. *International Journal of Chemical Kinetics* 39 (12): 672–680.

31 Xu, B., Jiang, W., Wang, Y. et al. (2014). Unprecedented reactivity of alkoxo/hydroxide-bridged dinuclear copper(II) complexes as artificial carboxylesterase: effects of gemini 16-6-16 micelles and bimetallic synergistic effect. *Colloids and Surfaces A: Physicochemical and Engineering Aspects* 456 (1): 222–230.

32 Mancin, F., Scrimin, P., and Tecilla, P. (2012). Progress in artificial metallonucleases. *Chemical Communications* 48 (45): 5545.

33 Gellman, S.H., Breslow, R., and Petter, R. (1986). Catalytic hydrolysis of a phosphate triester by tetracoordinated zinc complexes. *Journal of the American Chemical Society* 108 (9): 2388–2394.

34 Menger, F.M., Gan, L.H., Johnson, E., and Durst, D.H. (1987). Phosphate ester hydrolysis catalyzed by metallomicelles. *Journal of the American Chemical Society* 109 (9): 2800–2803.

35 Menger, F.M. and Tsuno, T. (1989). Organic reactions catalyzed by copper-loaded polymers. Reactivity vs polymer structure. *Journal of the American Chemical Society* 111 (13): 4903–4907.

36 Morrow, J.R. and Trogler, W.C. (1989). Hydrolysis of phosphate triesters with copper(II) catalysts. *Inorganic Chemistry* 28 (12): 2330–2333.

37 Chin, J. and Jubian, V. (1989). A highly efficient copper(II) complex catalysed hydrolysis of methyl acetate at pH 7.0 and 25 °C. *Journal of the Chemical Society, Chemical Communications* (13): 839–841.

38 Wadsworth, W.S. (1981). Lewis acid catalyzed methanolysis of a phosphate triester. *The Journal of Organic Chemistry* 46 (20): 4080–4082.

39 Scrimin, P., Ghirlanda, G., Tecilla, P., and Moss, R.A. (1996). Comparative reactivities of phosphate ester cleavages by metallomicelles. *Langmuir* 12 (26): 6235–6241.

40 Bunton, C.A. and Savelli, G. (1986). Organic reactivity in aqueous micelles and similar assemblies. *Advances in Physical Organic Chemistry* 22 (C): 213–309.

41 Bunton, C.A., Scrimin, P., and Tecilla, P. (1996). Source of catalysis of dephosphorylation of p-nitrophenyl-diphenylphosphate by metallomicelles. *Journal of the Chemical Society, Perkin Transactions 2* 3 (3): 419–425.

42 Bhattacharya, S., Snehalatha, K., and George, S.K. (1998). Synthesis of some copper(II)-chelating (dialkylamino)pyridine amphiphiles and evaluation of their esterolytic capacities in cationic micellar media. *The Journal of Organic Chemistry* 63 (1): 27–35.

43 Bhattacharya, S. and Snehalatha, K. (1995). Synthesis and esterolytic chemistry of some (dialkylamino)pyridine-functionalized micellar aggregates. Evidence of catalytic turnover. *Langmuir* 11 (12): 4653–4660.
44 Bhattacharya, S., Snehalatha, K., and Kumar, V.P. (2003). Synthesis of new Cu(II)-chelating ligand amphiphiles and their esterolytic properties in cationic micelles. *The Journal of Organic Chemistry* 68 (7): 2741–2747.
45 Jiang, F., Jiang, B., Cao, Y. et al. (2005). Metallomicellar catalysis: hydrolysis of phosphate monoester by Cu(II), Zn(II), Ni(II) and Co(II) complexes of pyridyl ligands in CTAB micellar solution. *Colloids and Surfaces A: Physicochemical and Engineering Aspects* 254 (1–3): 91–97.
46 Jiang, F., Huang, L., Meng, X. et al. (2006). Metallomicellar catalysis: hydrolysis of phosphate monoester and phosphodiester by Cu(II), Zn(II) complexes of pyridyl ligands in CTAB micellar solution. *Journal of Colloid and Interface Science* 303 (1): 236–242.
47 Fubin, J., Juan, D., Xiaoqi, Y. et al. (2004). Metallomicellar catalysis: accelerated hydrolysis of BNPP by copper(II), zinc(II), and nickel(II) complexes of long alkanol-imidazole in CTAB micellar solution. *Journal of Colloid and Interface Science* 273 (2): 497–504.
48 Ogino, K., Kashihara, N., Fujita, T. et al. (1987). Hydrolytic Metalloenzyme models. Metal ion dependent site-selective acylations of hydroxyl groups of Bis-imidazole ligands promoted by Zn^{2+} and Cu^{2+} ions in cationic micelles. *Chemistry Letters* 16 (7): 1303–1306.
49 Jiang, W., Xu, B., Lin, Q. et al. (2008). Metal-promoted hydrolysis of bis(*p*-nitrophenyl)phosphate by trivalent manganese complexes with Schiff base ligands in gemini micellar solution. *Colloids and Surfaces A: Physicochemical and Engineering Aspects* 315 (1–3): 103–109.
50 Liu, Y., Meng, X.G., Li, J.M., and Li, X.H. (2013). Hydrolysis of phosphodiester catalyzed by metallomicelles with histidine residue: kinetics and mechanism. *Colloids and Surfaces A: Physicochemical and Engineering Aspects* 436: 839–845.
51 Li, F.Z., Feng, F., Yu, L., and Xie, J.Q. (2014). The Metallomicelle of lanthanide metal (Ce, La) aza-macrocyclic complexes with a carboxyl branch: the catalytic activity and mechanism in the hydrolysis of a phosphate diester. *Journal of Solution Chemistry* 43 (8): 1331–1343.
52 Xiancheng, Z., Yuanqin, Z., Xiaoqi, Y., and Anmin, T. (1999). Metallomicellar catalysis. Cleavage of *p*-nitrophenyl picolinate in copper(II) coordinating *N*-myristoyl-*N*-(β-hydroxyethyl)ethylenediamine in CTAB micelles. *Langmuir* 15 (5): 1621–1624.
53 Hay, R.W., Govan, N., and Parchment, K.E. (1998). A metallomicelle catalysed hydrolysis of a phosphate triester, a phosphonate diester and O-isopropyl methylfluorophosphonate (sarin). *Inorganic Chemistry Communications* 1 (6): 228–231.

5

Metallosurfactants as Catalysts in Organic Reactions and Energy-Based Applications

Sakshi Goel and Supriya Rana

Panjab University, Department of Chemistry & Centre of Advanced Studies in Chemistry, Chandigarh 160014, India

5.1 Introduction

For the past few decades, metallosurfactants have been acting as an integral component to enhance the physio-chemical properties of surfactant systems compared to conventional surfactants in various ways [1]. The "metal ion" with hydrophobic chain makes them eligible candidates for various catalytic reactions. Traditional catalysis is carried out in organic solvents and, on the other hand, engineered micelles help in catalysis reactions, using water as a solvent. In accordance with green chemistry principles, the ideal solvent minimizes toxicity, energy demand, and pollution. Among all the possible solvent systems considered, water, undoubtedly, is the best choice as green solvent. However, strong polarity of water makes it least suitable solvent in organic reactions. To overcome this problem, surfactants are employed in the system, which help in the solubilization of substrates in water. In addition, the micellar aggregates of catalysts also facilitate their recycling and further extraction of products. The field of micellar catalysis in water has been extensively reviewed by Sorella et al. [2] and Lipshutz et al. [3]. Nonetheless, metallosurfactants possess an added advantage here, as the metal itself acts as a catalyst, along with retaining the surfactant properties. In water, metallosurfactant forms the micelles, which have hydrophobic pockets at the center. When the organic substrates are added, they tend to concentrate in those pockets, lying near to the metal ion, which facilitates the reaction. Metallosurfactants are currently being used for a number of organic reactions, including hydrogenation, oxidation, and coupling reactions, which will be discussed in detail in the present chapter. The esterolytic reactions involving metallosurfactants have already been presented in Chapter 4.

Apart from synthetic chemistry, metallosurfactants are being used in various energy-based applications. Nowadays, the unprecedented energy demands, due to modernization in our lifestyle and industrialization, require exponential growth in efforts to produce clean and green energy. At present, the increasing energy

Metallosurfactants: From Fundamentals to Catalytic and Biomedical Applications, First Edition.
Edited by Surinder K. Mehta and Ravneet Kaur.

demands are fulfilled by excessive usage of non-renewable resources such as fossil fuels. While utilizing these fossil fuels like coal, oil, natural gas, and coke, huge number of harmful gases are also produced, which have devastating side effects leading to serious environmental problems such as depletion of the earth crust and environmental pollution [4, 5]. So, there is a challenge for the world leaders to meet energy consumption demand and focus on the issue to provide clean and renewable energy resources at low cost [6–8]. The current scenario is aimed towards replacing the non-renewable energy resources with renewable materials and save the environment from resource depletion and climate change challenges [9, 10].

One of the most promising methods to produce clean energy is by splitting water into hydrogen and oxygen, either electrochemically or photochemically. Water splitting is an eco-friendly and clean process to meet the energy problem [11] and involves two half-cell reactions, i.e. hydrogen evolution reaction (HER) and oxygen evolution reaction (OER), which requires high overpotential and kinetic barriers [12]. On the other hand, the production of oxygen through water splitting during water electrolysis is a challenging process for energy storage and conversion as it is not feasible both thermodynamically and kinetically because of the complex charge transfer mechanism involved, which requires four-proton-coupled electron transfer and O=O bond formation [13, 14]. This process of water electrolysis leads to the generation of high-purity hydrogen and oxygen gas, which form the main constituents of the fuels.

To counter all these issues associated with water electrolysis reaction and meet up the high-scale requirements, the development of the advanced catalytic material is highly sought after. The presence of an efficient electrocatalyst for lowering these energy barriers during water oxidation is extremely important [15]. In case of OER, Ru, Ir, and their oxides show higher electrocatalytic activity [16] and, in case of HER, the use of noble metals like Pt and Pd fulfills the low kinetics and over potential requirement [17]. However, the commercialization of these materials is still a matter of concern because of their high cost, non-scalability due to their preciousness, scarcity, and instability with time in harsh conditions and less abundance in the earth crust [18, 19]. In the recent years, a lot of research is going on for the development of efficient electrocatalyst for water splitting to generate clean fuels and rechargeable metal air batteries [20]. Therefore, to tackle these energy problems efficiently, there is a need to develop efficient, durable, inexpensive, highly stable, and high catalytic activity materials that can work efficiently in any harsh and adverse conditions (acidic and alkaline solutions) for the catalysis and are more feasible than noble metals [21, 22].

Nowadays, the transition metal-based group elements such as iron, cobalt, nickel, and their oxides (perovskite [23], spinels [24], and oxyhydroxides and hydroxides [25]), chalcogenides, nitrides, carbides, borides, metal alloys, and carbon-based catalysts have been investigated as promising candidates for both photocatalysts and electrocatalysts toward water oxidation. Their low cost, great abundance, excellent chemical stability in broad pH range, resistance to corrosion, non-toxic nature, low processing cost, and biodegradability lead to higher economic viability than precious metals [26, 27]. Further, various methodologies have been developed to address the above-mentioned requirements for the development of the catalyst for water electrolysis in the harsh conditions. According to various studies, the use of the surfactants results in the exfoliation process, which leads to increase in stability of the dispersed material by preventing reaggregation and plays a role in monitoring the surface charge [28–30]. This phenomenon can help to regulate the catalytic properties. Metallosurfactant-aided catalysis is an emerging field in the area of water oxidation. The head group of a metallosurfactant usually contains the metal cation, which offers a pathway for the transportation of ions and acts as a redox center during catalytic performance. Therefore, the combination of the metal ion and the surfactants gives a new segment of advanced material that can be used as an electrocatalyst, which will be discussed in second part of this chapter.

5.2 Metallosurfactants as Catalysts in Organic Reactions

One of the initial reports of employing metallosurfactant in organic reaction was in 2009 by Lipshutz and Ghorai for ring-closing metathesis reaction in water at room temperature [31]. A Grubbs Hoveyda 1 Ru(II) catalyst was prepared for lipophilic dienes to form five- to seven-membered ring products. The reaction even worked in seawater and thus proved to be an excellent example of using metallosurfactant in organic reaction with excellent yield and recovery. Generally, a surfactant is composed of hydrophobic and hydrophilic parts. Apart from these two parts, a third component, which was free hydroxyl group, was present in this ruthenium surfactant. The ability of free hydroxyl group in the hydrophobic core, linked covalently with Ru carbine, is exploited to form nanomicelles in the organic reaction. The major tenets of green chemistry were accomplished in this reaction as it was performed in water, at room temperature, and the catalyst was continuously recycled without any workup. It was further extended to Grubbs Hoveyda 2 Ru(II) catalyst for ring-closing metathesis reaction of heterocyclics and carbocyclics [32]. Only 2% (w/w) of catalyst was found to be efficient for acrylates and alkenes as well. Lipshutz and Ghorai generated a newfound interest in the organic catalytic transformations. Since then, there have been several organic reactions catalyzed very efficiently and effectively using metallosurfactants or metallomicelles, including, but not limited to – hydrogenation, oxidation, C–C cross-coupling, hydration, epoxidation, sulfoxidation, etc. which are discussed below. The esterolytic reactions have already been reviewed quite in detail in Chapter 4 and will be excluded from the present chapter. Also, Chapter 3 reports an in-depth analysis of self-aggregating metalloaggregates in enhancing the reaction rate of organic transformations.

5.2.1 Types of Reactions

Metallomicellar catalysis can be categorized based on the type of reactions or metallosurfactants used. Going by the type of reactions, several groups have attempted to investigate different metallosurfactants catalyzing a specific type of reaction.

Transfer Hydrogenation Reaction

Transfer hydrogenation is a special class of hydrogenation reactions, which proceeds with the addition of hydrogen employing hydrogen transfer catalysts using a source other than gaseous hydrogen, which is quite expensive and inconvenient. A fine example of catalysis of organic reaction in aqueous medium was demonstrated by Li et al. with the help of an amphiphilic copolymer-based iridium catalyst for transfer hydrogenation of aldehydes [33]. The aldehydes were reduced in water where the emulsion droplets acted as microreactors. The catalyst, being amphiphilic, assembled both organic and $HCOO^-$ moieties around active sites, which accelerated the reaction and resulted in a turnover frequency of up to $3.0 \times 10^5\ h^{-1}$.

Kalsin et al. also investigated a similar hydrogenation reaction with a variety of ketones, depicting solubility from partially water soluble to hydrophobic, with formate in water [34]. The new complexes [(*p*-cymene)RuCl(L^n)] and [Cp*RhCl(L^n)] (Cp* = η^5-C_5Me_5) were prepared, where L represents water-soluble ligands functionalized with trialkyl ammonium side groups, i.e. $H_2N(CH_2)_2NHSO_2$-*p*-$C_6H_4CH_2[NMe_2$-$(C_nH_{2n+1})]^+$ ($[HL_n]^+$; $n = 8, 16$). The surface tension measurements showed micelle formation at very low concentration, i.e. 0.6 mM. The C_{16} tails were able to adsorb at the water–substrate interface and, thus, showed exceptionally high catalytic activity. This Ru-based catalyst doubled the hydrogenation reaction rate of hydrophobic ketones, as compared to non-micellar ones. In addition, the same authors studied the cooperative interaction of surface-active ruthenium analogue RuL^n ($n = 8, 16$) with other added surfactants (cationic, anionic, and zwitter-ionic), in transfer hydrogenation of hydrophobic ketones in water [35]. As expected, the added surfactants lowered the critical micellar concentration (CMC) of RuL^n, resulting in better emulsification. RuL^8, having shorter chain, were not capable of making micelles alone. Thus, after adding dodecyl sodium sulfate ($C_{12}H_{25}SO_3^-\ Na^+$), the performance matched with that of RuL^{16}. The authors concluded that the distribution of the catalyst/surfactant/substrate at the interface was affected by the solubility of the ketone in water.

Wang et al. found that the addition of commercially available cationic surfactant cetyl trimethylammonium bromide (CTAB), in presence of lipophilic [36] and hydrophilic [37] catalysts, improved the reactivity and enantioselectivity of asymmetric transfer hydrogenation (ATH) of aromatic ketones and imines. However, the same did not affect the enantiomeric excess (ee) for aliphatic ketones. To overcome this issue, Rh was complexed with the amphiphilic diamine ligand (Figure 5.1) [38]. The length of the catalyst chain (n) of ligand was varied from 1 to 16 and named as L1–L5. Out of these investigated ligands, a single-chain chiral metallosurfactant Rh-L4 gave incredible ee for α,β-unsaturated ketones, specifically octan-2-one, of up to 99%, which had $n = 11$. The ligands having 4–8 carbons in the tail drastically improved the efficiency of the reaction, which can be seen in Figure 5.1. However, further increase in the length did not affect the conversion and the ee value anymore.

Another classical example of performing hydrogenation reactions with water as a solvent was illustrated by Lin et al., who designed various amphiphilic ligands and complexed them with rhodium [39]. These chiral-surfactant-type catalysts

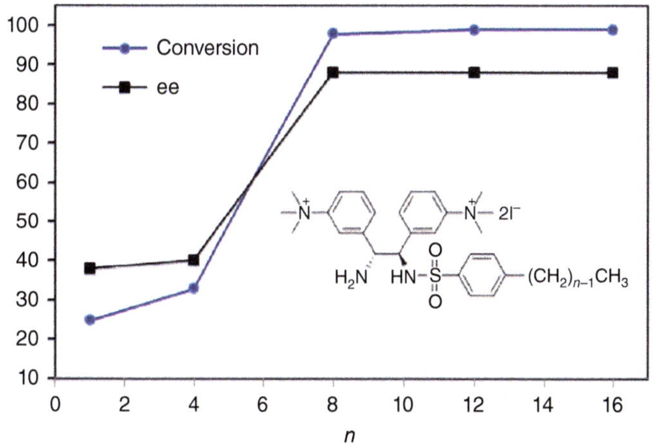

Figure 5.1 Effect of the length of catalyst chain on the ATH of aliphatical ketone octan-2-one. Reaction conditions: 0.004 mmol of ligand L1-L5, 0.002 mmol of $[Cp^*RhCl_2]_2$, 2 mmol of HCOONa, 0.4 mmol of octan-2-one, 2 ml of H_2O, 20 °C, 24 hours. Source: Li et al. [38].

were employed in the ATH of long-chained aliphatic ketoesters. The interactions between the catalyst and the substrate in the core of metallomicelles were found to be responsible for excellent enantioselectivity (up to 99% ee). The long alkyl chains of amphiphiles were ordered in a regular fashion in micelles, which gave advantage in stereoselectivity. Due to the use of metallomicellar catalyst in the reaction, the transition state formed was stabilized by CH-π interactions between methyl group of catalyst and the aromatic ring of substrate.

Oxidation Reaction

Oxidation reactions are some of the most important organic transformations generating quite useful products. When we compare the micellar and non-micellar medium, the concentration of dissolved molecular oxygen is more in the latter. This property gives an advantage of using metallosurfactants in the organic reactions as catalysts. Thiruvengetam et al. [40] reported the use of molybdenum-based metallomicellar catalyst, $(C_{19}H_{42}N)[MoO(O_2)_2(C_6H_4NO_2)]$ (Mo1), in oxidation of unsaturated alcohols. The mentioned Mo complex activated the molecular oxygen in aqueous medium, and its mildness led to controlled and selective oxidation. The same catalyst was found to be versatile and chemoselective, as it performed excellently in presence of other functionalities as well, like cyano, sulfide, aryl-hydroxyl, tertiary amine, internal/terminal alkene, internal/terminal alkyne, and acetal. The recyclability was investigated through one of the model substrates, which demonstrated yields between 92% and 97% till eight catalytic runs (Figure 5.2).

Cycloaddition – Diels–Alder Reaction

Diels–Alder reaction has been utilized as a powerful tool to introduce chemical complexity, particularly six-membered rings in the synthesis of new materials. Otto Diels and Kurt Alder were even awarded Nobel Prize for the discovery of

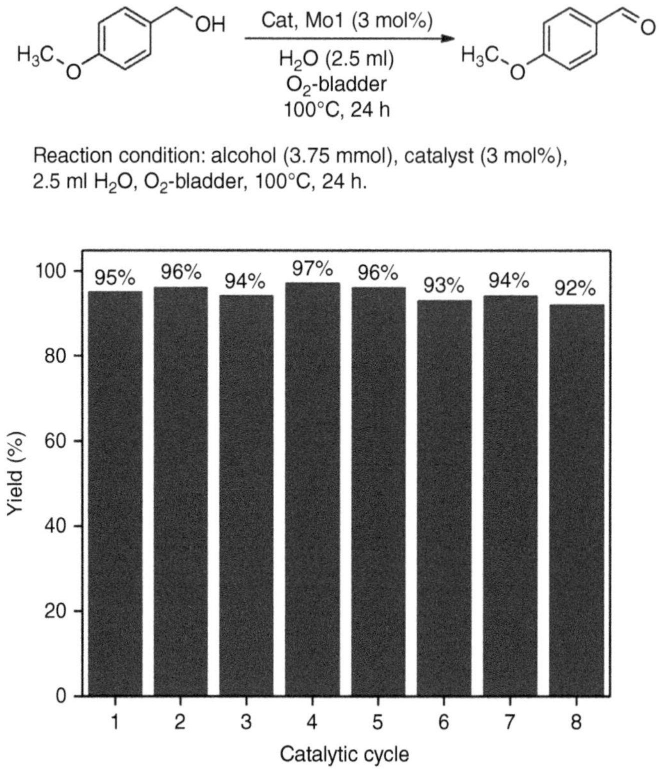

Figure 5.2 Recycling test of the catalyst Mo1.

Diels–Alder reaction. The advantage of using metallosurfactants over conventional surfactants like sodium dodecyl sulfate (SDS), CTAB, and dodecyl heptaoxyethylene ether ($C_{12}E_7$) for the Diels–Alder reaction has been investigated by Engberts and coworkers [41]. The reaction was slowed down by micelles of SDS, CTAB, and $C_{12}E_7$ [42], while the metallosurfactant, $Cu(DS)_2$ micelles, enhanced the rate by up to 1.8 million times making the reaction very efficient. The diene and dienophile used in the reaction were in close proximity when $Cu(DS)_2$ was used, because the dienophile complexed with Cu ions, resulting in favorable orientation, which accelerated the overall reaction. Moreover, Engberts also proved vesicular catalysis to be more interesting than micellar catalysis, taking green chemistry into consideration [43]. Even at very low concentration (0.1 mM), vesicles formed from a cyclic phosphate ester (5,5-di-*n*-dodecyl-2-hydroxy-1,3,2-dioxaphosphorinan-2-one) with copper(II) counterions ($Cu(dDP)_2$) were found to be very efficient. The rate of the reaction depended a lot on the preparation technique of vesicles.

5.2.2 NHC Metallosurfactant-Catalyzed Reactions

When N-heterocyclic carbenes (NHC) are compared to phosphanes in using them as ligands in catalysis, it has been found that NHC perform better under stress

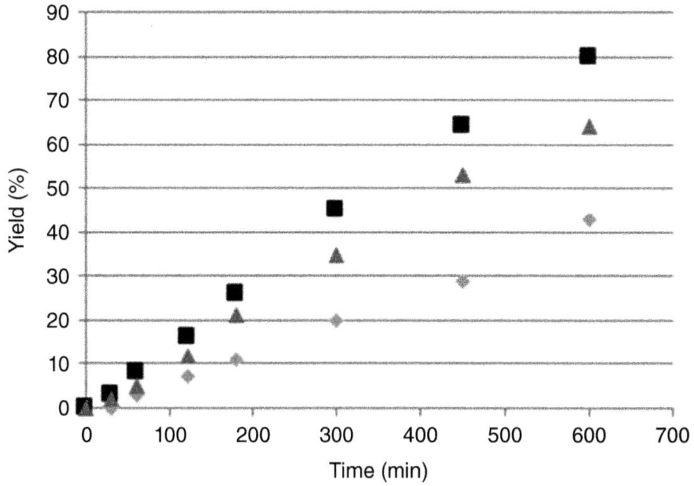

Figure 5.3 Kinetics of our authors' model reaction with different alkyl chain lengths. ♦: C7-IPr; ■: C11-IPr; ▲: C15-IPr. Source: Rühling et al. [47].

and heat conditions; therefore, several groups have attempted to synthesize and utilize metal-coordinated NHC for various applications [44]. Polarz and coworkers have reported the synthesis of Pd–NHC metallosurfactant to use as catalyst in C–C cross-coupling reactions [45]. To compare the results, another non-amphiphilic coordination compound of same catalytic center was also prepared and employed as catalyst under identical reaction conditions. The micellar catalysis improved the turnover frequency by onefold. Similarly, two more novel complexes were prepared using NHC with Cu(I) and Fe(II) as head for polymerization of methyl methacrylate [46]. The colloids of polymethyl methacrylate formed in the reaction were stabilized by catalyst located at their surface.

Another hybrid surfactant was synthesized using NHC ligand (IPr) and a structurally simple surfactant, which was further complexed with gold to employ it as a catalyst in hydration reactions [47]. IPr is one of the *N,N′*-diarylimidazolium carbenes, which has hydrophilic head group and can form robust and catalytically active complexes with gold. When the simple surfactants were used, the anionic surfactant displayed better results than cationic and neutral surfactants because of effective complexation with Au(I). The effect of the different alkyl chain lengths on the catalytic activity was also studied, and it was deduced that the chain length should be optimal (C11-IPr) to place the gold atom on the boundary of the micelle and increase the rate of the reaction (Figure 5.3). If the chain is short (C7-IPr), then the gold atom will be immersed in the micelle, while the long chain (C15-IPr) forces the gold atom out of the micelle.

5.2.3 Stimuli-Responsive Metallosurfactant-Catalyzed Reactions

To overcome the difficulties associated with the organic reactions, it is pertinent to look into synthesizing surfactants with desired properties. One of the major issues was separation of catalyst from the medium during work-up, for which a stimuli-responsive catalyst was synthesized, which undergoes hydrophilic-to-hydrophobic changes in response to temperature. A "smart" poly(*N*-isopropylacrylamide) (PNIPAAm) was grafted on the metal center of Mn(III) complex to develop a novel thermoresponsive surfactant-type catalyst (denoted as catalyst I) (Scheme 5.1) [48]. The nanoreactor, used as catalyst in the asymmetric epoxidation of alkenes in water, had hydrophobic core for reaction and hydrophilic surface for solubilization in aqueous medium. The thermoresponsivity of catalyst helped to recover and reuse the catalyst for five cycles by only changing the temperature.

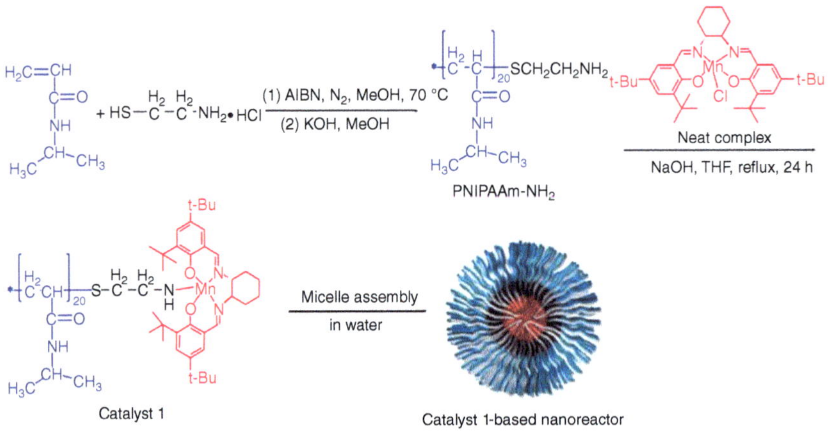

Scheme 5.1 Schematic representation of synthesis and self-assembly of surfactant-type catalyst 1. Source: Zhang et al. [48].

Another thermoresponsive catalyst using Ti(IV) was synthesized, by the same group, for asymmetric sulfoxidation in water [49]. The transmission electron microscope (TEM) images of the synthesized surfactant system proved the formation of micelles in water at room temperature. These metallomicelles acted as catalyst in the sulfoxidation reaction using H_2O_2 as an oxidant. High chemoselectivity and enantioselectivity were obtained with the synthesized metallosurfactants and thermo-controlled separation aided in easy recovery.

5.2.4 Miscellaneous Reactions

An outstanding example of enantioselective addition in water was given by Kobayashi and coworkers by using metallosurfactant, scandium tris(dodecyl sulfate) [$Sc(DS)_3$] [50]. Asymmetric ring opening of epoxides was carried out with the metallosurfactant as catalyst. When the same reaction was performed with $Sc(OTf)_3$, the reaction was very slow and even after increasing the concentration of nucleophile or increasing reaction time, the yield did not improve. Moreover, $Sc(DS)_3$ was also found to catalyze the synthesis of various polysubstituted and polycyclic quinolines [51]. Table 5.1 shows the reaction of 2-aminobenzophenone (1a) and with ethyl acetoacetate (2a), which is catalyzed by various Lewis

Table 5.1 Reaction of 2-aminobenzophenone 1a with ethyl acetoacetate 2a catalyzed by various Lewis acid-surfactant combined catalysts (10 mol%) in water at room temperature.

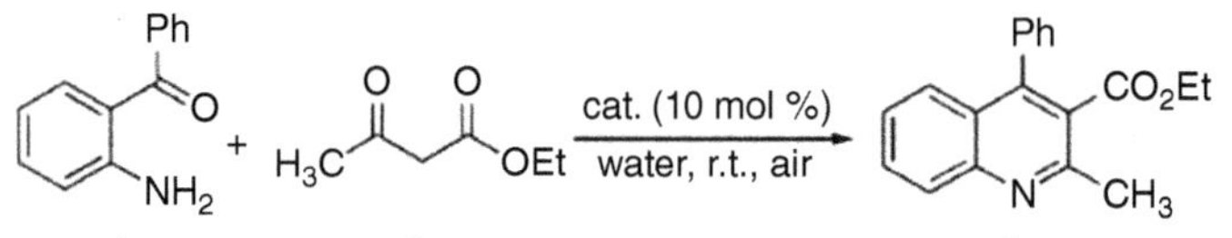

1a 2a 3a

Entry	LASC (catalyst)	Yield [%][a]
1	$In(O_3SOC_{12}H_{25})_3$	67
2	$In(O_3SC_{12}H_{25})_3$	50
3	$Sm(O_3SOC_{12}H_{25})_3$	82
4	$Gd(O_3SOC_{12}H_{25})_3$	56
5	$Dy(O_3SOC_{12}H_{25})_3$	45
6	$Fe(O_3SOC_{12}H_{25})_3$	58
7	*$Sc(O_3SOC_{12}H_{25})_3$*	*90*

a) Isolated yield based on 2-aminobenzophenone **1a**.

acid-surfactant-combined (LASC) catalysts (employing various metals such as In, Sm, Gd, Fe, and Dy) in aqueous medium at room temperature. Among the LASC investigated, $[Sc(DS)_3]$ displayed the finest result. Same reaction was carried out to produce other quinolones and no side reactions were observed, which are otherwise present in basic conditions.

As discussed previously, metallosurfactants are particularly efficient in vesicular catalysis. One such example of vesicular catalysis is hydroformylation of 1-octene [52]. Rhodium complexes of a series of xantphos derivatives with surface-active pendant groups were used as catalysts in this reaction. When the ligand was large, it formed vesicles in aqueous solution. These vesicles were stable at high temperature, which was an advantage over micellar catalysis. The high reaction rates were result of the hydrophobicity of large ligands, which ultimately increased the solubility of organic substrates in water.

Another cross-coupling reaction is reported in which metallomicelles of palladium(II) complex were efficient catalysts [53]. Suzuki–Miyaura reactions of aryl bromides substituted with a long alkyl chain and arylboronic acids at 80 °C were carried out in neat water. The reaction was found efficient for synthesis of biphenyl derivatives, which are used as liquid-crystal compounds.

5.3 Metallosurfactants as Catalyst in Water Oxidation

Exploiting the catalytic activity of metallosurfactants, their applicability was extended to water oxidation reactions. Metallosurfactants can act as excellent

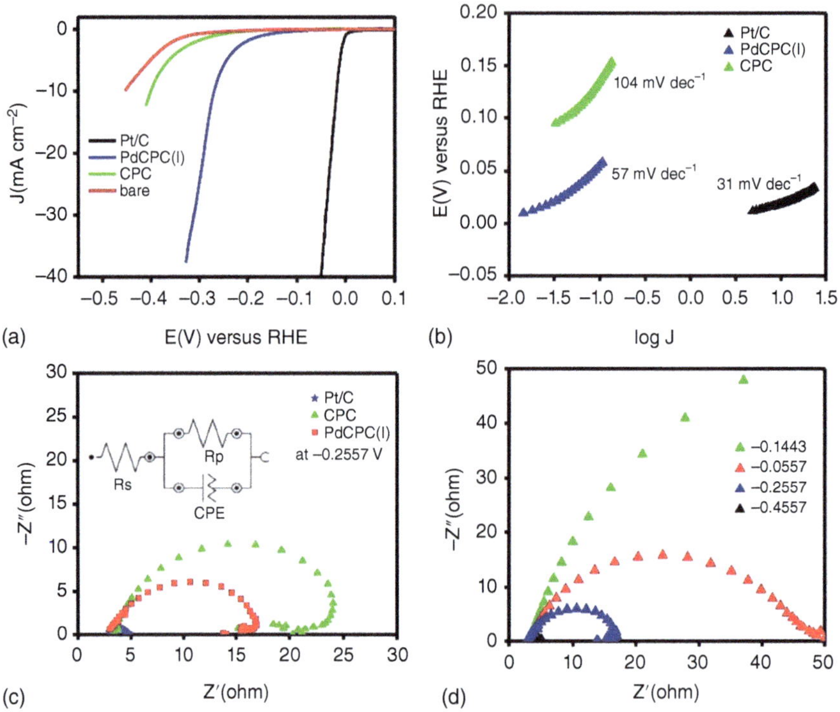

Figure 5.4 (a) Polarization curves of CPC, PdCPC(I), and Pt/C, (b) Tafel slopes of CPC, PdCPC(I), and Pt/C; their respective electrochemical impedence spectroscopy (EIS) studies, (c) comparative impedance spectra of CPC, PdCPC(I), and commercially available Pt/C at 0.2557 V (versus RHE), (d) effect of change in potential from 0.1443, 0.0557, 0.2557, 0.4557 V (versus RHE) on PdCPC(I)-coated surface in 0.5 N H_2SO_4. Source: Kaushik et al. [54].

electrocatalysts aiding water splitting, leading to fuel cell generation. Pradeep et al. synthesized the palladium-based metallosurfactants, i.e. Pd(CPC) using cetylpyridinium chloride (CPC) as a surfactant [54]. CPC surfactants are amphiphilic in nature, which retard the particle aggregation and act as a linker between the electrode surface and the electrolyte. Further, the HER performance was measured in 0.5 N H_2SO_4 using metallosurfactants coated over stainless steel as working electrode, and the linear sweep voltammetry (LSV) polarization curve shows the catalytic performance where onset potential and overpotential correspond to 15 and 269 mV versus RHE (Figure 5.4). The lesser Tafel slope value of 57 mV dec^{-1} confirms the faster kinetics of the metallosurfactant during electrocatalysis. The electrochemical impedance studies were measured at different potentials of 0.1443, 0.0557, 0.2557, 0.4557 V versus RHE, which indicated the decrease in charge transfer resistance with increase in potential. This work gives new design to develop Pd-based metallosurfactants for energy-based applications.

Pradeep et al. also designed a new redox active, cost-effective Co(II)-based metallosurfactant, i.e. hexadecylpyridinium trichloro cobaltite [CoCPC(I)], which was used as an electrocatalyst in HER [55]. The activity of the electrocatalyst was

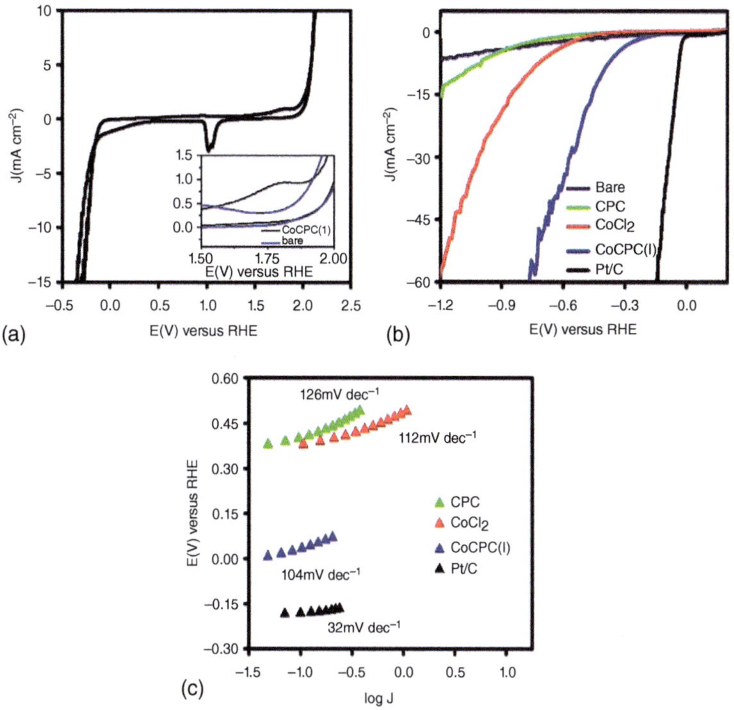

Figure 5.5 (a) Cyclic voltammetry of CoCPC(I) in 0.5 N H_2SO_4 at 100 mV s^{-1} and 298 K. Inset shows a comparison of a bare and coated electrode (b) Polarization curve of CPC, $CoCl_2$, CoCPC(I), and commercial Pt/C catalyst (C) Tafel slope of CPC, $CoCl_2$, CoCPC(I), and commercial Pt/C in 0.5 N H_2SO_4. Source: Pradeep et al. [55].

evaluated in acidic medium (0.5 N H_2SO_4). During catalysis, the electrocatalyst CoCPC(I) showed an onset potential of −30 mV, an overpotential of −307 mV, and the Tafel slope corresponding to 104 mV dec^{-1} as illustrated in Figure 5.5. The authors have studied the detailed mechanism for the electrocatalytic activity by evaluating various factors that include mass activity, turn over frequency (TOF), exchange current density, contact angle measurements, stability studies, and poisoning effect on activity. According to the studies, cobalt metal present as counter ion with surfactant leads to the faster formation of the micelles than surfactant CPC. This metallomicellar structure modifies the interfacial property of the surfactants and is responsible for the better catalytic performance.

Further, Gonawala et al. developed cobalt-based phenolate metallosurfactants $[Co^{III}(L^{N2O3})H_2O]$, which was deposited on the fluorine-doped tin oxide glass (FTO) substrate using Langmuir–Blodgett films (LB) technique [56]. This LB technique provided the control to optimize the films thickness and its uniformity toward heterogeneous water oxidation. It was found that the phenolate-rich complexes are more active for proton reduction in acidic media. The uniformity and amphiphilic behavior of the LB films was analyzed and confirmed that high pressure leads to the formation of homogeneous films. Further, they evaluated the catalytic activity of the

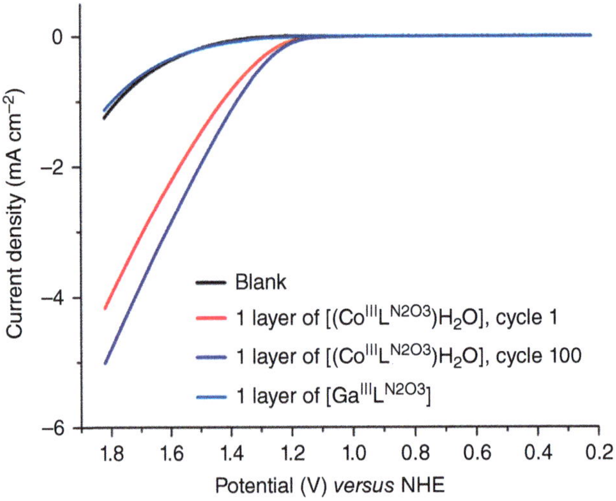

Figure 5.6 Polarization curves, initial and after 100 cycles at pH 11 for 1 layer of [$Co^{III}(L^{N2O3})H_2O$] and initial cycle for 1 layer of [$Ga^{III}L^{N2O3}$]on an FTO electrode at pH 11. Source: Gonawala et al. [56].

films toward water oxidation in the sodium borate solution at different pH values. As per the results obtained, the single LB monolayer of the metallosurfactants showed an overpotential of 0.5 V. Figure 5.6 shows the correlation between the number of layers deposited and its electrocatalytic efficiency. The optimal point of the LB films shows better activity up to 9th layer with a constant increment in the catalytic performance from 1st to 9th LB layer. Further, the catalytic efficacy of the LB films was estimated by evaluating different factors like turn over number, stability studies, post-morphological and compositional analysis by scanning electron microscope (SEM) and X-ray photoelectron spectroscopy (XPS) studies.

Kunkel et al. generated a new concept that mainly focuses on the phenomenon of self-assembly in catalysis where surfactants with functional head group aggregates lead to better catalysis [57]. Herein, vesicular-shaped Co(II)-based metallosurfactants were synthesized and head group consisted of diethylenetriamine fullerenes, which were used as electrocatalysts toward the water oxidation. The electrochemical catalysis was studied at concentration below and above critical aggregation concentration (cac). Further, the amphiphilic properties were investigated by evaluating surface tension and dynamic light scattering (DLS) parameters. It was shown that there is no aggregation at lower concentration, but the aggregates formed at 0.2 M concentration, which have hydrodynamic diameter of 150 nm.

The morphology shown by TEM confirmed the formation of vesicular-type structure, not the micelles. The electrocatalytic activity measurements were done by using cyclic voltammetry (CV) as shown in Figure 5.7. It has been found that Co-based metallosurfactants show the onset potential for OER to be +1180 mV and for HER the value corresponds to −1260 mV in comparison to pure organic surfactant, which is catalytically inactive. This clearly established the dependency

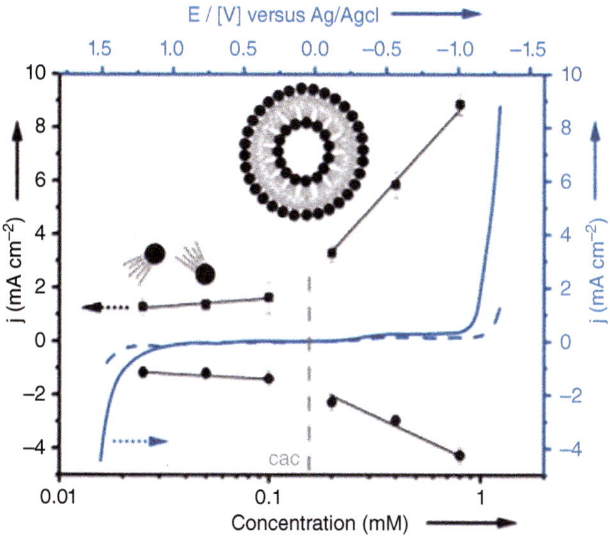

Figure 5.7 CV measurement (black) and Randles–Sevcik plot (blue); OER = circles; HER = squares of (4) at 0.025 mM compared to the CV of the organic ligand (3; dashed line).

of the catalytic properties on the level of aggregation. Further, they found an increase in the current density when concentration is greater than cac value.

5.4 Light-Driven Hydrogen Generation

The urge to replace the conventional energy resources with the solar fuels has gained immense attention in the last few decades. The mechanism of photocatalytic water splitting has complex and tedious kinetics in homogeneous solution. The light-driven H_2 evolution or water reduction has been done by means of various constituents like photosensitizer (PS), electron mediator, sacrificial reductant (SR) or electron donor, and proton-reducing catalyst [58–60]. The efficiency of the H_2 evolution can be improved by using linked donor–acceptor molecules with a large electron transfer rate [61]. The photoexcitation of the donor–acceptor system facilitates the direct transfer of electrons to the catalyst without mediator. Further, a lot of investigation in the field of photocatalysis used for water splitting has been carried out by using sacrificial electron donor to achieve higher H_2 evolution efficiency [62]. H_2 evolution process is usually done in both aqueous and organic solvents-based catalytic systems [63]. Gratzel et al. had investigated the effect of CTAB micelles on the catalytic activity of e_{aq}^- and CO_2^- by evaluating the rate constant, and they concluded that the kinetics of the reaction increased in the presence of the micelle solution due to efficient electron transfer [64]. Basically, the use of micelles helps in modifying the light-induced electron transfer [65].

Kagalwala et al. demonstrated the photocatalytic H_2 evolution using amphiphilic Ir metallosurfactants [Ir(ppy)$_2$(dhpdbpy)]Cl (where ppy = 2-phenylpyridine and

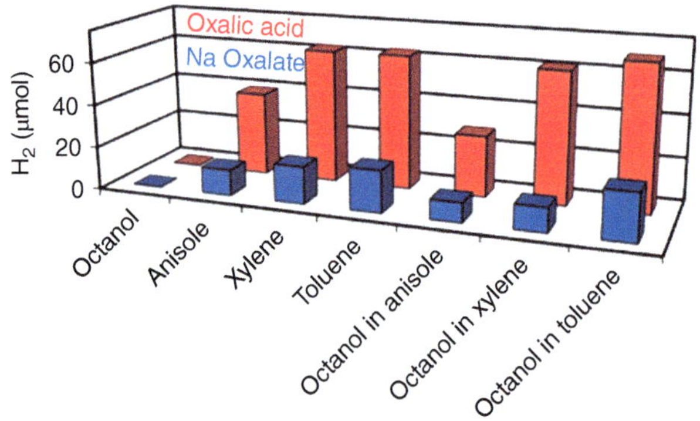

Figure 5.8 Solvent study. Variation of organic solvent, with the final solution containing 4 ml of the respective organic solvent, 4 ml of water, and 1 ml of 1-octylamine. Toluene with 5% 1-octanol (v/v) was found to be the most optimal solvent. Oxalic acid (red bars) was observed to be a better SR than sodium oxalate (blue bars). Source: Kagalwala et al. [66].

dhpdbpy = 4,4′-diheptadecyl-2,2′-bipyridine) in equimolar toluene and water system. This biphasic catalytic study is only possible by using amphiphilic molecules such as metallosurfactants due to their capability of forming microemulsions [66]. The microemulsion system leads to the photochemical conversions through photoinduced electron transfer and charge separation in microscopic entities. Herein, the Rh-based amphiphile $[Rh(dhpdbpy)_2Cl_2]Cl$ was used as reducing catalyst and oxalic acid as sacrificial electron donor. The electrochemical behavior of the metallosurfactants was determined from CV using glassy carbon electrode as working electrode. The addition of 1-octylamine results in an ion pair formation with oxalic acid, which interacts with the metallosurfactants and commences the catalytic process. Herein, 1-octylamine plays an important role for better solubility of the two layers. Figure 5.8 explains the effect of the solvent system on the catalytic performance. Here, 1-octylamine is crucial to generate H_2 as it facilitates considerable mixing of the biphasic solvent and in situ ion formation between 1-octylamine and oxalic acid. This system acts as surfactant and strongly interacts with metallosurfactants. But no such ion formation takes place for sodium oxalate and catalytic activity was reduced. Furthermore, the effect of concentration of the metallosurfactants on catalytic activity was also evaluated. Their study established the importance of the microemulsion formation in the biphasic system that leads to the improved photosensitization process.

Adams et al. have studied the effect of cationic micelle on the light-induced electron transfer, charge separation, and back electron transfer between electron donor $[Ru(NH_3)_6]^{2+}$ and series of other Ru(II) diamine complexes, which act as chromophores having general formula $[(bpy)_2Ru(LL)]^{2+}$ (LL = bpy; 4-R-4′-methyl-2,2′-bpy) [67]. The processes of quenching and back electron transfer were strongly influenced in the presence of CTAB as surfactant for MC13, MC17, and DC17 solution (MCx = 4-methyl, 4′-x-2,2′-bipyridine where x = pentyl (MC5), tridecyl (MC13), or heptadecyl (MC17) or DC17 = 4,4′-diheptadecyl-2,2′-bipyridine). Figure 5.9 signifies that in the presence of CTAB micelles, the charge separation efficiencies of chromophores MC13 and MC17 increase drastically. The rate of back

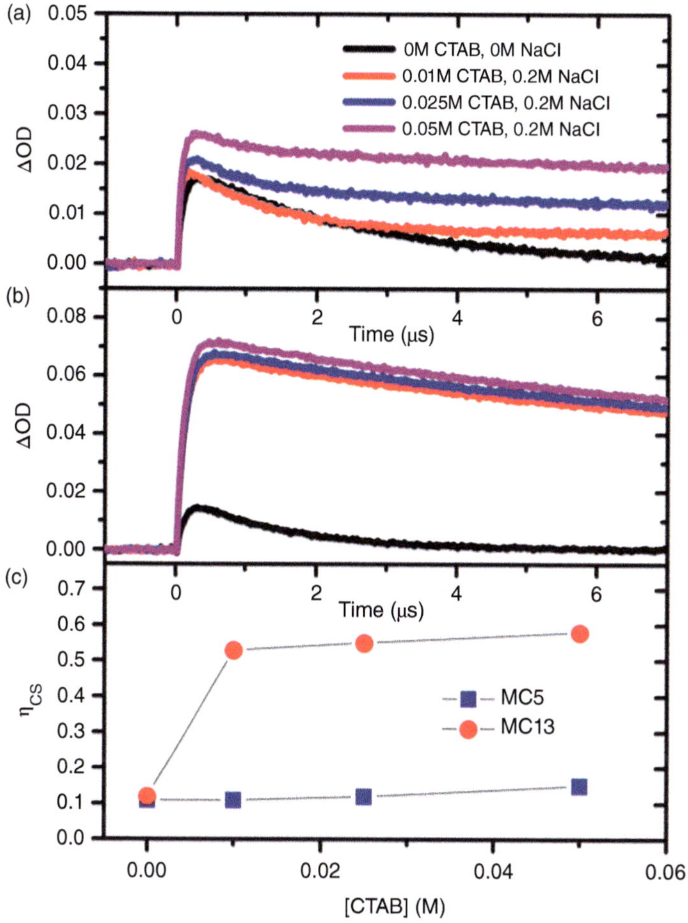

Figure 5.9 Kinetic traces at 510 nm for the back reactions of (a) $[Ru(bpy)_2MC5]^{1+}$ and (b) $[Ru(bpy)_2MC13]^{1+}$ with $[Ru(NH_3)_6]^{3+}$ in the absence and presence of CTAB micelles (0, 0.01, 0.025, and 0.05 M); 0.2 M NaCl added to solutions with CTAB to diminish ionic strength effects. (c) Charge separation yields as a function of CTAB concentration.

electron transfer by chromophores MC13, MC17, DC17 complexes decreased in the micellar solution under high ionic concentration. Basically, the micellar system leads to the inhibition of back electron transfer, which is an important photo redox reaction. This study shows the advantage of micellar system, which leads to the charge separation of the reactants due to hydrophobic and electrostatic forces and consequently influences the catalytic H_2 and O_2 production from water [63, 68].

5.5 Conclusions

Metallosurfactants can be very effectively employed as nanoreactor containing active catalytic centers to carry out various organic transformations. The chapter particularly highlighted their applicability in organic reactions and several energy-based systems. This could be a potential answer to the rising energy demands of the world, which needs a clean and green fuel alternative that works to mitigate the climate change issues as well. In addition to improvement in yields and selectivity in organic reactions, the major advantage of using metallosurfactants is their ability to switch from organic solvent into aqueous medium to carry out reaction conversions, thus working in accordance with the principles of green chemistry. This also leads to minimal toxicity, easy product extraction, and recyclability of the catalyst, which is most desired in any catalytic conversion. Metallosurfactants have garnered a lot of attention with their potential to serve as excellent electrocatalysts and photocatalysts employing the easily available transition metal ions as compared to the precious Rh, Ir candidates in terms of commercial availability, cost, and stability. The field of metallosurfactants is an emerging area as surfactants act as a linker and reduce the particle aggregation, the interfacial properties between electrode and electrolyte, which is important for charge transfer catalysis. However, there is still a lot of scope of research in using metallosurfactants as electrocatalysts to meet up the renewable energy resources requirements, which provides an open and exciting area to researchers to work on, related to energy-based applications.

References

1 Clint, J.H. (1992). *Surfactant Aggregation*. New York: Chapman and Hall.

2 Sorella, G.L., Strukul, G., and Scarso, A. (2015). Recent advances in catalysis in micellar media. *Green Chemistry* 17: 644–683.

3 Lipshutz, B.H., Ghorai, S., and Cortes-Clerget, M. (2018). The hydrophobic effect applied to organic synthesis: recent synthetic chemistry "in water". *Chemistry A European Journal* 24: 6672–6695.

4 Iwakura, C., Honji, A., and Tamura, H. (1981). The anodic evolution of oxygen on Co_3O_4 film electrodes in alkaline solutions. *Electrochimica Acta* 26: 1319–1326.

5 Wang, H.F., Chen, L., Pang, H. et al. (2020). MOF-derived electrocatalysts for oxygen reduction, oxygen evolution and hydrogen evolution reactions. *Chemical Society Review* 49: 1414–1448.

6 Hosseini, S.E. and Wahid, M.A. (2016). Hydrogen production from renewable and sustainable energy resources: promising green energy carrier for clean development. *Renewable Sustainable Energy Reviews* 57: 850–866.

7 Montoya, J.H., Seitz, L.C., Chakthranont, P. et al. (2017). Materials for solar fuels and chemicals. *Nature Materials* 16: 70–81.

8 Yu, L., Yang, J.F., Guan, B.Y. et al. (2018). Hierarchical hollow nanoprisms based on ultrathin Ni–Fe layered double hydroxide nanosheets with enhanced electrocatalytic activity towards oxygen evolution. *Angewandte Chemie International Edition* 57: 172–176.

9 Kauffman, D.R., Deng, X., Sorescu, D.C. et al. (2019). Edge-enhanced oxygen evolution reactivity at ultrathin, Au-supported Fe_2O_3 electrocatalysts. *ACS Catalysis* 9: 5375–5382.

10 Qi, J., Zhang, W., and Cao, R. (2018). Porous materials as highly efficient electrocatalysts for the oxygen evolution reaction. *ChemCatChem* 10: 1206–1220.

11 Zhang, B., Wang, H., Zuo, Z. et al. (2018). Tunable CoFe-based active sites on 3D heteroatom doped graphene aerogel electrocatalysts via annealing gas regulation for efficient water splitting. *Journal of Material Chemistry A* 6: 15728–15737.

12 Liu, X., Cui, S., Sun, Z. et al. (2016). Self-supported copper oxide electrocatalyst for water oxidation at low overpotential and confirmation of its robustness by Cu K-Edge X-ray absorption spectroscopy. *Journal of Physical Chemistry C* 120: 831–840.

13 Dou, Y., Liao, T., Ma, Z. et al. (2016). Graphene-like holey Co_3O_4 nanosheets as a highly efficient catalyst for oxygen evolution reaction. *Nano Energy* 30: 267–275.

14 Sui, C., Chen, K., Zhao, L. et al. (2018). MoS_2 -modified porous gas diffusion layer with air–solid–liquid interface for efficient electrocatalytic water splitting. *Nanoscale* 10: 15324–15331.

15 Esswein, A.J., Mc Murdo, M.J., Ross, P.N. et al. (2009). Size-dependent activity of Co_3O_4 nanoparticle anodes for alkaline water electrolysis. *Journal of Physical Chemistry C* 113: 15068–15072.

16 Mamaca, N., Mayousse, E., Arrii-Clacens, S. et al. (2012). Electrochemical activity of ruthenium and iridium based catalysts for oxygen evolution reaction. *Applied Cataysis B: Environment* 111, 112: 376–380.

17 Yu, F.Y., Lang, Z.L., Yin, L.Y. et al. (2020). Pt-O bond as an active site superior to Pt^0 in hydrogen evolution reaction. *Nature Communications* 11: 490.

18 Antolini, E. (2014). Iridium as catalyst and cocatalyst for oxygen evolution/reduction in acidic polymer electrolyte membrane electrolyzers and fuel cells. *ACS Catalysis* 4: 1426–1440.

19 Geiger, S., Kasian, O., Shrestha, B.R. et al. (2016). Activity and stability of electrochemically and thermally treated iridium for the oxygen evolution reaction. *Journal of Electrochemical Society* 163: F3132–F3138.

20 Zhang, Z., Zhou, D., Bao, X. et al. (2019). One-pot synthesis of Fe_2O_3/C by urea combustion method as an efficient electrocatalyst for oxygen evolution reaction. *International Journal of Hydrogen Energy* 44: 2877–2882.

21 Koza, J.A., He, Z., Miller, A.S., and Switzer, J.A. (2012). Electrodeposition of crystalline Co_3O_4 – a catalyst for the oxygen evolution reaction. *Chemistry of Materials* 24: 3567–3573.

22 Yang, H., Liu, Y., Luo, S. et al. (2017). Lateral-size-mediated efficient oxygen evolution reaction: insights into the atomically thin quantum dot structure of $NiFe_2O_4$. *ACS Catalysis* 7: 5557–5567.

23 Vojvodic, A. and Norskov, J.K. (2011). Optimizing perovskites for the water-splitting reaction. *Science* 334: 1355–1356.

24 Li, M., Xiong, Y., Liu, X. et al. (2015). Facile synthesis of electrospun MFe_2O_4 (M = Co, Ni, Cu, Mn) spinel nanofibers with excellent electrocatalytic properties for oxygen evolution and hydrogen peroxide reduction. *Nanoscale* 7: 8920–8930.

25 Jin, H., Mao, S., Zhan, G. et al. (2017). Fe incorporated α-$Co(OH)_2$ nanosheets with remarkably improved activity towards the oxygen evolution reaction. *Journal of Materials Chemistry A* 5: 1078–1084.

26 Haschke, S., Zhuo, Y., Schlicht, S. et al. (2019). Enhanced oxygen evolution reaction activity of nanoporous $SnO_2/Fe_2O_3/IrO_2$ thin film composite electrodes with ultralow noble metalloading. *Advance Materials Interfaces* 6: 1801432.

27 Wang, S., Min, Y., and Yu, S. (2007). Synthesis and magnetic properties of uniform hematite nanocubes. *The Journal of Physical Chemistry C* 111: 3551–3554.

28 Backes, C., Higgins, T.M., Kelly, A. et al. (2017). Guidelines for exfoliation, characterization and processing of layered materials produced by liquid exfoliation. *Chemistry of Materials* 29: 243–255.

29 Gupta, A., Arunachalam, V., and Vasudevan, S. (2015). Water dispersible, positively and negatively charged MoS_2 nanosheets: surface chemistry and the role of surfactant binding. *Journal of Physical Chemistry Letters* 6: 739–744.

30 Smith, R.J., King, P.J., Lotya, M. et al. (2011). Large-scale exfoliation of inorganic layered compounds in aqueous surfactant solutions. *Advanced Materials* 23: 3944–3948.

31 Lipshutz, B.H. and Ghorai, S. (2009). PQS: a new platform for micellar catalysis. RCM reactions in water, with catalyst recycling. *Organic Letters* 11: 705–708.

32 Lipshutz, B.H. and Ghorai, S. (2010). PQS-2: ring-closing- and cross-metathesis reactions on lipophilic substrates; in water only at room temperature, with in-flask catalyst recycling. *Tetrahedron* 66: 1057–1063.

33 Li, J., Zhang, Y., Han, D. et al. (2008). Transfer hydrogenation of aldehydes on amphiphilic catalyst assembled at the interface of emulsion droplets. *Green Chemistry* 10: 608–611.

34 Kalsin, A.M., Peganova, T.A., Novikov, V.V. et al. (2014). Transfer hydrogenation of ketones catalyzed by surface-active ruthenium and rhodium complexes in water. *Chemistry A European Journal* 20: 846–854.

35 Kalsin, A.M., Peganova, T.A., Novikov, V.V. et al. (2015). Cooperative effects of ruthenium micellar catalysts and added surfactants in transfer hydrogenation of ketones in water. *Catalysis Science and Technology* 5: 4458–4465.

36 Wang, F., Liu, H., Cun, L. et al. (2005). Asymmetric transfer hydrogenation of ketones catalyzed by hydrophobic metal–amido complexes in aqueous micelles and vesicles. *The Journal of Organic Chemistry* 70: 9424–9429.

37 Wu, J., Wang, F., Ma, Y. et al. (2006). Asymmetric transfer hydrogenation of imines and iminiums catalyzed by a water-soluble catalyst in water. *Chemical Communications* 16: 1766–1768.

38 Li, J., Han, J., Lin, Z. et al. (2019). Influence factors studies on the Rh-catalyzed asymmetric transfer hydrogenation of ketones with surfactant-type ligand in water. *Tetrahedron* 75: 422–428.

39 Lin, Z., Li, J., Huang, Q. et al. (2015). Chiral surfactant-type catalyst: enantioselective reduction of long-chain aliphatic ketoesters in water. *The Journal of Organic Chemistry* 80: 4419–4429.

40 Thiruvengetam, P., Chakravarthy, R.D., and Chand, D.K. (2019). A molybdenum based metallomicellar catalyst for controlled and chemoselective oxidation of activated alcohols in aqueous medium. *Journal of Catalysis* 376: 123–133.

41 Otto, S., Engberts, J.B.F.N., and Kwak, J.C.T. (1998). Million-fold acceleration of a Diels–Alder reaction due to combined Lewis acid and micellar catalysis in water. *Journal of the American Chemical Society* 120: 9517–9525.

42 Wijnen, J.W. and Engberts, J.B.F.N. (1997). Retro-Diels–Alder reaction in aqueous solution: toward a better understanding of organic reactivity in water. *The Journal of Organic Chemistry* 62: 2039–2044.

43 Rispens, T. and Engberts, J.B.F.N. (2001). Efficient catalysis of a Diels–Alder reaction by metallo-vesicles in aqueous solution. *Organic Letters* 3: 941–943.

44 Kantchev, E.A.B., O'Brien, C.J., and Organ, M.G. (2007). Palladium complexes of N-heterocyclic carbenes as catalysts for cross-coupling reactions – a synthetic chemist's perspective. *Angewandte Chemie International Edition* 46: 2768–2813.

45 Donner, A., Hagedorn, K., Mattes, L. et al. (2017). Hybrid surfactants with N-heterocyclic Carbene heads as a multifunctional platform for interfacial catalysis. *Chemistry A European Journal* 23: 18129–18133.

46 Donner, A., Trepka, B., Theiss, S. et al. (2019). NHC-metallosurfactants as active polymerization catalysts. *Langmuir* 35: 16514–16520.

47 Rühling, A., Galla, H.J., and Glorius, F. (2015). A remarkably simple hybrid surfactant–NHC ligand, its gold-complex, and application in micellar catalysis. *Chemistry A European Journal* 21: 12291–12294.

48 Zhang, Y., Tan, R., Zhao, G. et al. (2016). Asymmetric epoxidation of unfunctionalized olefins accelerated by thermoresponsive self-assemblies in aqueous systems. *Catalysis Science and Technology* 6: 488–496.

49 Zhang, Y., Tan, R., Zhao, G. et al. (2016). Thermo-responsive self-assembled metallomicelles accelerate asymmetric sulfoxidation in water. *Journal of Catalysis* 335: 62–71.

50 Boudou, M., Ogawa, C., and Kobayashi, S. (2006). Chiral scandium-catalysed enantioselective ring-opening of *meso*-epoxides with N-heterocycle, alcohol and thiol derivatives in water. *Advanced Synthesis and Catalysis* 348: 2585–2589.

51 Zhang, L. and Wu, J. (2007). Friedländer synthesis of quinolines using a Lewis acid-surfactant-combined catalyst in water. *Advanced Synthesis and Catalysis* 349: 1047–1051.

52 Goedheijt, M.S., Hanson, B.E., Reek, J.N.H. et al. (2000). Accelerated biphasic hydroformylation by vesicle formation of amphiphilic diphosphines. *Journal of the American Chemical Society* 122: 1650–1657.

53 Liu, Y., Ma, X., Xie, J. et al. (2013). Metallomicelles of palladium(II) complexes as efficient catalysts for the Suzuki–Miyaura reaction in neat water. *Applied Organometallic Chemistry* 27: 494–498.

54 Kaushik, P., Kaur, G., Chaudhary, G.R., and Batra, U. (2021). Tuning the surface using palladium based metallosurfactant for hydrogen evolution reaction. *Journal of Colloid and Interface Science* 582: 894–905.

55 Pradeep, Kaur, G., Chaudhary, G.R., and Batra, U. (2020). Investigating affordable cobalt based metallosurfactant as an efficient electrocatalyst for hydrogen evolution reaction. *Journal of Colloid and Interface Science* 562: 598–607.
56 Gonawala, S., Baydoun, H., Wickramasinghe, L., and Verani, C.N. (2016). Efficient water oxidation with electromodified Langmuir-Blodgett films of procatalytic [$Co^{III}(N_2O_3)$] metallosurfactants on electrodes. *Chemical Communication* 52: 8440–8443.
57 Kunkel, M., Bitter, S., Sailer, F. et al. (2020). Aggregation-induced improvement of catalytic activity of metalfullerene-based surfactants used in water splitting reaction. *ChemCatChem* 12: 2726–2731.
58 Bernhard, S. and Cline, E.D. (2009). The transformation and storage of solar energy: Progress towards visible-light induced water splitting. *Chimia Inernational Journal of Chemistry* 63: 709–713.
59 Windisch, J., Orazietti, M., Hamm, P. et al. (2016). General scheme for oxidative quenching of a copper bis-phenanthroline photosensitizer for light-driven hydrogen production. *ChemSusChem* 9: 1719–1726.
60 Han, Z. and Eisenberg, R. (2014). Fuel from water: the photochemical generation of hydrogen from water. *Accounts of Chemical Research* 47: 2537–2544.
61 Borchardt, D., Pool, K., and Wherland, S. (1982). Electron transfer between a cobalt clathrochelate and ferrocene in acetonitrile. *Inorganic Chemistry* 21: 93–97.
62 Mancin, F., Scrimin, P., Tecilla, P., and Tonellato, U. (2009). Amphiphilic metalloaggregates: catalysis, transport, and sensing. *Coordination Chemistry Reviews* 253: 2150–2165.
63 Alkaitis, S.A. and Gratzel, M. (1976). Laser photoionization and light-initiated redox reactions of tetramethylbenzidine in organic solvents and aqueous micellar solution. *Journal of American Chemical Society* 98: 3549–3554.
64 Gratzel, M., Kozak, J.J., and Thomas, J.K. (1975). Electron reactions and electron transfer reactions catalyzed by micellar systems. *The Journal of Chemical Physics* 62: 1632–1640.
65 Schmehl, R.H. and Whitten, D.G. (1981). Photoinduced redox reactions in aqueous micelles. Quenching, back-reaction, and competing processes for tetraanionic porphyrins with alkylviologens. *The Journal of Physical Chemistry* 85: 3473–3480.
66 Kagalwala, H.N., Chirdon, D.N., Mills, I.N. et al. (2017). Light-driven hydrogen generation from microemulsions using metallosurfactant catalysts and oxalic acid. *Inorganic Chemistry* 56: 10162–10171.
67 Adams, R.E. and Schmehl, R.H. (2016). Micellar effects on photoinduced electron transfer in aqueous solutions revisited: dramatic enhancement of cage escape yields in surfactant Ru(II) diimine complex/[$Ru(NH_3)_6]^{2+}$ systems. *Langmuir* 32: 8598–8607.
68 Brugger, P.A. and Gratzel, M. (1980). Light-induced charge separation by functional micellar assemblies. *Journal of American Chemical Society* 102: 2461–2463.

6

Metallosurfactants as Drug-Delivery Vehicles

Rohini Kanwar[1,2], Amit Kumar[1], Jyoti Rathee[1], and Surinder K. Mehta[1]*

[1] *Panjab University, Department of Chemistry and Centre of Advanced Studies in Chemistry, Chandigarh, 160014, India*
[2] *Mehr Chand Mahajan DAV. College for Women, Department of Chemistry, Chandigarh, 160036, India*

6.1 Introduction

Surfactants are amphiphilic compounds comprising a hydrophilic head group and a hydrophobic tail that provides an ability to solubilize polar and nonpolar components, respectively. Since the head group of a surfactant is polar, the transition metal ions upon incorporation in the surfactant moiety interact with the headgroup, resulting in the formation of a surfactant metal complex called "metallosurfactant." With the introduction of a transition metal into the conventional surfactant system, the interaction and aggregation behavior of fabricated metallosurfactant becomes of great interest owing to its multidimensional applications. Catalysis, sensing, energy-transfer devices, antimicrobial agents, drug delivery, magnetic resonance imaging (MRI) contrast agents (CAs), optoelectronics, and template agents for the synthesis of mesoporous nanomaterials, are a few well-explored applications of metallosurfactants [1–10].

Distinct metal ions like cobalt, copper, ruthenium, platinum, and gadolinium can coordinate with the metallosurfactant [11]. Veeralakshmi et al. [12] studied the influence of hydrophobicity of cobalt(III)-based metallosurfactant on the interaction with nucleic acids and proteins. They selected single- and double-chain amine-based surfactants, that is, diethylenediamine and tetradecylamine, respectively, for studying the interaction with calf thymus DNA and human serum albumin (HSA). A strong interaction of HSA and DNA with the double-chain metallosurfactant was reported in comparison to single-chain surfactant. Figure 6.1 shows the structure of a typical cobalt(III)-based metallosurfactant [13].

Like conventional surfactants, metallosurfactants retain the property of self-aggregation, vesicle, or micelle formation. Various aggregates with different geometries such as metallomicelles, inverted metallomicelles, bilayers or vesicles, and metal-liposome (metallosomes) have been formulated in different types of solvents. Zha et al. [7] evaluated the aggregation behavior of C_n–Cu–C_n-based metallovesicles and employed it for the encapsulation and

Metallosurfactants: From Fundamentals to Catalytic and Biomedical Applications, First Edition.
Edited by Surinder K. Mehta and Ravneet Kaur.

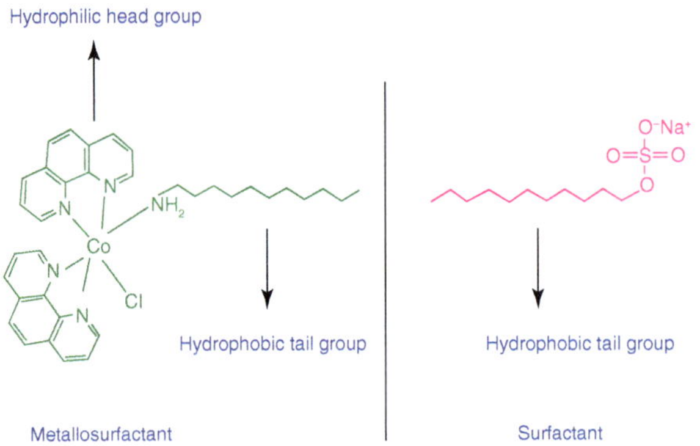

Figure 6.1 Representation of a cobalt-based metallosurfactant complex. Source: Reproduced with permission from Vignesh et al. [13].

release of doxorubicin. During this study, numerous metal-based complexes with the formula $Na_2\{Cu[C_nH_{2n+1}N(CH_2COO)_2]_2\}$ (n = 8, 12, 16) were synthesized by using ultrasonication method. The Cu(II) coordinates with the carboxylic acid moiety of the surfactant's head group, overcoming the electrostatic repulsion between the head groups and thereby directing the headgroups in proximity to each other to form C_n–Cu–C_n metallovesicles. The fabricated metallovesicles showed 70% encapsulation efficiency for doxorubicin hydrochloride and exhibited Fickian diffusion release. Zhiltsova et al. [14] synthesized copper-based metallosurfactant complex by employing a bicyclic amphiphilic ligand, 1-hexadecyl-4-aza-1-azonibicyclo[2.2.2]octane bromide. The micellar solution of the prepared complex has been employed for the solubilization of quercetin. The metallomicellar complex exhibited good solubilization of quercetin in both pre-micellar and post-micellar region. The interaction between polyphenol moieties and metallomicelles was found to be purely electrostatic and the aqueous solubility of drug was enhanced.

Dominguez-Gutierrez et al. [15] formulated Ru(III)-based metallosurfactant, that is, $[Ru(bipy)(4,4'\text{-diheptadecyl-2,2'-bipyridyl})_2]Cl_2$, by reacting 4,4′-diheptadecyl-2,2′-bipyridine and $RuCl_3$ in dimethylformamide (DMF). Figure 6.2 shows the structure of the formed metallosurfactant. Since the synthesized complex, $[Ru(bipy)_3]Cl_2$ was insoluble in toluene, its apolar carbon chains get directed toward the apolar solvent while the cationic headgroups get oriented into the core of the bilayer, forming inverted metallo-micelles in apolar solvent.

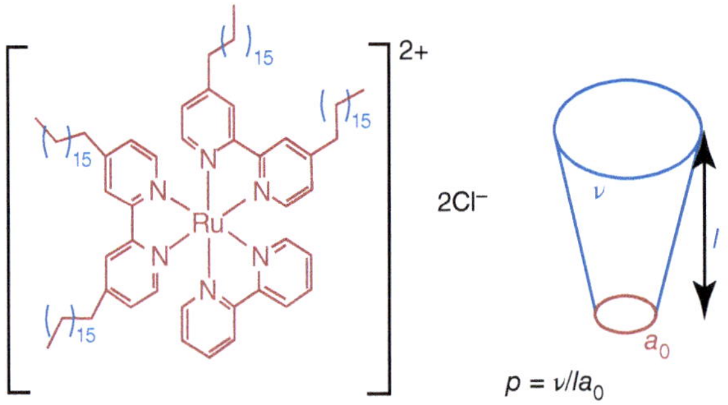

Figure 6.2 $[Ru(bipy)(4,4'\text{-diheptadecyl-2,2'-bipyridine})_2]Cl_2$ metallosurfactant with the ability to aggregate in an inverted micelle. Source: Reproduced with permission from Domínguez-Gutiérrez et al. [15].

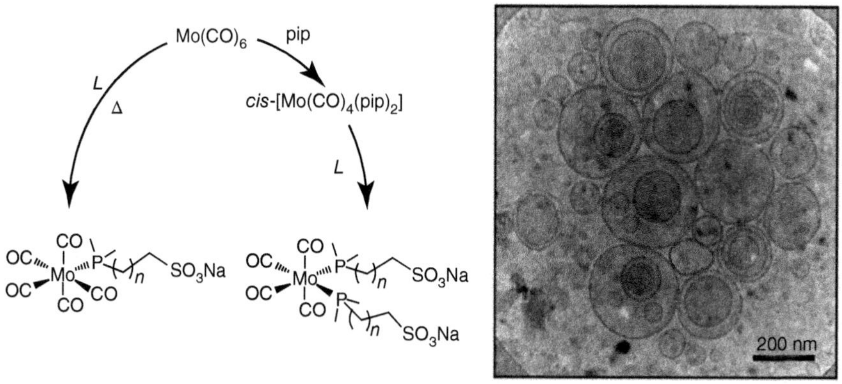

Figure 6.3 Structure of molybdenum metallosurfactant (left) and TEM image of its vesicles (right). Source: From Parera et al. [17]/with permission of Royal Society of Chemistry.

The formed inverted micellar aggregates were 92 nm sized and had an aggregation number of 1.1×10^5.

Sharma et al. [16] formulated supramolecular metallosurfactant of platinum using cationic amphiphilic ligands (*N,N,N*-cetyltrimethylammonium bromide) of various chain lengths (C-8, C-10, C-12, C-14, and C-16). The authors reported a decrease in the self-aggregation behavior of the prepared metallosurfactants with respect to an increase in alkyl chain length, and in addition, demonstrated the use of prepared metallosurfactants in therapeutic applications. Parera et al. [17] fabricated metallovesicles using supramolecular rearrangement of metal carbonyl metallosurfactants. Molybdenum metallosurfactants were synthesized (Figure 6.3) and their aggregation behavior was investigated. Cryo-transmission electron microscopy (TEM) analysis of metallovesicles revealed the presence of small unilamellar vesicles (size <100 nm), large unilamellar vesicles (size >100 nm), multilamellar (onion-like concentric vesicle inside) vesicles, and multivesicular (non-concentric small vesicles inside large vesicle) vesicles.

Based on the booming research in the field of metallosurfactant-based self-assemblies such as metallo-vesicles, metallo-micelles, metallosomes, along with their wide application arena, the present chapter focuses on the utility of metallosurfactants as nanovectors, drug-delivery agents, and MRI contrast agents, which has been discussed in detail.

6.2 Distinct Assemblies of Metallosurfactants in Drug Delivery

6.2.1 Metallosomes as Drug-Delivery Agents

Nanomaterial-based assemblies proffer to be a promising scaffold in the field of drug delivery that increases the pharmacokinetics, potency, and safety profile of potential drugs. In this scenario, the best way to utilize these nanostructures is to develop supramolecular self-assemblies comprising metallosurfactants and phospholipids called metallosomes (or metallovesicles) [18]. Metallosomes can mimic biological membranes and are utilized as carriers for targeted and controlled drug delivery. Till date, numerous bioactive and agrochemical agents have been encapsulated within the metallosomes using two-tailed or single-tailed surfactant. However, mostly, double-chain surfactants are exploited to fabricate metallosomes.

Marín-García et al. [19] fabricated metallosomes by mixing soybean phosphatidylcholine (a natural non-toxic phospholipid) with metallosurfactant, that is, molybdenum pentacarbonyl ($Mo(CO)_5$), or molybdenum tetracarbonyl ($Mo(CO)_4$). The toxicity of developed metallosomes was found to be 10-fold lower than that of the metallosurfactants. The formed assembly remained stable even after dilution. Following the approach, Marín-Garcia et al. [20] reported a new supramolecular system, prepared by mixing molybdenum metallosurfactant with phospholipid phosphatidylcholine. The investigation of formed metallosomes using microfluidics revealed that the aggregation tendency of metallosomes is solvent dependent.

Wang et al. [21] reported higher encapsulation of carboxyfluorescein in anionic and cationic surfactant-based molybdenum vesicles. Nehru et al. [22] focused on the fabrication of fluorescent metallosomes by employing double-chain surfactant-based ruthenium(II) complexes. The self-assembled complex was found to be of spherical morphology and showed excellent stability in an aqueous medium. An increase in size of the metallovesicles was noted with an increase of head size and chain length of surfactant. Ruthenium complexes greatly decreased α-helix's percentage content in bovine serum albumin (BSA)/HSA.

Further, Sakla et al. [23] developed CO-releasing metallosomes via incorporation of a carbonyl complex of manganese into 1,2-stearoyl-*sn*-glycerol-3-phosphocholine (DSPC) lipid-based system. Prepared metallosomes were stable under low temperature and dark environment. The presence of a large surface area of the vesicles made them an efficacious CO delivery vehicle. Further, cytotoxicity studies demonstrated the biosafety of the metallosomes in a dark environment. Kidani et al. [24] formulated metallosomes using the complex (1,2-diaminocyclohexane)platinum(II) (DACHPt) and employed them for the delivery of oxaliplatin (an anticancer drug). Similarly, Osada et al. [25] provide a promising strategy for the development of the metallosomes using DACHPt with the help of copolymers (Figure 6.4).

Ionic surfactants or lipids interact easily with metal due to the presence of electrostatic interactions. Following this approach, Pal et al. [26] reported the fabrication of metallovesicles using amphiphilic bis-(4-pyridylmethyls hexadecanoate)(1,4,7-triazacyclononane)copper(II) triflate. Electronic microscopy revealed that the fabricated metallo-vesicles have a particle size of 33 nm. The cytotoxic concentration (C_{50}) of metallosomes in embryonic kidney cells (human embryonic kidney [HEK] 293-T) was found to be 50% at 1.25 mM concentration. Arroyo et al. [27] synthesized the metallosomes containing copper(II) and zinc(II) complexes of 1-alkyl-1,4,7-triazacyclonoanane and explored their DNA transport studies *in vitro* and *in vivo* environment. HEK 293-T-cells were utilized for the DNA transfection efficiency studies. Obtained results demonstrated the good transfection efficiency of Cu-based metallosomes in delivering of DNA vaccine in murine

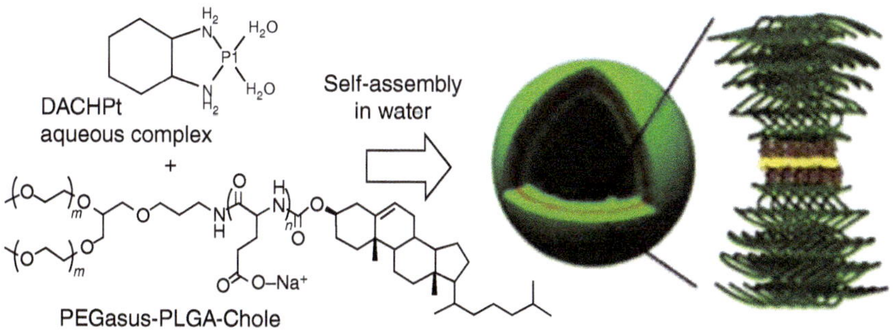

Figure 6.4 Schematic diagram of metallosomes comprising a metal complex (DACHPt) with the carboxylic moiety of poly lactic-co-glycolic acid (PLGA) segment. Source: From Osada et al. [25]/with permission of American Chemical Society.

experimental models. The tunable properties of metallosomes make them the smart carrier in drug-delivery purposes by reducing the complexity of multidrug administration.

6.2.2 Metallosurfactants and Its Self-Assembled Structures as Nanovectors

During the past decade, nanovectors have been extensively studied for the encapsulation, transportation, and delivery of pharmaceutical drugs. Nanovectors not only enhance the absorption of drug in the body but also provide controlled target-specific release, improve the efficacy of drug delivery, and prevent drug degradation. Metallosurfactant acts as a pharmaceutical agent and possesses the ability to act as a nanovector for drugs.

Lebrón et al. [9] designed metallosurfactant-based nanovectors from (2,2′-bipyridine)ruthenium(II) complex, $[Ru(bpy)_3]^{2+}$ as a possible anti-cancer drug candidate. In this study, the metallomicelles fabricated from the cytotoxic metal complex showed promising anticancer activity. Beside anticancer activity, the $[Ru(bpy)_3]$ metallomicellar system showed an encapsulation efficiency of 60% and 10% for doxorubicin and levofloxacin, respectively. Moreover, ruthenium(II) polypyridyl complexes were toxic to cancer lines and demonstrated a strong bacterial and bacteriostatic activity.

Badawi et al. [28] evaluated *in vivo* antitumor, antibacterial, and antimicrobial activity of copper (Cu)-cetyl trimethyl ammonium bromide (CTAB) metallosurfactant-based nanovector. Cardiotoxicity investigation revealed a significant effect of Cu-CTAB metallosurfactant on serum enzymes. Cu-CTAB exhibited a promising anticancer effect by decreasing the tumor volume and raising the means survival rate in an ethyl actetate (EAC)-inoculated group. Apart from this, Cu-CTAB showed excellent antimicrobial activity against *Desulfonamonas pigra, Escherichia coli, Staphylococcus aureus*, and *Candida albicans*.

Zheng et al. [29] synthesized metal-surfactant-based host–metal–guest coordination-bonding architecture in a mesostructured silica hybrid for

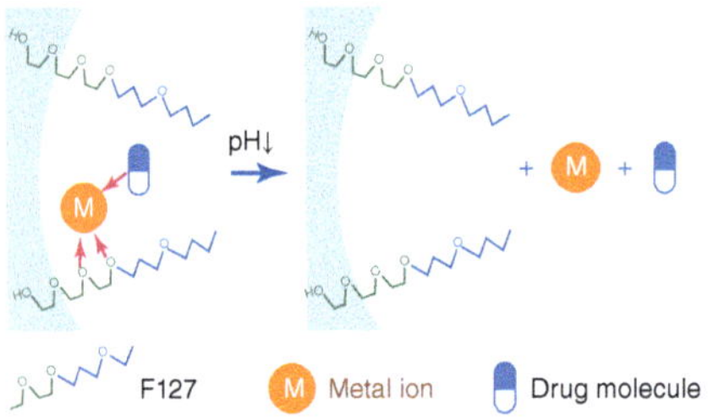

Figure 6.5 Construction of pH-responsive mesostructured surfactant silica hybrid. Source: Reproduced with permission from Zheng et al. [29].

pH-responsive drug delivery. Anticancer drugs – mitoxantrone, doxorubicin, and a conjugated complex of fluorescein isothiocyanate with octaarginine (R8-fluorescein isothiocyanate [FITC]) – were used as model drugs. The Pluronic F127-Zn-R8-FITC coordination complex exhibited successful pH-responsive release under weakly acidic conditions (Figure 6.5). *In vivo* tumor imaging and cell assay (SPCA-1) confirmed the pH-responsive release in the pH range of 6.2–6.8 using fluorescent confocal micrographs.

6.2.3 Metallosurfactant Co-polymer-Based Micellar Systems for Drug Delivery

Besides using metallosurfactant or metallo-vesicles for drug entrapment, the micellar solution of metallosurfactant has garnered attention for solubilization of hydrophobic drugs. Polymeric micelles have been extensively explored for the entrapment of drugs. For such systems, amphiphilic block copolymers such as Pluronics are more commonly used. The hydrophobic segments of the co-block polymers self-assemble to avoid contact with the aqueous media and get associated together by hydrophobic interactions. The interaction between the metal and co-block polymer controls the fabrication and application of polymer–metal complex micelles (PMCM) for the encapsulation and release of drugs. Kim et al. [30] synthesized PMCMs by mixing *cis*-dichlorodiammineplatinum II (CDDP), triblock poly(lactic acid)-poly(ethylene glycolpoly(ethylene glycol) (PEG–PLA), and dendrimer phthalocyanine (DPc) in phosphate buffer saline solution via polymer–metal coordination interactions. The sustained release of anti-cancer drug, CDDP, makes PMCM a promising biomedical nano-imaging device.

Furthermore, supramolecular-based copolymer micelles have been explored as a potential nanocarrier for the successful entrapment and release of drugs. The self-assembly of these metallo-supramolecule-based block copolymers has been investigated by researchers for application in drug delivery. Liang et al. [31] developed ultrasound responsive metallo-supramolecular block copolymer micelle using diblock poly(propylene glycol) (PPG) and PEG copolymer (Figure 6.6). PPG and PEG copolymers were bonded together by a Tpy bond at the junction point PPG-[Cu]-PEG. The fabricated metallo-micelles showed high-intensity focused ultrasound (HIFU)-responsive release for hydrophobic molecules such as Nile red and pyrene.

In another study [32], authors investigated the micellar behavior and properties of metallo copolymers with different transition metals such as ruthenium. The

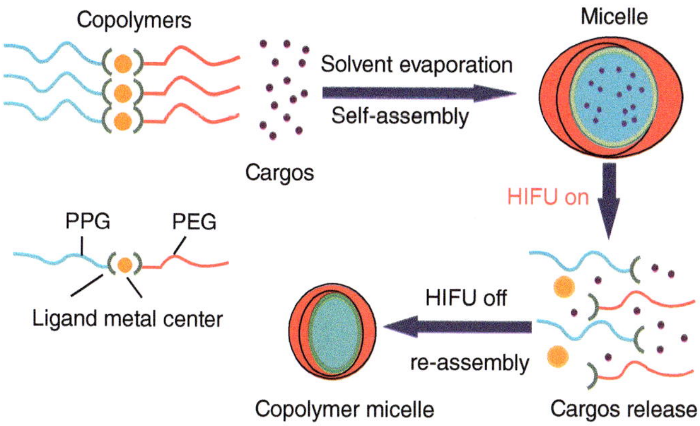

Figure 6.6 HIFU-responsive process of Cu(II)-terpyridine bonds containing block copolymer micelles. Source: Reproduced with permission from Liang et al. [31].

HIFU-responsive release of hydrophobic cargo from PPG-[Cu]-PEG·PPG-[Cu]-PEG with site-specific cleavage allowed the control of drug-release behavior of such systems from the polymeric micelles.

Ge and coworkers [33] synthesized metal-supramolecular core cross-linked micelles from poly(2-(2-methoxyethoxy)ethyl methacrylate)-*b*-poly(2-(diethylamino) ethyl methacrylate-*co*-4′-(6-methacryloxyhexyloxy)-2,2′:6′,2″-terpyridine) [PMEO 2MA-*b*-P(DEA-*co*-TPHMA)] block copolymers. The synthesis route involving the formation of bis(terpyridine)ruthenium(II) complexes is represented in Figure 6.7. The metallosupramolecular micelle exhibited good structural integrity in the aqueous solutions at varying temperatures and pH and showed multiresponsive (pH-responsive and thermoresponsive) release of the model drug (the Nile red).

6.3 Metallosurfactants as MRI Contrast Agents

MRI is a non-invasive biomedical diagnostic imaging procedure for three-dimensional tomographic imaging of soft tissue of an organism. In MRI, contrast agents (CAs) are typically used to enhance sensitivity and to improve the contrast between healthy and diseased tissue. The faster the relaxation rate of tissue, the greater the intensity of the signal is, and brighter the image it produces. Thus, metallosurfactant-based micellar structures and metallosomes have gained much attention as MRI contrast agents.

For instance, the high magnetic moment and the long electron spin relaxation time of gadolinium (Gd, III) make it an ideal candidate for CAs. Note that to circumvent the toxicity of free Gd(III), it is employed in the stable chelate complex form. Generally, the Gd(III) chelates are attached to the self-assembled structures such as micelles and liposomes to increase the relaxation rate. Gong et al. [5] synthesized high relaxivity MRI contrast agents using Gd(III)-based metallosurfactants via miniemulsion polymerization. The synthesized metallo-colloids showed 240%

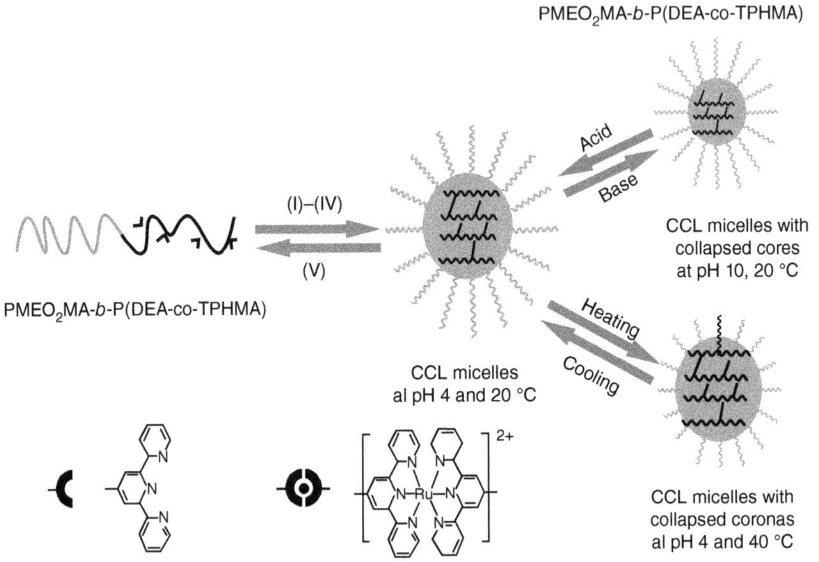

Figure 6.7 Fabrication of reversible metallo-supramolecular pH and thermoresponsive micelles from terpyridine-containing double hydrophilic diblock copolymer. Source: Reproduced with permission from Ge and Liu [33].

of enhancement in the relaxivity. The metallo-colloids were able to produce contrast much brighter than the deionized water in the presence of Gd(III) metallosurfactant complex owing to the faster relaxation of proton of deionized water in the presence of metallosurfactant.

Chen et al. [34] synthesized Gd(III)-based MRI CA using quaternary ammonium long alkyl chain surfactant via miniemulsion polymerization. The fabricated metallo-colloids were reported in the size range of 50–110 nm with a relaxivity of up to 22.77 mM^{-1} s^{-1}. *In vivo* cytotoxicity study showed only 10% decrease in cell viabilities after 48 hours incubation of Gd(III) metallosurfactants solution of concentration up to 200 μM.

Zhu et al. [35] developed multimodal imaging probes using carbon dots to form metallosurfactant-based CAs as bimodal contrasting probes (Figure 6.8). Negatively charged carbon dots formed the metallo-micelles of the size 50 nm through ionic interactions. A high magnetic resonance (MR) relaxivity value of 19.73 mM^{-1} s^{-1} was observed. The cytotoxicity results showed that the carbon dot-metallosurfactant CA is benign to cells with good photoluminescence.

Xu et al. [36] developed bimodal MRI and fluorescence imaging agents using co-assembly of Gd(III) metallosurfactant, conjugated polymeric fluorescent nanoparticles, and Pluronic F127 cross-linked via organosilica (Figure 6.9). Fabricated micelles of size 90–100 nm with a relaxivity of 105.37 mM^{-1} s^{-1} were observed. The obtained relaxivity was found to be 20 times higher than commercial contrast agents. *In vivo* cytotoxicity study revealed biocompatibility of the synthesized CAs. MRI imaging of kidney tumor showed significant enhancement in MR signal intensity. The signal was detectable even after 24 hours administration because of the long circulation of CAs in blood.

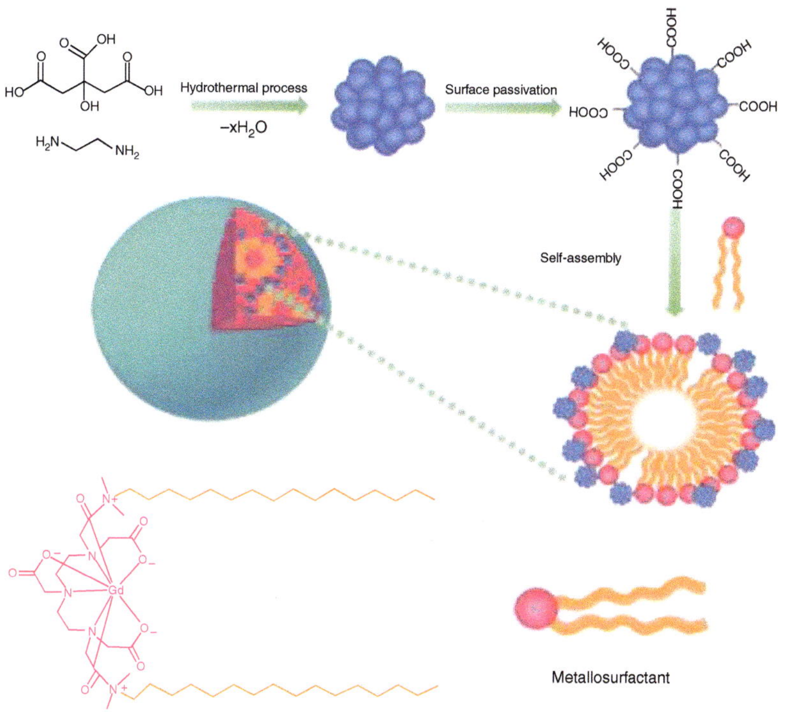

Figure 6.8 An illustration of synthetic route for carbon dots (CDs) and self-assembly process of MS–CDs. Source: Reproduced with permission from Zhu et al. [35].

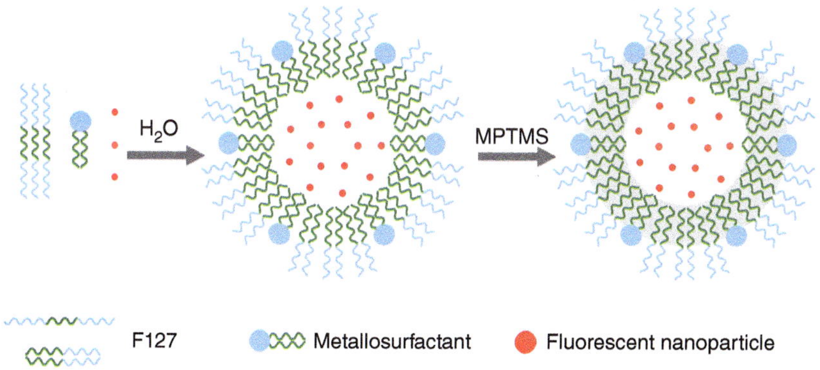

Figure 6.9 Illustration of fabrication of bimodal nanoparticular contrast agent. Source: Reproduced with permission from Xu et al. [36].

6.4 Conclusion

The present chapter has highlighted the use of metallosurfactant as a promising agent in the field of drug delivery. Metallosurfactants possess self-aggregation properties and form distinct assemblies such as metallomicelles, inverted metallomicelles, nanovectors, metallovesicles, metallosomes, and MRI contrast agents. Briefly, cobalt, copper, platinum, ruthenium, and gadolinium-based metallosurfactants, including their synthesis, characteristic features, and potential applications, have been discussed elaborately. For instance, gadolinium(III) metallosurfactants have been extensively studied as MRI agents due to the higher magnetic moment and longer spin relaxation time of gadolinium. However, more *in vivo* studies related to the safety profile of these metal-based complexes are required for their practical applicability.

Acknowledgments

Amit Kumar is thankful to Department of Science and Technology (DST) for the INSPIRE-SRF(IF180772) award. Jyoti Rathee acknowledges University Grant Commission (UGC) for the Senior Research Fellow (SRF). The authors are thankful to Panjab University, Chandigarh, for its support.

References

1 Kagalwala, H.N., Chirdon, D.N., Mills, I.N. et al. (2017). Light-driven hydrogen generation from microemulsions using metallosurfactant catalysts and oxalic acid. *Inorganic Chemistry* 56: 10162–10171.
2 Guerrero-Martínez, A., Vida, Y., Domínguez-Gutiérrez, D. et al. (2008). Tuning emission properties of iridium and ruthenium metallosurfactants in micellar systems. *Inorganic Chemistry* 47: 9131–9133.
3 Sharma, N.K. and Singh, M. (2018). Multifunctional supramolecular ionic metallosurfactants (SMIMSs) for antimicrobial, anticancer and serum albumins binding. *Journal of Molecular Liquids* 263: 463–471. https://doi.org/10.1016/j.molliq.2018.04.138.
4 Kashapov, R., Razuvayeva, Y., Ziganshina, A. et al. (2020). Supraamphiphilic systems based on metallosurfactant and calix[4]resorcinol: self-assembly and drug delivery potential. *Inorganic Chemistry* 59 (24): 18276–16286. https://doi.org/10.1021/acs.inorgchem.0c02833.
5 Gong, P., Chen, Z., Chen, Y. et al. (2011). High-relaxivity MRI contrast agents prepared from miniemulsion polymerization using gadolinium(III)-based metallosurfactants. *Chemical Communications* 47: 4240–4242. https://doi.org/10.1039/c0cc05746k.

6 Griffiths, P.C., Fallis, I.A., Chuenpratoom, T., and Watanesk, R. (2006). Metallosurfactants: interfaces and micelles. *Advances in Colloid and Interface Science* 122: 107–117. https://doi.org/10.1016/j.cis.2006.06.010.

7 Zha, Q., Xie, Q., Hu, Y. et al. (2016). Metallosurfactants C_n–Cu–C_n: vesicle formation and its drug-controlled release properties. *Colloid & Polymer Science* 294: 841–849. https://doi.org/10.1007/s00396-016-3841-7.

8 Wang, R., Liang, Y., and Schmehl, R.H. (1994). Light-induced exchange energy transfer reactions of micelle forming Ru(II) diimine complexes. *Inorganica Chimica Acta* 225: 275–283. https://doi.org/10.1016/0020-1693(94)04058-3.

9 Lebrón, J.A., Ostos, F.J., López-López, M. et al. (2019). Preparation and characterization of metallomicelles of Ru(II). Cytotoxic activity and use as vector. *Colloids and Surfaces B: Biointerfaces* 175: 116–125. https://doi.org/10.1016/j.colsurfb.2018.11.081.

10 Berro, Y., Gueddida, S., Bouizi, Y. et al. (2020). Imprinting isolated single iron atoms onto mesoporous silica by templating with metallosurfactants. *Journal of Colloid and Interface Science* 573: 193–203. https://doi.org/10.1016/j.jcis.2020.03.095.

11 Owen, T. and Butler, A. (2011). Metallosurfactants of bioinorganic interest: coordination-induced self assembly. *Coordination Chemistry Reviews* 225: 678–687. https://doi.org/10.1016/j.ccr.2010.12.009.

12 Veeralakshmi, S., Sabapathi, G., Nehru, S. et al. (2017). Surfactant–cobalt(III) complexes: the impact of hydrophobicity on interaction with HSA and DNA – insights from experimental and theoretical approach. *Colloids Surfaces B: Biointerfaces* 153: 85–94.

13 Vignesh, G., Sugumar, K., Arunachalam, S. et al. (2013). A comparative study on the binding of single and double chain surfactant-cobalt(III) complexes with bovine serum albumin. *Spectrochimica Acta Part A: Molecular and Biomolecular Spectroscopy* 113: 415–422. https://doi.org/10.1016/j.saa.2013.04.123.

14 Zhiltsova, E.P., Ibatullina, M.R., Lukashenko, S.S. et al. (2018). Spectrophotometric study of quercetin in metallomicellar solutions of 1-hexadecyl-4-aza-1-azoniabicyclo[2.2.2]octane bromide complex with copper dibromide. *Journal of Molecular Liquids* 249: 716–722. https://doi.org/10.1016/j.molliq.2017.11.091.

15 Domínguez-Gutiérrez, D., Surtchev, M., Eiser, E., and Elsevier, C.J. (2006). Ru(II)-based metallosurfactant forming inverted aggregates. *Nano Letters* 6: 145–147. https://doi.org/10.1021/nl051944v.

16 Sharma, N.K., Singh, M., and Bhattarai, A. (2016). Hydrophobic study of increasing alkyl chain length of platinum surfactant complexes: synthesis, characterization, micellization, thermodynamics, thermogravimetrics and surface morphology. *RSC Advances* 6: 90607–90623. https://doi.org/10.1039/c6ra20330b.

17 Parera, E., Comelles, F., Barnadas, R., and Suades, J. (2011). Formation of vesicles with an organometallic amphiphile bilayer by supramolecular arrangement of metal carbonyl metallosurfactants. *Chemical Communications* 47: 4460–4462. https://doi.org/10.1039/c0cc05493c.

18 Wang, J., Wang, A.Z., Lv, P. et al. (2018). Advancing the pharmaceutical potential of bioinorganic hybrid lipid-based assemblies. *Advanced Science* 5. https://doi.org/10.1002/advs.201800564.

19 Marín-García, M., Benseny-Cases, N., Camacho, M. et al. (2017). Low-toxicity metallosomes for biomedical applications by self-assembly of organometallic metallosurfactants and phospholipids. *Chemical Communications* 53: 8455–8458. https://doi.org/10.1039/c7cc04945e.

20 Marín-García, M., Benseny-Cases, N., Camacho, M. et al. (2018). Metallosomes for biomedical applications by mixing molybdenum carbonyl metallosurfactants and phospholipids. *Dalton Transactions* 47: 14293–14303. https://doi.org/10.1039/c8dt01584h.

21 Wang, W., Chen, X., and Efrima, S. (1999). Fabrication of semiconductor nanoparticles in a three-dimensional organic-layered solid crystal. *Chemistry of Materials* 11: 1883–1889. https://doi.org/10.1021/cm990079a.

22 Nehru, S., Veeralakshmi, S., Kalaiselvam, S. et al. (2020). Protein binding and antioxidant studies of diimine based emissive surfactant–ruthenium(II) complexes. *Journal of Biomolecular Structure and Dynamics* 1102. https://doi.org/10.1080/07391102.2020.1733664.

23 Sakla, R. and Jose, D.A. (2018). Vesicles functionalized with a CO-releasing molecule for light-induced CO delivery. *ACS Applied Materials & Interfaces* 10: 14214–14220. https://doi.org/10.1021/acsami.8b03310.

24 Kidani, Y., Inagaki, K., Iigo, M. et al. (1978). Antitumor activity of 1,2-diaminocyclohexane-platinum complexes against Sarcoma-180 ascites form. *Journal of Medicinal Chemistry* 21: 1315–1318. https://doi.org/10.1021/jm00210a029.

25 Osada, K., Cabral, H., Mochida, Y. et al. (2012). Bioactive polymeric metallosomes self-assembled through block copolymer-metal complexation. *Journal of the American Chemical Society* 134: 13172–13175. https://doi.org/10.1021/ja304615y.

26 Pal, S., Islam, M.T., Moore, J.T. et al. (2017). Self-assembly of a novel Cu(II) coordination complex forms metallo-vesicles that are able to transfect mammalian cells. *New Journal of Chemistry* 41: 11230–11237. https://doi.org/10.1039/c7nj02161e.

27 Arroyo, I.Z., Gomez, C., Alarcon, H. et al. (2018). Alkyl length effects on the DNA transport properties of Cu(II) and Zn(II) metallovesicles: an in vitro and in vivo study. *Journal of Drug Delivery* 2018: 1–11. https://doi.org/10.1155/2018/2851579.

28 Badawi, A.M., Zakhary, N.I., Morsy, S.M.I. et al. (2012). Biochemical study on the effect of metallo-surfactant and its loaded nano-analogue as anticancer drug. *Journal of American Science* 8: 280–288. http://www.jofamericanscience.org/journals/amsci/am0805/036_8998am0805_280_288.pdf.

29 Zheng, H., Gao, C., Peng, B. et al. (2011). pH-responsive drug delivery system based on coordination bonding in a mesostructured surfactant/silica hybrid. *Journal of Physical Chemistry C* 115: 7230–7237. https://doi.org/10.1021/jp110808f.

30 Kim, J., Yoon, H.J., Kim, S. et al. (2009). Polymer-metal complex micelles for the combination of sustained drug releasing and photodynamic therapy. *Journal of Materials Chemistry* 19: 4627–4631. https://doi.org/10.1039/b904224e.

31 Liang, B., Tong, R., Wang, Z. et al. (2014). High intensity focused ultrasound responsive metallo-supramolecular block copolymer micelles. *Langmuir* 30: 9524–9532. https://doi.org/10.1021/la500841x.

32 Liang, B., Wang, Z., and Xia, H. (2020). High intensity focused ultrasound responsive release behavior of metallo-supramolecular block PPG–PEG copolymer micelles. *Ultrasonics Sonochemistry* 68: 105217. https://doi.org/10.1016/j.ultsonch.2020.105217.

33 Ge, Z. and Liu, S. (2013). Facile fabrication of multistimuli-responsive metallo-supramolecular core cross-linked block copolymer micelles. *Macromolecular Rapid Communications* 34: 922–930. https://doi.org/10.1002/marc.201300072.

34 Chen, Y., Yang, H., Tang, W. et al. (2013). Attaching double chain cationic Gd(III)-containing surfactants on nanosized colloids for highly efficient MRI contrast agents. *Journal of Materials Chemistry B* 1: 5443–5449. https://doi.org/10.1039/c3tb20807a.

35 Zhu, Q., Pan, F., Tian, Y. et al. (2016). Facile synthesis of Gd(III) metallosurfactant-functionalized carbon nanodots with high relaxivity as bimodal imaging probes. *RSC Advances* 6: 29441–29447. https://doi.org/10.1039/c6ra02654k.

36 Xu, K., Xu, N., Zhu, Y. et al. (2020). Co-assembly of Gd(III)-based metallosurfactant and conjugated polymer nanoparticles in organosilica cross-linked block copolymer micelles for highly efficient MRI and fluorescent bimodal imaging. *Particle and Particle Systems Characterization* 37: 1–8. https://doi.org/10.1002/ppsc.202000044.

7

Metallosurfactants as Molecular Machines for the Preparation of Hybrid Silica-Based Porous Material

Rekha Bhar and Surinder K. Mehta

Panjab University, Department of Chemistry and centre of Advanced studies in Chemistry, Sector-14, Chandigarh 160014, India

7.1 Introduction

Surfactants have emerged as compounds of utmost scientific and technological importance, witnessing an escalating progress in their structure as well as functionality. Surfactant systems with inorganic constituents (metal ions) form a subclass of hybrid surfactants having properties corresponding to both interfacial and metal ion chemistry. Metal ion-containing surfactants are known to the scientific world with different names like metallosurfactants, metalloamphiphiles, metal–hybrid surfactants, metal ion-decorated surfactants, and organic–inorganic hybrid surfactants. Characteristically, metallosurfactants are a means of localizing many physicochemical properties of the metal ions at air/water, oil/water interface. Thus, it can be said that metallosurfactants are perfect combination of co-ordination and surface chemistry. Metals introduced chemical reactivity to the otherwise naive surfactant molecules. Metal ions also carry their redox behavior in metallosurfactant as well owing to their variable oxidation states, while the conventional surfactants include carbon, hydrogen, oxygen, and sulfur-like elements, which have fixed covalence and cannot show redox activity. It can be said that conventional surfactants do not possess any chemical reactivity of their own. But the embodiment of metal ion into their structure can overcome this limitation. Such remarkable features of these designer amphiphilic molecules attract enormous attention owing to several advantages over conventional surfactants. The most fascinating property of the surfactants and amphiphiles is to self-assemble above a particular threshold concentration [1]. Metallosurfactants possess a variety of fascinating applications in the fields of magnetic resonance imaging [2], template for porous materials [3], template for fabricating nanomaterials [4], membrane mimetics [5], thin-film optoelectronics [6], interfacial photo physics [7], various separation methods [8], catalysis [9], etc. Dual functionality of metallosurfactants and ability to act as two-in-one precursor (i.e. metal ion source as well as particle protector) for nanomaterial fabrication and for generation of metal-containing mesoporous

Metallosurfactants: From Fundamentals to Catalytic and Biomedical Applications, First Edition.
Edited by Surinder K. Mehta and Ravneet Kaur.

silica material (metal ion source and micellar template) made them promising novel precursor materials. Potential of metallosurfactants as nanoreactor for the fabrication of nanomaterials is discussed in Chapter 9. In the present chapter, a comprehensive discussion on the role of metal hybrid amphiphiles as molecular machine for generating porous silica material has been outlined. Also, a brief introduction on molecular machines (MM) and porous silica-based materials has also been provided herein.

7.2 Porous Silica-Based Materials

Generation of porous silica material gained momentum after the discovery of surfactant-based templating approach for its synthesis by Mobil Corporation laboratories [10]. Since then, porous silica material has garnered vast attention and found application in innumerable fields like sensing [11], storage [12], heavy metal ion removal [13], enhancing the bioavailability of poorly soluble drugs [14], and as excellent adsorbent material [15]. Besides manmade porous materials like ceramics and cements, nature is also a good reservoir of pervious material. For example soil, zeolites, bones, wood are excellent porous materials, which exist in nature. Characteristically, porous silica is a low-density and high surface area material and, depending on the pore size, can be classified into three categories, *viz.* microporous, mesoporous, and macroporous. Microporous materials have pore size of less than 2 nm, while mesoporous have their pore size in the range of 2–50 nm. Materials with pore size more than 30 nm fall into the category of macroporous materials (Figure 7.1).

Among various types of porous materials, mesoporous silica-based materials are most extensively studied because of their ordered pore nature, tunable pore size, and wide range of morphologies. Due to the very high specific surface area, porous silica material can be further exploited in catalysis; however, catalytically active agent (metal ions) needs to be incorporated. To do so, metal salt is usually added

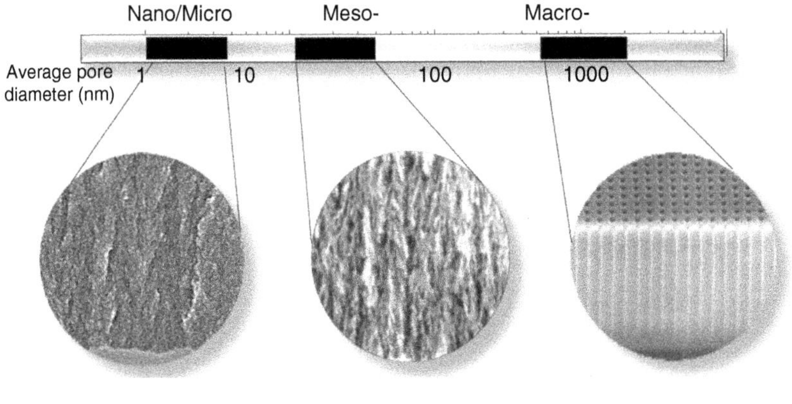

Figure 7.1 Porous materials with different size of pores.

to the synthesized porous silica material, but it leads to non-uniform distribution of metal ion and the harsh treatment during calcination leads to the structural deformation of the composite. Along with structural deformation, leaching of metal ion can also occur during the course of reaction, which suggests that metal ions are not firmly attached with the silica, hence losing their purpose. An alternate method is to use metallosurfactant, which potentially produces homogeneously distributed metal nanocolloids onto the surface of porous silica material. A detailed description about the same has been provided in Section 7.3 of this chapter, but before proceeding further, let us have a glimpse at the molecular machines.

7.2.1 Molecular Machine

In simple words, machine is a tool created by human beings to make their work easier. The molecular machines (MM) are the example of machines at molecular level comprising certain set of molecules, which can respond to an external stimuli. MM came into focus in 1980, after the breakthrough work by Jean-Pierre Sauvage, a French chemist. He designed a molecule named "CATENANE" consisting of two entangled molecular rings, which can rotate around each other when energy is supplied (Figure 7.2).

Catenanes are termed as first-generation molecular machines [16]. Another ground-breaking discovery in this field was done by James Fraser Stoddart and Bernard Feringa in 1990s [17, 18]. The three shared the 2016 Nobel Prize in Chemistry. It is surprising to realize that molecular machines are omnipresent and vital for living beings. All the activities happening in a living body are controlled by the biological molecular machine. Inspired by the nature, manmade molecular machines are the area of most intensive research nowadays.

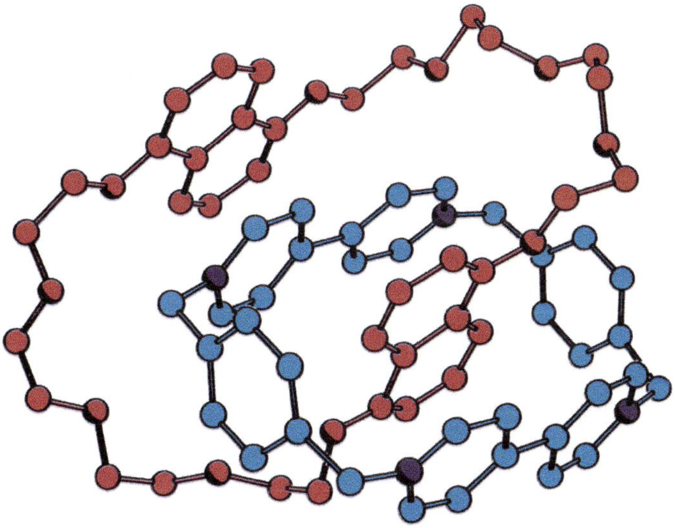

Figure 7.2 Structural representation of a Catenanes.

Micro/nanoscale molecular machines have revolutionary applications especially in biomedical field. From drug delivery to tumor diagnosis and biopsies, MM have made things easier and approachable, which otherwise seems impossible. After two decades of basic research in the field of MM designing and functioning, it is illustrated that MM can write, image, and repair nanoscale structures. Scientists believe that designing and potential of molecular machines can only be limited by human imagination.

To design MM with precise control over its translational and rotary motion is still a challenge. Self-assembly, which is a characteristic feature of amphiphilic moieties, can play a significant role and can be exploited as molecular machines, specifically metal ion-directed self-assembly as exemplified by copper-based catenanes developed by Sauvage. Since then, metal-based molecular assemblies have been continuously explored and discussed. Major contributors to the field have been discussed in detail herein.

7.2.2 General Synthesis Procedure of Porous Silica Material

Porous silica material can be prepared by two approaches, i.e. template-free and template-based synthesis [19]. Both the approaches have their own advantages. However, template-based synthesis can provide a better control over size, distribution, and interconnectivity of the pores. The most common template-based synthesis involves the use of a cationic surfactant [20, 21]. Usually, an acidic or basic aqueous solution of surfactant is taken at or above critical micellar concentration (Figure 7.3).

To the mixture, a silica source (orthosilicates, fumed silica, sodium silicates, etc.) is added for silicalization. Presence of surfactants leads to the polymerization of silica source around the surfactants aggregates. Finally, the template is removed either by extraction or calcination to get the porous material. Template word can be related to

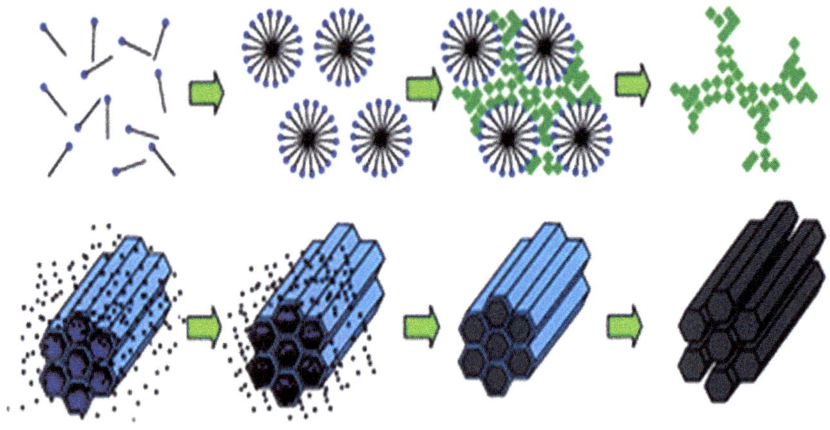

Figure 7.3 Pictorial representation of surfactant-based template synthesis of ordered mesoporous silica.

the molecular machine as it takes input in form of chemicals and gives off porous silica as output.

7.3 Different Types of Porous Silica-Based Material

7.3.1 Copper@Surfactants-Based Porous Material

Metal–surfactant complexes having both the templating agent (surfactant) and catalytic agent (metal ion) are suited best to the production of metal ion containing porous silica material. Dual nature of metallosurfactants made them favorite precursor for the synthesis of metal ion-embedded porous silica material. In this respect, the very first experimental work was reported by Poppl et al. in 1995 [22]. A macrobicyclic cage compound (C_{12}CESTO) having surfactant properties has been utilized for capturing cupric ion.

A solution of 5-dodecyl-12,17-dimethyl-1,5,9,12,17 pentaazobicyclo-[7.5.5] nonadecane (C_{12}CESTO) was mixed with cupric nitrate solution to make complexation of Cu(II) with C_{12}CESTO. Cu(II)@C12CESTO complex was then mixed with micellar solution of cetyltrimethylammonium chloride (CTAC) to form CTAC-Cu(II) C_{12}CESTO mixed micelles. Mixed micellar solution and tetrabutylammonium silicate were added together. Resulting gel was subjected to hydrothermal treatment to generate mesoporous silica materials designated as CESTO-CuMCM-41. To compare the role of Cu(II) complexation with C_{12}CESTO, the authors also prepared CuMCM-41 without C_{12}CESTO. The term MCM-41 stands for Mobil Composition of Matter no. 41 with a pore size within in the range 2–6.5 nm. Thus, the prepared material, i.e. CESTO-CuMCM-41, has a pore size of 2–6.5 nm. Further, it was confirmed that complexation of copper ion with surfactant macrobicyclic cage compound leads to the firm association of metal ions with silica framework. Cupric ions occupy ion-exchange sites on silica walls by interacting with four framework oxygens. In conclusion, it was made clear that location of cupric ion into the silica material can be easily controlled by metallosurfactant.

In another study [3], sarcophagines were used to form metallosurfactants. Sarcophagines are macrobicyclic hexamines, which acts as metal binder to form copper metallosurfactants (Figure 7.4). To prepare porous silica, TEOS was added to a mixture of conventional organic surfactant, i.e. CTAB and prepared copper metallosurfactants with constant stirring.

Resulting reaction mixture was then subjected to hydrothermal treatment followed by calcination. It was observed that calcination process resulted in extensive aggregation of copper in the form of copper crystallite inside the porous silica as evident from the TEM image (Figure 7.4). Temperature of calcination process also affects copper crystallite size. At 300 °C, smaller aggregates were formed, while 500 °C resulted in bigger aggregates. The authors concluded that different metal-doped mesoporous silica can be prepared using different metal-caged sarcophagine. Additionally, by altering the ratio of metallosurfactants to conventional surfactants, pore size and homogeneity of pores can be easily tuned. However, findings did not throw any light on the application part of the prepared material.

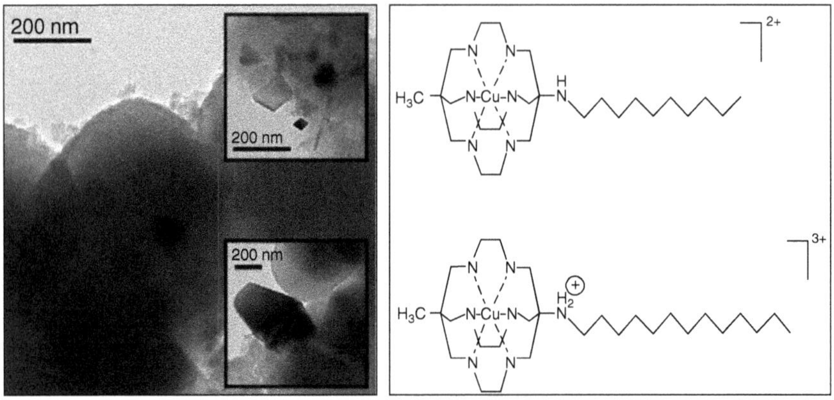

Figure 7.4 Copper surfactants based on sarcophagine and TEM image of mesoporous silica; upper inset shows copper crystallite at 330 °C calcination and lower inset show bigger crystallite at 500 °C calcination temperature.

Pim et al. [23, 24] reported fabrication of meso- and microporous silica without using conventional surfactants at all. Bistable copper metallosurfactants-based molecular machine were prepared by authors. Characteristic feature of the synthesized material is that it responds to the pH change, i.e. complex can exist in deprotonated (pH 12) and protonated (pH 5) forms (Figure 7.5).

Hydrolysis of TEOS in the presence of deprotonated metallosurfactant template results in mesoporous silica, while protonated metallosurfactant template resulted in microporous silica. Till now, porous material with uniform pore size had been discussed stating that creating uniformity in pore size is the key factor that is to be controlled. However, material with dual porosity, i.e. two different size of pores within the same material, have their own advantages. Slowly thrust is shifted toward preparation of porous material with pore size of two and three scale length simultaneously in a single step. In this regard, Kanwar et al. [25] reported the generation of meso-macroporous silica material embedded with copper oxide nanoparticles (NPs). Solid lipid NPs and copper–cetylpyridinium chloride surfactant were used as dual template. Copper surfactant micelle acts as template for mesopore material and copper ion source and also as a stabilizing agent or co-surfactant for solid lipid NPs, which are indeed the building blocks for macropores. Close proximity of metal ion with hydrophobic part leads to the uniform distribution of copper oxide NPs into the mesopores of silica matrix as seen in electron microscopy. Prepared material was then explored as heterogeneous catalyst for the reduction of nitro aromatic compounds and hexacyanoferrate(III). Presence of pores of different length in the single matrix provides extra benefit to the diffusion controlled reaction. It was stated that mesopores are the active for catalysis, while interconnected macropores act as a gateway to incoming reactant and outgoing product. In this manner, saturation of active can be easily avoided as proven by the product yield. The product yield of 99.9% for nitrophenol reduction and 98.7% for hexacyanoferrate reduction verified that prepared composite acts as efficient catalyst. In a separate study, the prepared material was then explored as adsorbent for genomic DNA by Bhar et al. [26].

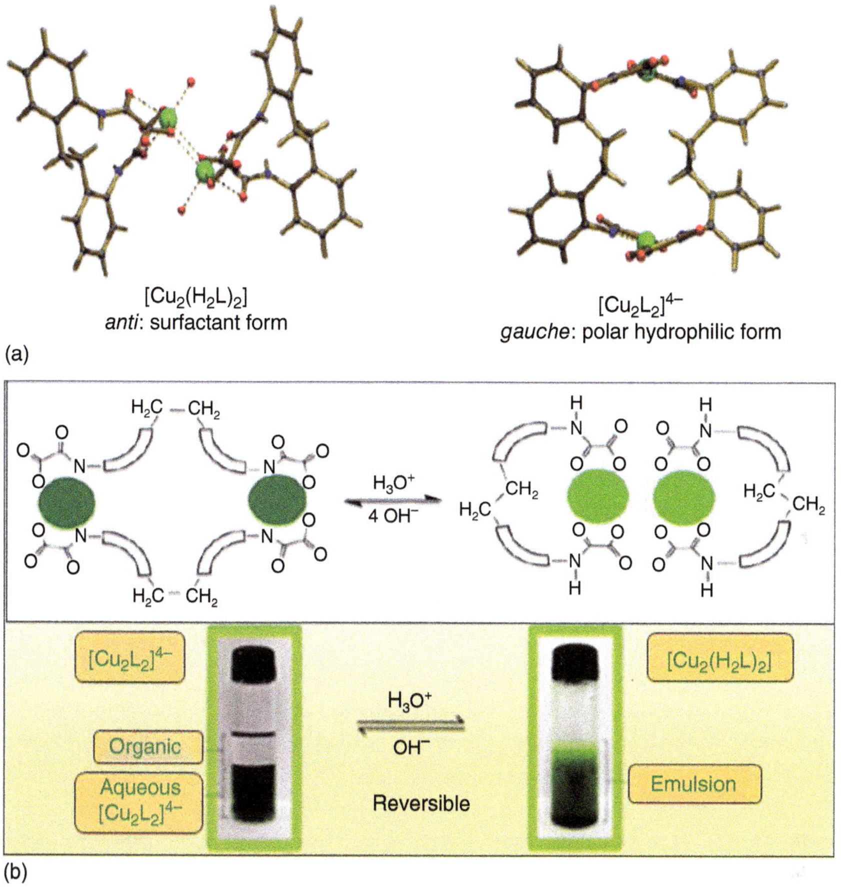

Figure 7.5 Crystal structure of deprotonated and protonated bistable copper metallosurfactant (a), schematic representation of mechanism involved in pH-triggered conformation (b).

To do so, meso-macroporous silica was first surface passivated by aminopropyltriethoxysilane. Porous material accompanied by nanoparticle deliver extra surface area for attachment to the adsorbate. It was the first ever report describing the adsorption of whole genomic DNA onto the silica surface.

Keeping up the pace, Rehman et al. [27] developed metallic and bimetallic oxide/SiO_2 material using corresponding metallosurfactants as template. In this case, metal ion is not present as an integral part of the head group of the surfactant but as modified counter ion. Counter ion of CTAB, i.e. bromide, was modified by Fe, Cu, Ni, Zn metal ions. Metallosurfactants of the type $CTA^+[CuCl_2Br]$, $CTA^+[NiCl_2Br]$, and similarly for other metals were synthesized. But the hydrolysis of TEOS was carried out only in the presence of $CTA^+[CuCl_2Br]$, $CTA^+[NiCl_2Br]$ and equimolar ratio of these two surfactants as a template. Hydrolyzed product was then calcined at 550 °C to get the final product, i.e. CuO/SiO_2, NiO/SiO_2, and $CuO–NiO/SiO_2$. Prepared material was tested as a potential catalyst for nitrophenol reduction. It was found that bimetallic material is way more an efficient catalyst as compared to their monometallic counterparts. It was argued that synergy between copper and nickel increases the activity of catalyst, hence, higher efficiency.

7.3.2 Ruthenium@Surfactants-Based Porous Material

The study made by Jervis et al. [28] in this field was a huge development. The authors did excellent work on ruthenium metallosurfactant as a precursor for mesoporous silica material. The authors proposed metallosurfactant-based route as a direct approach for the fabrication of heterogeneous catalyst. To an acidic solution of metallosurfactant, silica precursor tetramethylorthosilicate was added. After two hours of stirring, red-colored gel possessing a hexagonal phase was produced as observed by optical microscopy. The final product was obtained by calcination of the gel at 600 °C for six hours and viewed under TEM (Figure 7.6). The material obtained was then tested for the hydrogenation reaction of hex-1-ene to hexane with an excellent turnover number, i.e. 877 mol hexane (mol Ru h)$^{-1}$ Furthermore, the role of calcination temperature and other conditions on the mesoporous structure was also checked. It was observed that at 400 °C, the material obtained was of high homogeneity and the size of RuO_2 NPs was no larger than 10 Å while at 500 °C, irregular-sized and uneven distribution of RuO_2 particles were observed.

BET surface area and turnover number for the conversion of 1-hexene to hexane was also found to be higher for the material calcined at 400 °C as compared to the 500 °C sample. Also, a significant weight loss was observed upon calcination in oxygen atmosphere rather than nitrogen atmosphere.

After realizing the catalytic potential of metal-embedded mesoporous silica material, Amos et al. [29] carried out in-depth investigation of the synthesis protocol. Ruthenium metallosurfactant templated synthesis was also compared to that of conventional surfactant (CTAB). The authors adopted both the liquid-crystal template (LCT) and true liquid-crystal templating (TLCT) approach to prepare porous material. Conclusively, the role of TLCT and metallosurfactant was found

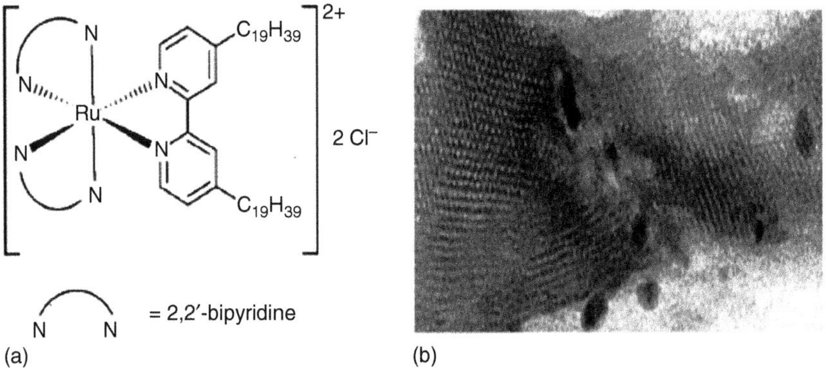

Figure 7.6 Ruthenium metallosurfactant used as template for porous silicate (a), TEM image of as-obtained silicate (b).

to be superior in terms of specific surface area and homogeneity. On similar lines, Co-surfactant complexes having tetradentate amine, triethylenetetramine, alkylated pyridine ligands have also been explored as templating precursors [30]. Similarly, Yam et al. [31] also explored the LCT efficacy of ruthenium-based surfactants. They prepared nine ruthenium metallosurfactants with different chain length of hydrophobic part to control porosity. Several noticeable features were decoded by this study. First one is that head group of the prepared metal–surfactant complexes is quite bigger than the head group of the conventional surfactants. But the size of head group does not affect the mesophase behavior. Also, these complexes were stable under acidic conditions (pH 2), which is a determining factor while doing hydrolysis of silica source to generate ordered structure. Further, it was observed that RuO_2 NPs are evenly distributed throughout the porous material. To summarize, the authors concluded that by altering the chain length and head group of the metallosurfactants, it is feasible to tune the pore size.

In a breakthrough experiment by King et al. [32], oxidation of water to oxygen gas in the presence of cerium(IV) was successfully achieved by catalyst using RuO_2 NPs anchored mesoporous silica obtained from ruthenium metallosurfactant. Oxidation of water has a great significance in splitting of water in electrochemical cell, but it is a difficult redox reaction to carry out. Due to the requirement of $4e^-$ simultaneously the reaction suffers very large over-potential. This difficulty was easily surpassed using the above-mentioned catalyst. Silica-supported RuO_2 NPs catalyze the oxidation of water even in anhydrous conditions. The authors [33] further synthesized and characterized mesoporous silica with uniformly distributed Au, Pd, Pt, Ir NPs by utilizing the TLCT feature of polyoxyethylene-10-lauryl ether.

7.3.3 Cobalt@Surfactants-Based Porous Material

Along with copper surfactants, Hondow et al. [3] fabricated cobalt-based sarcophagine surfactants as well. Size of pores/porosity in the final product was found to be dependent on the shape of template. Cobalt surfactant forms long worm-like micelle, which aids in the formation of M41S type of porous material; also, at much higher concentration it forms lamellar structure. Due to this, ordered pore arrangement was not observed for cobalt surfactant. However, regular strips were visible throughout the material pointing toward the multilamellar nature of template or aggregates of cobalt metallosurfactant, though they have less surface area as compared to copper counterpart but much higher than aqueous impregnation of cobalt salt and also found to be more promising template for mesoporous silica as compared to copper surfactants of the same type.

Cobalt soap was used as co-template for the preparation of cobalt integrated with mesoporous silica [34]. $[CoCH_3(CH_2)_{10}COO]_2 \cdot 2H_2O$ was used as a co-template along with the micellar solution of C_{16}TMABr, which leads to the formation of Co-MCM-41 material (Figure 7.7), which may be exploited for catalysis.

Co soap alone was not able to form micelles but in the presence of micellar concentration of CTAB, it can directly form mixed micelles, thus, resulting in Co NPs-embedded mesoporous silica. The synthesized material may be exploited in

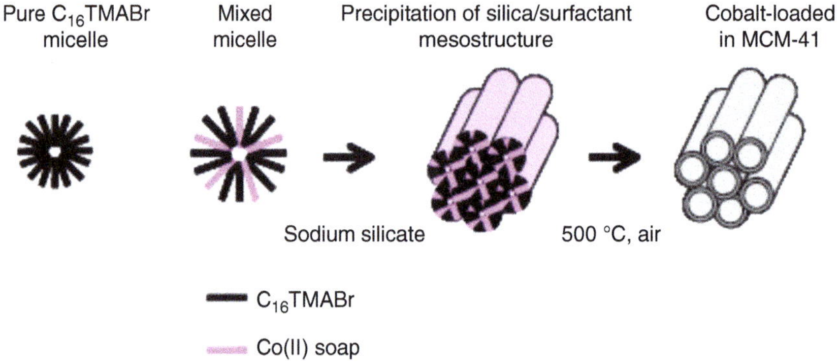

Figure 7.7 Representation of templating phenomena of Co soap.

catalysis as demonstrated by Corberán et al. [35] in the selective oxidative dehydrogenation of isobutane. Co-containing mesoporous silica acts as a homogeneous catalyst.

Activation of the alkane molecules was achieved at the inner pore volume, which resulted in generation of active free radicals. Thus, formed free radical can either act inside the void volume, which is termed as nanoreactor, or interact with another center present on the mesopore wall. Also, UV–vis diffuse reflectance showed unique tetrahedral coordination of cobalt(II) in the present material, therefore, making it an active and selective catalyst for the said reaction. Active centers available for catalysis were ascribed to the Co NPs present within the porous material.

7.3.4 Iron@Surfactants-Based Porous Material

Magneto-responsive mesoporous silica can also be prepared by magneto-responsive surfactant. Magneto-responsive surfactant, i.e. cetyltrimethylammonium ferrate (CTAF), can act as a structure directing agent for the synthesis of porous material having magnetic character [36]. Synthetic methodology adopted was simple, but the appearance of magnetic character in the end product surprised the authors because iron NPs were absent in the end product (Figure 7.8). Magnetism was ascribed to the molecular and electronic spin of the iron ion present within the porous material.

The major breakthrough in iron surfactant chemistry came in 2015, when the synthesis of porous material with dual porosity, i.e. meso-macroporous silica, was reported [37].

Along with the solid lipid NPs, which were synthesized in the presence of CTAF (responsible for macroporosity), micelles of Pluronic P-123 (responsible for mesoporosity) were employed as template. Thus, the formed composite was characterized as Fe_2O_3 NPs-impregnated meso-macroporous silica by different techniques, *viz.* SAXS, BET, TEM, etc. SAXS pattern revealed the presence of hexagonally arranged mesopores interconnected with the macropores.

Evidence of macroporosity was given by TEM, BET analysis, and mercury intrusion porosimetry. Fabricated composite was then tested for catalytic activity in the Fenton-like reaction involved in the degradation of methylene blue (MB). According

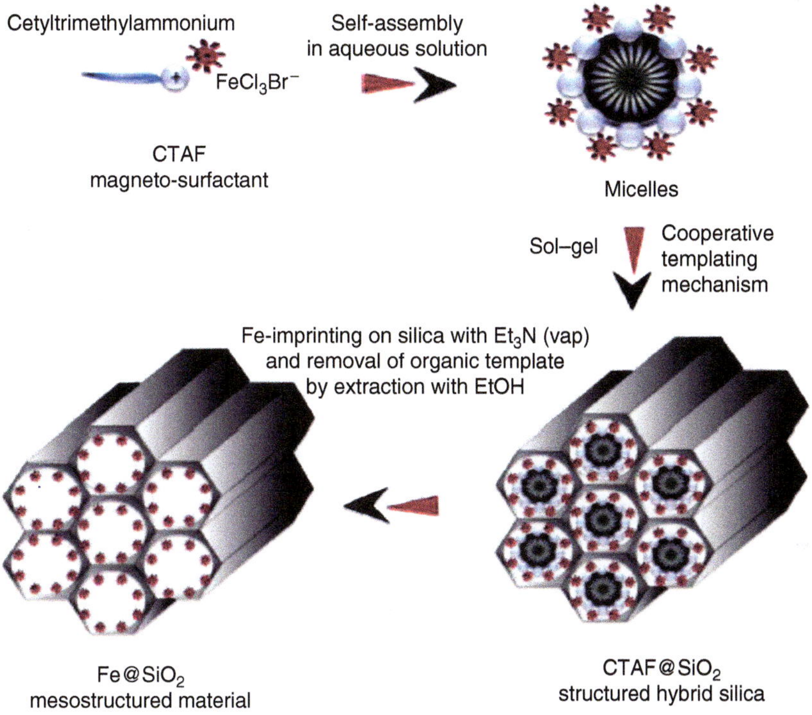

Figure 7.8 Schematic illustration of synthetic pathway for magnetic mesoporous silica through a self-assembly mechanism with CTAF.

to the report, degradation of MB is a diffusion-controlled reaction and macropores easily channelize the diffusion of reactants as well as the product making the catalyst highly efficient in terms of yield and time consumption. Catalyst was found to be quite active even after the fifth repetitive use.

An advancement to phenomena was reported by Berro et al. [38]. Instead of NPs, they devised a methodology to impregnate single metal atom into the mesoporous silica using metallosurfactants. Single-atom catalyst popularly known as SAC is in demand nowadays. SACs are even more efficient than NPs as catalyst. To make non-precious metal similar or even better catalyst than precious metals (Au, Pt, Ag), SAC strategy was developed for such metals. For example, Pt catalyst can be replaced with Co or Fe for oxygen reduction reaction. Keeping this in mind, single iron atom containing mesoporous silica was prepared using cetyltrimethylammonium trichloromonobromoferrate (CTAF) surfactant as template. Mixed micelle of Pluronic P-123 (non-ionic surfactant) and CTAF was taken as template, while tetramethylorthosilicate was used as a silica source (Figure 7.9).

Non-ionic surfactant plays an important role of controlling structural and textural characteristics of the framework and CTAF controls the Fe loading into the silica matric as well as homogeneous distribution of iron atom. During the calcination process, metal ion usually gathers and forms clusters in the form of metal/metal oxide nanoparticle. However, in the present case, no agglomeration was observed

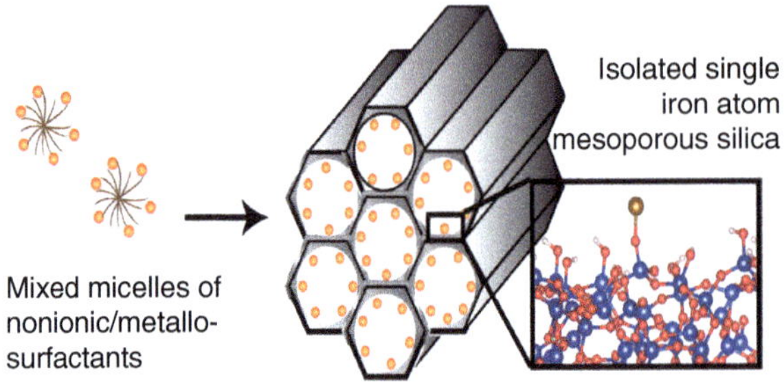

Figure 7.9 Graphical representation of single iron atom-loaded mesoporous silica.

as confirmed by ICP–AES and UV–vis diffuse reflectance spectroscopy. One can determine nature of metal ion present by understanding the coordination of metal. Spectra band at 235 nm suggested that tetrahedral iron atom remains insulated from each other. Also, lack of any peak above 350 nm confirmed the absence of octahedral iron and, hence, absence of iron NPs in the prepared material. The authors concluded that by regulating the ratio of CTAF to non-ionic surfactant, loading of iron atom into the silica framework can be easily controlled, minimum being 0.15 Fe atom nm^{-2}. This is a remarkable achievement in the field of metal ion-embedded porous silica materials.

7.3.5 Other Metallosurfactant-Based Porous Material

Metallosurfactant containing metal ions other than copper, ruthenium, cobalt, and iron are less explored, but their chemistry is well established. Yuan et al. [39] synthesized manganese-modified mesoporous silica (Mn@MCM-41) using cetylpyridinium bromide complexed with $[Mn(CH_3COO)_2 \cdot 4H_2O]$ in alkaline medium. Interestingly, it was observed that the shape of Mn@MCM-41 depends upon the stirring rate of the reaction medium.

Higher stirring rate tends to form vesicles of the surfactant, while lower stirring rate corresponds to the typical spherical-shaped micelles arranged in hexagonal manner. Thus, produced composite material was successfully characterized by various techniques, *viz.* ESR, NMR, TEM, and BET.

In another report [7], iridium surfactant was used as template for the production of iridium ion-containing mesoporous silica (Figure 7.10).

Being highly luminescent, optical property of the synthesized complex was tuned instead of catalysis. The authors argued that by using iridium metallosurfactant, it was easy to minimize the aggregation of active centers responsible for luminescence as templating could lead to tailoring of interatomic distance. Another problem overcome by using this methodology was avoiding the unlimited excess of molecular oxygen, which otherwise caused optical quenching. By incorporating iridium inside the mesoporous silica, the excess of molecular oxygen got limited, thus proving it to be excellent optical material with high quantum yield.

Recently, Chang et al. [40] realized the potential of metal-chelating surfactant as a structure-directing agent for the fabrication of mesoporous silica and reported the formation of alkali metal (Na^+, K^+, Cs^+), alkaline earth metal (Mg^{2+}, Ca^{2+},

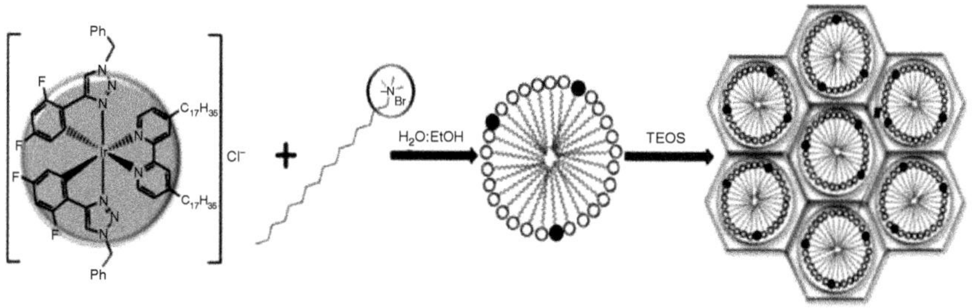

Figure 7.10 Schematic representation of the strategy used for the encapsulation of the iridium–surfactant complex, the metallosurfactant is mixed with CTAB to form micelles, which are then used for the synthesis of MCM-41 via sol–gel process.

Sr^{2+}, Ba^{2+}), and transition metal ion (Co^{2+}, Ni^{2+}, Cu^{2+}, Zn^{2+}) containing mesoporous silica particles. Metal ions form salt by coupling with Pluronic P-123 in aqueous solution. Micellar solution of metal chelated surfactant complex produced corresponding metal ion containing mesoporous silica particles. Complexation of metal ion takes place within the hydrophilic domain of the surfactant. Strength of bonding between metal ion and surfactant played a major role in deciding the size of the mesoporous particles formed. Stability constant of the complex followed the order $Co^{2+} < Ni^{2+} \leq Cu^{2+} > Zn^{2+}$, which is in accordance with the Irving–Williams stability constant series.

Besides porous silica material, porous alumina loaded with metal NPs was also explored as catalyst by Hackett et al. [41]. Atomically dispersed Pd(II) centers in the porous alumina support act as highly efficient and selective catalyst for the aerobic oxidation of certain allylic alcohols like benzyl, cinnamyl, crotyl alcohols. The authors gave a new term selox for selective oxidation as no evidence of solvent oxidation/isomerization/polymerization was observed and only pure benzaldehyde, cinnamaldehyde, crotonaldehyde were formed as a principal reaction product.

7.4 Future Perspectives

Till now, metallosurfactants are more explored in terms of structural and texture direct agents for porous silica. Though generated porous material is well explored as a catalyst, the potential or capability in other field is kept at bay. Metal ion-containing mesoporous silica can be further exploited as a compartmentalized drug-delivery agent, for selective and efficient adsorption/desorption of biomolecules, metal ion detection and sensing, pesticide removal from waste water, optoelectronics, etc. Furthermore, only few transition metal ion-containing metallosurfactants have been tested as molecular machines for mesoporous silica. Each metal has its own characteristic features, which can be employed for specific purpose. Employing metallosurfactants with different types of metal ion can open gateway for numerous other applications. As a future prospective, maybe porous material with pores of three different length scales can be designed. Likewise, metal ions/or metallic/bimetallic NPs-impregnated mesoporous silica seems a common outcome of metallosurfactant-based template synthesis. However, porous material containing single atom can also be designed from metallosurfactants. Not only silica-based porous material but also alumina-based porous material can be designed and fabricated using metal hybrid surfactants. All these studies signify that modified surfactants hold great potential in terms of structural diversity and their applicability.

7.5 Conclusion

Modified surfactants are excellent molecular machines for generating porous silica material with tailor made properties. Employing metallosurfactants as soft template, problem of leaching active centers or metal ion from the silica sieve can be easily avoided. Metal ion or NPs can be firmly and orderly arranged within the silica matrix because of close association of metal ion with the surfactant template itself. Porous material impregnated with nanomaterials can be easily engineered using metallosurfactant alone or a mixture of metallosurfactant and conventional surfactant. By regulating the ratio of metallosurfactant to conventional surfactant during the synthesis process, it is possible to get pores of desirable size. In other words, metallosurfactants give researchers more freedom to play with porosity, structural and textural diversity of the silica matrix. Prepared material found tremendous application in selective catalysis, while applications in other fields are yet to be explored and synchronized.

References

1 Griffiths, P.C., Fallis, I.A., Chuenpratoom, T., and Watanesk, R. (2006). Metallosurfactants: interfaces and micelles. *Advances in Colloid and Interface Science* 122 (1–3): 107–117.

2 Brown, P., Alan Hatton, T., and Eastoe, J. (2015). Magnetic surfactants. *Current Opinion in Colloid and Interface Science* 20 (3): 140–150.

3 Hondow, N., Harowfield, J., Koutsantonis, G. et al. (2012). Metallosurfactants in the preparation of mesoporous silicas. *Microporous and Mesoporous Materials* 151: 264–270.

4 Kaur, R. and Mehta, S.K. (2014). Self aggregating metal surfactant complexes: precursors for nanostructures. *Coordination Chemistry Reviews* 262 (1): 37–54.

5 Gokel, G.W. and Negin, S. (2012). Synthetic membrane active amphiphiles. *Advanced Drug Delivery Reviews* 64 (9): 784–796.

6 Guerrero-Martinez, A., Vida, Y., Dominguez-Gutierrez, D. et al. (2008). Tuning emission properties of iridium and ruthenium metallosurfactants in micellar systems. *Inorganic Chemistry* 47 (20): 9131–9133.

7 Botelho, M., Fernandez-Hernandez, J.M., De Queiroz, T.B. et al. (2011). Iridium(III)-surfactant complex immobilized in mesoporous silica via templated synthesis: a new route to optical materials. *Journal of Materials Chemistry* 21 (24): 8829.

8 Luk, Y.Y. and Abbott, N.L. (2002). Applications of functional surfactants. *Current Opinion in Colloid and Interface Science* 7 (5, 6): 267–275.

9 Polyzos, A., Hughes, A.B., and Christie, J.R. (2007). Catalysis of aryl ester hydrolysis in the presence of metallomicelles containing a copper(II) diethylenetriamine derivative catalysis of aryl ester hydrolysis in the presence of metallomicelles containing a copper(II) diethylenetriamine derivative. *Langmuir* 23 (4): 1872–1879.

10 Beck, J.S., Vartuli, J.C., Roth, W.J. et al. (1992). A new family of mesoporous molecular sieves prepared with liquid crystal templates. *Journal of the American Chemical Society* 114 (27): 10834–10843.

11 Sancenón, F., Pascual, L., Oroval, M., and Aznar, E. (2015). Gated silica mesoporous materials in sensing applications. *Chemistry Open* 4: 418–437.

12 Thomas, K.M. (2007). Hydrogen adsorption and storage on porous materials. *Catalysis Today* 120: 389–398.

13 Schroden, R.C., Al-daous, M., Sokolov, S. et al. (2002). Hybrid macroporous materials for heavy metal ion adsorption. *Journal of Materials Chemistry* 12 (5): 22–25.

14 Arruebo, M. (2012). Drug delivery from structured porous inorganic materials. *WIREs Nanomedicine and Nanobiotechnology* 4: 16–30.

15 Hoong, C.W., Noor, M.M., Ahmad, Z.A. et al. (2013). Sustainable porous materials for gas adsorption applications: a concise review. *Advances in Materials Research* 795: 96–101.

16 Reviriego, F. and Sauvage, J. (2010). Copper-complexed catenanes and rotaxanes in motion: 15 years of molecular machines. *Dalton Transactions* 39: 10557–10570.

17 Liu, Y., Flood, A.H., and Stoddart, J.F. (2004). Thermally and electrochemically controllable self-complexing molecular switches. *Journal of the American Chemical Society* 126: 9150–9151.

18 Wiel, K.J. and Feringa, B.L. (2005). Synthesis of functionalized molecular motors. *Synthesis* 11: 1789–1796.

19 Zhao, X.S., Su, F., Yan, Q. et al. (2006). Templating methods for preparation of porous structures. *Journal of Materials Chemistry* 16: 637–648.

20 Cheng, C. and Luan, Z. (2000). The role of surfactant micelles in the synthesis of the mesoporous molecular sieve MCM-41. *Langmuir* 11 (7): 2815–2819.

21 Huo, Q., Margolese, D.I., and Stucky, G.D. (1996). Surfactant control of phases in the synthesis of mesoporous silica-based materials. *Chemistry of Materials* 8 (5): 1147–1160.

22 Poppl, A., Baglioni, P., and Kevan, L. (1995). Electron spin resonance and electron spin echo modulation studies of the incorporation of macrocyclic-complexed cupric ions into siliceous MCM-41. *The Journal of Physical Chemistry* 99 (38): 14156–14160.

23 Walace, D., Ribeiro-santos, T.A., Jardim, I.S. et al. (2018). Bistable copper(II) metallosurfactant as molecular machine for the preparation of hybrid silica-based porous materials. *Materials and Design* 160: 876–885.

24 do Pim, W.D., Oliveira, W.X.C., Ribeiro, M.A. et al. (2013). A pH-triggered bistable copper(II) metallacycle as a reversible emulsion switch for biphasic processes. *Chemical Communications* 49: 10778–10780.

25 Kanwar, R., Bhar, R., and Mehta, S.K. (2018). Designed meso-macroporous silica framework impregnated with copper oxide nanoparticles for enhanced catalytic performance. *ChemCatChem* 10 (9): 2087–2095.

26 Bhar, R., Kanwar, R., and Mehta, S.K. (2020). Surface engineering of nanoparticles anchored meso-macroporous silica heterostructure: an efficient adsorbent for DNA. *Materials Chemistry and Physics* 255: 123541.

27 Rehman, A.U., Khan, S.A., Nazar, S. et al. (2019). Counterion engineered surfactants for the novel synthesis of colloidal metal and bimetal oxide/SiO_2 materials with catalytic applications. *Colloids Surfaces A* 571: 80–85.

28 Jervis, H.B., Raimondi, M.E., Raja, R. et al. (1999). Templating mesoporous silicates on surfactant ruthenium complexes: a direct approach to heterogeneous catalysts. *Chemical Communications* 2 (20): 2031–2032.

29 Amos, K.E., Brooks, N.J., King, N.C. et al. (2008). A systematic study of the formation of mesostructured silica using surfactant ruthenium complexes in high- and low-concentration regimes. *Journal of Materials Chemistry* 18 (43): 5282–5292.

30 Danks, M.J., Jervis, H.B., Nowotny, M. et al. (2002). High-activity heterogeneous catalysts prepared in one step from the mesophases of metallosurfactants. *Catalaysis Letters* 82 (1, 2): 95–98.

31 Yam, V.W.W., Li, B., and Zhu, N. (2002) Synthesis of mesoporous silicates with controllable pore size using surfactant ruthenium(II) complexes as templates. Advance Materials 14 (10); 719–722.

32 King, N.C., Blackley, R.A., Zhou, W., and Bruce, D.W. (2006). The preparation by true liquid crystal templating of mesoporous silicates containing nanoparticulate metals. *Chemical Communications* 32: 3411–3413.

33 King, N.C., Dickinson, C., Zhou, W., and Bruce, D.W. (2005). The oxidation of water by cerium(IV) catalysed by nanoparticulate RuO_2 on mesoporous silica. *Dalton Transactions*: 1027–1032.

34 Grosshans-Vièles, S., Tihay-Schweyer, F., Rabu, P. et al. (2007). Direct synthesis of mesoporous silica containing cobalt: a new strategy using a cobalt soap as a co-template. *Microporous and Mesoporous Materials* 106 (1–3): 17–27.

35 Corberán, V.C., Jia, M.J., El-Haskouri, J. et al. (2004). Oxidative dehydrogenation of isobutane over Co-MCM-41 catalysts. *Catalysis Today* 91, 92: 127–130.

36 Kim, S., Bellouard, C., Pasc, A. et al. (2013). Nanoparticle-free magnetic mesoporous silica with magneto-responsive surfactants. *Journal of Materials Chemistry C* 1 (42): 6930–6934.

37 Kim, S., Durand, P., Roques-Carmes, T. et al. (2015). Metallo-solid lipid nanoparticles as colloidal tools for meso-macroporous supported catalysts. *Langmuir* 31 (5): 1842–1849.

38 Berro, Y., Gueddida, S., Bouizi, Y. et al. (2020). Imprinting isolated single iron atoms onto mesoporous silica by templating with metallosurfactants. *Journal of Colloid and Interface Science* 573: 193–203.

39 Yuan, Z., Ma, H., Luo, Q., and Zhou, W. (2002). Synthesis and characterization of manganese-modified MCM-41. *Materials Chemistry and Physics* 77: 299–303.

40 Chang, J.H., Lee, H., Jang, S., and Kim, J. (2016). Facile preparation of size-controlled mesoporous silica particles by metal-chelating surfactant micelle complexes. *Materials Letters* 173: 50–54.

41 Hackett, S.F.J., Brydson, R.M., Gass, M.H. et al. (2007). High-activity, single-site mesoporous Pd/Al_2O_3 catalysts for selective aerobic oxidation of allylic alcohols. *Angewandte Chemie International Edition* 46 (45): 8593–8596.

8

Metallosurfactants as Non-viral Vectors in Transfection

Pilar López-Cornejo[1], José A. Lebrón[1], Francisco J. Ostos[1], Manuel López-López[2], María L. Moyá[1], Eva Bernal[1], and Carmen Martín[1]

[1] *University of Seville, Department of Physical Chemistry, Faculty of Chemistry, c/Prof. García González n° 1, Seville 41012, Spain*
[2] *Campus de El Carmen, Department of Chemical Engineering, Physical Chemistry and Materials Science, Faculty of Experimental Sciences, Avenida de las Fuerzas Armadas s/n 21071, Huelva, Spain*

8.1 Introduction

Gene therapy is an experimental technique that uses genes in the treatment and/or cure of a disease. The replacement of defective genes or the cure of some genetic disorders in a patient is included in such a technique. In general, this therapy is being employed in incurable diseases or without known treatments, always with the goal of saving lives.

Up to now, most trials carried out were related to diseases caused by recessive gene disorders, such as hemophilia, sickle cell anemia, cystic fibrosis, or muscular dystrophy; with acquired genetic diseases, as cancer; or to viral infections as AIDS [1–9]. Results obtained in recent years seem promising. In fact, the FDA approved the first gene therapy in 2017 (Tisagenlecleucel, commercialized by Novartech and used in patients with acute lymphoblastic leukemia), and more studies are expected to be approved in 2022 [10]. Although there have been seven marked products of gene therapy authorized in EU, only six are still available [11].

The most used procedure in this therapy consists of the insertion of healthy genes into target cells to repair the damaged nucleic acids. The size of the gene material, the repulsive forces between the phosphate groups of the DNA backbone, the negatively charged heparin sulfate proteoglycan expressed on the external cellular membrane, as well as the innate immune system of the cells, make the transfection process difficult. Therefore, some nanostructures, called vectors (also known as nanocarriers or nanovehicles), help to get a successful insertion.

The administration of genetic material into cells can be carried out by physical or chemical methods. The former does not need any vector, the gene transfer happens by an electrical or mechanical stimulus provoked on the cells such as electroporation [12], microinjection [13], or electrotransfection [14]. All physical methods show

Metallosurfactants: From Fundamentals to Catalytic and Biomedical Applications, First Edition.
Edited by Surinder K. Mehta and Ravneet Kaur.

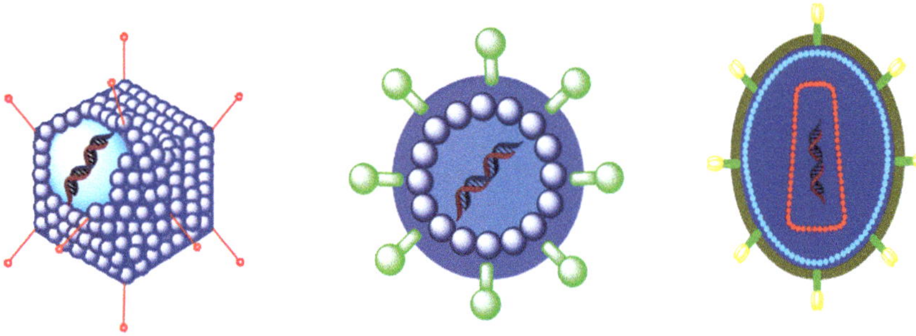

Figure 8.1 Structures of different viral vectors.

great disadvantages since they put cells under great stress and it takes a long time to obtain an effective delivery.

The chemical methods require the use of vectors, which can be viral or non-viral:

- **Viral vectors** are habitually retrovirus, lentivirus, adenovirus, etc. (Figure 8.1). They have demonstrated a high transfection efficiency, ease of production, a low toxicity, and a large stability. However, their use also shows disadvantages. Although they are usually modified to render them safe, sometimes they can generate negative immune responses and provoke serious side effects in the patients [15–17].
- **Non-viral vectors** are usually cyclic molecules or aggregates of amphipathic molecules. These vehicles are normally characterized by having regions with different polarity. Some examples are nanoparticles, micelles, cyclodextrins, calixarenes, or liposomes; all of them cationic in nature (Figure 8.2) [18, 19]. They show lower transfection efficiency compared to viral vectors, but their side effects do not appear to be so harmful as those provoked when a strange agent enters into the organism.

There are numerous studies in the literature about the synthesis and characterization of surfactants, and about the use of monomers or their self-assemblies as vectors of nucleic acid (DNA or RNA) in gene therapy [20–23]. However, even if their use for the incorporation of genetic material into cells has become a frequent and effective treatment, there are still aspects that require in-depth study to achieve scientific advances. One of them is related with the synthesis and optimization of new vectors with a greater efficacy.

Some decades ago, several amphiphilic molecules containing metal ions were prepared and used in several applications such as electron storage devices in photoredox processes, potential precursors in new synthetic methods, catalysts in heterogeneous processes, antimicrobial or antibacterial agents, nanocarriers of drugs, etc. [24–28]. These amphipathic molecules, with properties of metal complexes as well as of surfactants, are known today as metallosurfactants. The metal ions are usually located in the polar head groups of the molecules, although in some cases they can act as counter ions or even they can be part of organometallic fragments in the hydrophobic region of the surfactants [29–31].

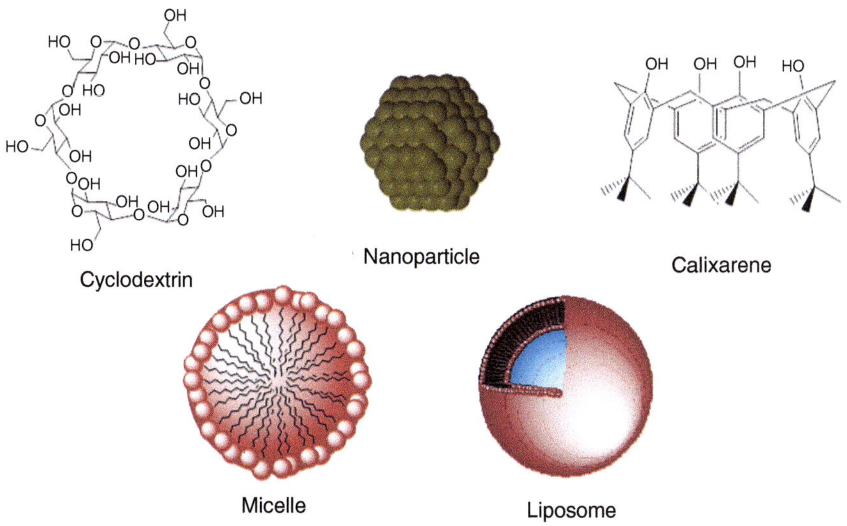

Figure 8.2 Structures of different non-viral vectors.

Given the importance of the use of metals in both the therapeutics and the diagnostic of medicinal chemistry [32], some vectors containing metals have been prepared as vehicles of genetic material. The incorporation of metal ions into the nanocarriers seems to improve the transfection processes of several aggregates, leaving an interesting open door in the advance of the research in gene therapy. In this chapter, we will explore the use of metallosurfactants, and of diverse structures obtained from their assemblies, in human gene delivery studies.

8.2 Metallosurfactant Monomers

In vitro studies about DNA condensation have become important working tools to understand the processes that occur in the cells *in vivo* when a gene therapy treatment is going to be performed. In fact, the condensation represents an important process by which the genetic information is packaged and protected to be released later in the task cell.

Most of the studies on DNA condensation have been carried out with cationic surfactants, single- or double-chained, or gemini surfactants, with different polar head groups [33–39]. When cationic surfactant molecules are added to a DNA solution, electrostatic interactions occur between them. The monomers are located at the surface of the DNA backbone forming premicellar aggregates or hemimicelles. Due to the hydrophobic character of the surfactant tails, an increase in the surfactant concentration causes a conformational change in the double-stranded helix of the polynucleotide from a more extended structure to a globular one, more compact, that could allow the passage of genetic material through the cell and nuclear membranes (Figure 8.3).

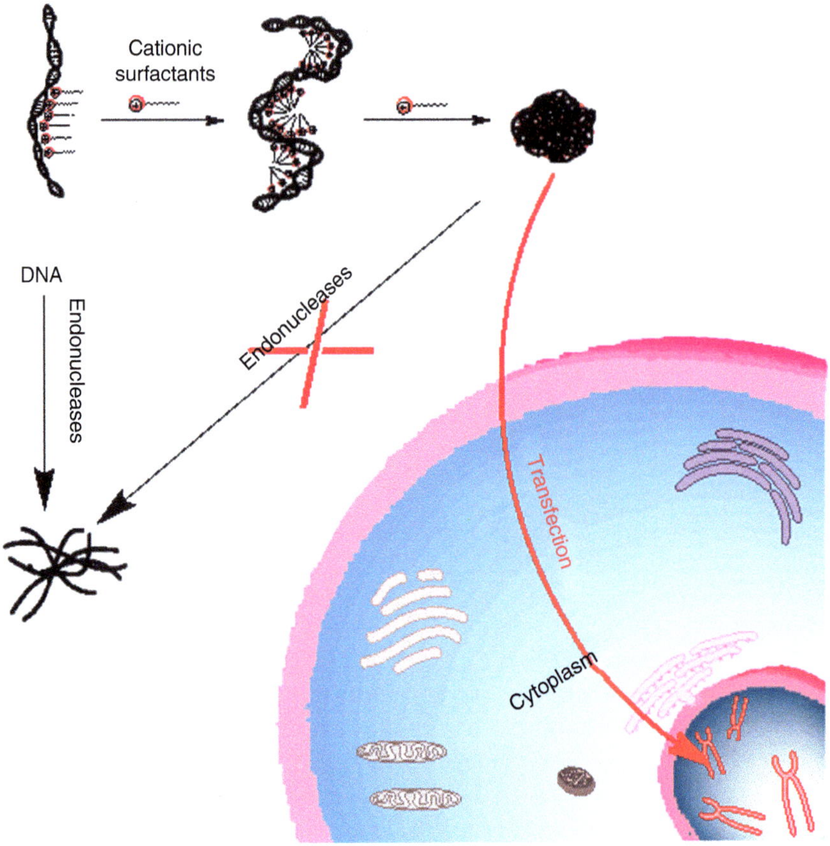

Figure 8.3 Stages involved in gene transfection process using surfactant monomers as vectors.

Recently, metallosurfactants have caught the attention of a lot of researchers. These molecules have the properties of surfactants, but also those corresponding to the metal contained in their structure. These metals, generally from the transition series, provide the amphiphilic molecules properties such as paramagnetism, color, catalytic and redox properties, or sensitivity to pH [40, 41]. Bearing this in mind, diverse surfactant-cobalt(III) complexes of the type *cis*-$[Co(X)_2(C_{14}H_{29}NH_2)Cl](ClO_4)_2$ (X = ethylenediamine, 2,2′-bipiridyl or 1,10-phenanthroline) and *cis*-$[Co(X)(C_{14}H_{29}NH_2)Cl](ClO_4)_2$ (X = triethylenetetramine) were synthesized and characterized by using techniques such as IR, NMR, UV–vis absorption spectra, and elemental analysis [42]. These metallosurfactants demonstrated a higher capacity of self-assembly to form micelles than their surfactant analogues, without the presence of the metal complexes in the polar head group. The authors found that all these metallosurfactants interacted with DNA. According to the characteristics of the different cobalt complexes present in the surfactant, electrostatic interactions, van der Waals forces, and intercalation binding modes drive the formation of DNA–surfactant complexes. In this regard, phenanthroline has been shown to make the association more favorable than other ligands. Viscosity measurements demonstrated that the binding of this planar ligand with the polynucleotide was not fully intercalative, there was a slight steric hindrance. The interaction modes observed with the other ligands were only external (surface binding) and/or groove binding.

The cytotoxic activity of the metallosurfactant *cis*-[Co(trien)($C_{14}H_{29}NH_2$)Cl]$(ClO_4)_2$ on human breast cancer cells (MCF-7) has been measured [43]. Different *in vitro* studies showed time-dependent toxicity effects on the cells due to the existence of hydrophobic forces in the binding surfactant/DNA. This metallosurfactant kills the cells mainly by apoptosis, a process of programmed cell death characterized by chromatin marginalization, nuclear fragmentation, and apoptotic body formation. The apoptosis occurs in cells, which are meant to be removed without affecting the neighboring cells [44]. Some necrosis was also observed in the p53-positive breast cancer cells. According to the authors, the Co(III)-based surfactant studied could be of application in cancer therapy. It was also pointed out that the ability to modulate the death of a cell must be recognized as a potential item in any therapeutic treatment.

It is known that transition metal complexes display important physical properties. In particular, Ru complexes show photochemical and photophysical properties that can be exploited in medical applications [45]. For example, some Ru compounds are able to penetrate tumors, binding to nucleic acids in a different way than cisplatin, an antineoplastic drug frequently used in chemotherapy, and decreasing the toxic effects on the cells. Also, improvements in cancer treatments can be experimentally observed thanks to the photochemical properties of Ru complexes containing planar aromatic nitrogen ligands. For example, they can suppress the immune response inhibiting the proliferation of T cells. Accordingly, two Ru-based metallosurfactants, both derived from the 2,2′-trisbipiridyl Ru(II) complex, with different number of hydrophobic tails of the same length (the double-tailed $[Ru(2,2'\text{-bipy})_2(4,4'\text{-}(C_{11}H_{23})_2\text{-}2,2'\text{-bipy})]^{2+}$ complex and the monocatenary $[Ru(2,2'\text{-bipy})_2(4\text{-}(CH_3)\text{-}4'\text{-}(C_{11}H_{23})_2\text{–}2,2'\text{-bipy})]^{2+}$ complex – RuC11C11 and RuC1C11 – Figure 8.4) were synthesized and their interactions with calf thymus DNA (ct-DNA) were quantified to improve and advance the design of non-viral vectors for gene therapy [46]. The metal center in the polar head group possesses measurable light-responsive (redox and spectroscopic) properties [47]. This permitted the surfactant/ct-DNA binding to be investigated without the use of any external dye (note that the latter could modify the structural characteristics of the polynucleotide). The addition of RuC1C11 and RuC11C11 monomers to a DNA solution provoked the formation of premicellar aggregates (hemimicelles) on the surface of the polynucleotide (surfactant concentrations lower than critical micellar concentration – CMC – values were used). This assembly is due to two types of interactions: an electrostatic interaction between the cationic head groups of the surfactant and the negative phosphate groups of the ct-DNA and the hydrophobic forces among the hydrocarbon tails. Binding constant (K_b) values at different nucleic acid concentrations were obtained from fluorescence emission intensity values of the Ru surfactants, by using a quantitative method developed in such a manuscript.

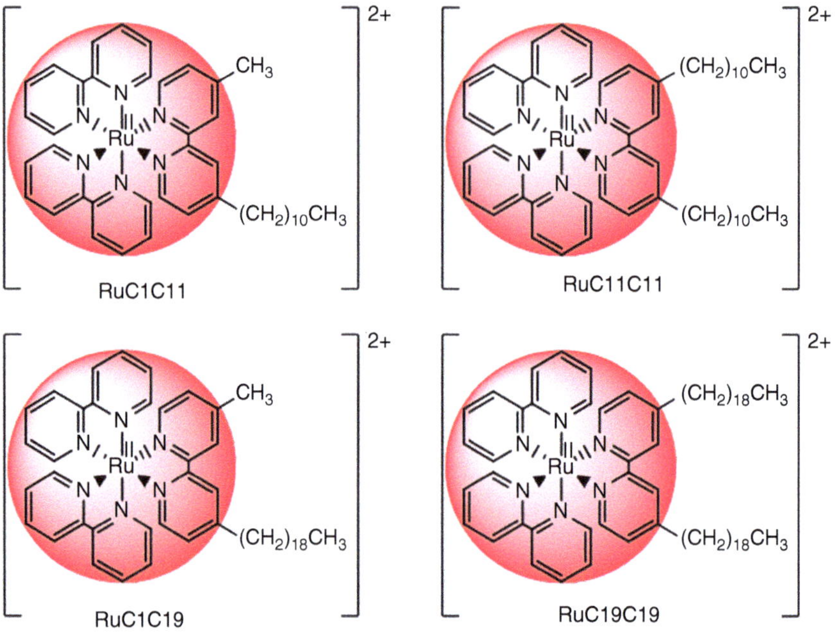

Figure 8.4 Structures of the Ru metallosurfactants.

K_b values demonstrated a sigmoidal dependence with the ct-DNA concentration. This behavior confirmed the cooperative character of the binding process, that is, the interaction of a first metallosurfactant molecule to a receptor (DNA) favors the binding of a second ligand to the same receptor and so on. This interaction finally resulted in a conformational change in the polynucleotide from an extended double-helix structure to a more globular form. AFM images demonstrated such a conformational change. The metallosurfactant monomer concentration necessary to produce this change was lower for the double-tailed surfactant than for the single-chain surfactant due to the stronger hydrophobic forces.

Results revealed the importance of the hydrophobic forces on the binding between surfactants and DNA. This was also confirmed when cyclodextrin molecules were added to the surfactant/DNA solutions. The formation of inclusion complexes of the surfactants with cyclodextrin molecules (with stoichiometries 1 : 1 and 1 : 2) [48] provoked the decompaction process of the surfactant/DNA globular complex (Figure 8.5). The cyclodextrin concentration necessary to de-condense the nucleic acid was lower for the RuC1C11/DNA complex than for the RuC11C11/DNA complex due to the lower hydrophobic character of the former [46, 48]. Comparison of the binding equilibrium constants corresponding to the formation of 1 : 1 and 1 : 2 metallosurfactant : cyclodextrin complexes showed that for the 1 : 2 complex both cyclodextrin molecules are placed on the same hydrophobic chain of the amphiphilic molecule.

The main role of the surfactant hydrophobic character in the interaction of the amphiphilic molecules with DNA has also been highlighted with longer-chained

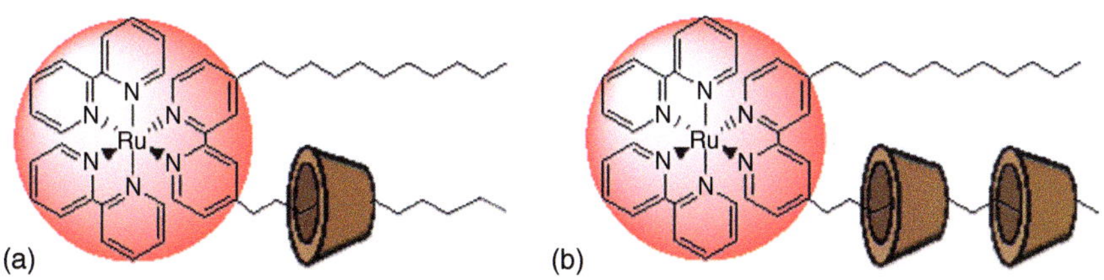

Figure 8.5 Structural representations of RuC11C11/cyclodextrin complexes with different stoichiometries: 1 : 1 (a) and 1 : 2 (b).

Ru metallosurfactants: the monocatenary [Ru(2,2′-bipy)$_2$(4-(CH)$_3$-4′-(C$_{19}$H$_{39}$)-2,2′-bipy)]$^{2+}$ complex (RuC1C19) and the double-chained [Ru(2,2′-bipy)$_2$-4,4′-(C$_{19}$H$_{39}$)$_2$-2,2′-bipy]$^{2+}$ complex (RuC19C19); both with the same cationic polar head group than those studied previously (Figure 8.4). Results obtained are now shown in this chapter.

A simple route to get information about the interaction surfactant/DNA is by using the competitive assay with the dye ethidium bromide. This probe binds with the base pairs of the polynucleotide through an intercalative mode [49, 50] and shows a high emission intensity when it is intercalated in the nucleic acid. A decrease of this emission intensity was observed when both RuC1C19 and RuC19C19 were added to an ethydium bromide–DNA solution. This behavior, due to the displacement of ethydium bromide molecules from its intercalative position into the base pair to the aqueous solution, was also observed for the analogous metallosurfactants RuC1C11 and RuC11C11. Comparing results in Figure 8.6, a sharper decrease for the double-chained surfactants than for the monocatenary ones is observed. This is due to the greater hydrophobic character of the former, which facilitates the formation of hemimicelles on the surface of the polynucleotide and its subsequent condensation. An analogous difference between single- and double-tailed surfactants was also observed from circular dichroism (CD) measurements. Figure 8.6 shows the dependence of the relative molar ellipticity of ethydium bromide obtained at a λ value of 308 nm (maximum wavelength corresponding to the probe band in CD) [51] on the

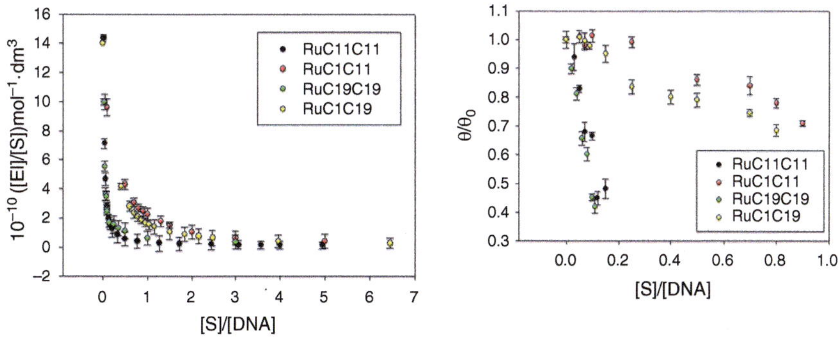

Figure 8.6 Dependence of both the apparent emission intensity values (EI) and relative molar ellipticity (θ/θ$_0$) of ethidium bromide with the molar ratio [S]/[DNA] for the different Ru(II)-based surfactants (S).

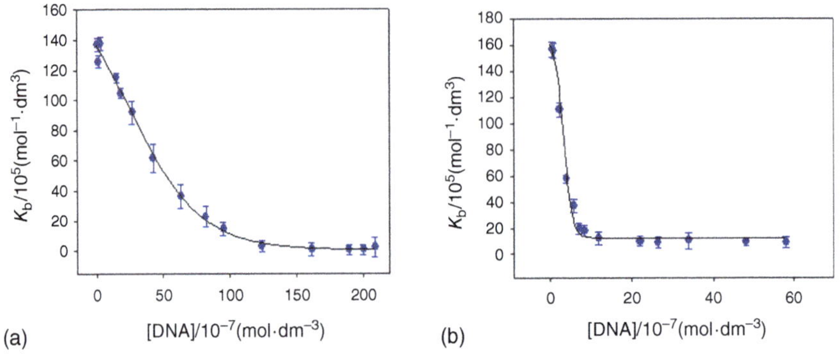

Figure 8.7 Sigmoidal dependence of the binding constant K_b with the [DNA] for the RuC11C11 (a) and RuC19C19 (b) metallosurfactants.

molar ratio [surfactant]/[DNA]. A much more pronounced decrease in ellipticity is clearly observed for the double-chain surfactants.

Binding constants K_b corresponding to the interaction between both the RuC11C11 and the RuC19C19 surfactants and ct-DNA were estimated from emission intensity values of such Ru surfactants, in the presence of different polynucleotide concentrations, and by using the method developed in [46]. A sigmoidal dependence of K_b on the DNA concentration is observed. The same trend was found for both the RuC11C11 and the RuC19C19 surfactants (Figure 8.7). This confirms the cooperative character of the interaction between surfactants and DNA.

K_b contains two different contributions: electrostatic and non-electrostatic (it must be noted that non-electrostatic interactions such as hydrogen bonds or π–π stacking interactions could also occur between the bipyridyl ligands of the cationic heads of the surfactants and the polynucleotide). When the DNA concentration is high, the electrostatic interactions between the polar heads of the surfactant and the phosphate groups of the DNA are the driving forces in the association process. Therefore, the minimum value of the binding constants obtained for high DNA concentrations will be mainly due to these electrostatic interactions and, consequently, this value must not depend on the hydrophobic interactions of the aliphatic chains of cationic surfactants (single and double chained). To verify the validity of the method used to obtain the binding constants [46], the authors compared the minimum values of K_b for all the Ru surfactants studied (RuC1C11, RuC11C11, RuC1C19, and RuC19C19). The minimum values of K_b obtained were $8.7 \times 10^5\ M^{-1}$ for RuC1C11, $8.6 \times 10^5\ M^{-1}$ for RuC11C11, $9.4 \times 10^5\ M^{-1}$ for RuC1C19, and $9.7 \times 10^5\ M^{-1}$ for RuC19C19. As can be seen, all these values are similar because the cationic head groups of these surfactants are the same. Besides, this value agrees with the binding constant between the $[Ru(bpy)_3]^{2+}$ complex (without any hydrocarbon tail) and ct-DNA, $8.0 \times 10^5\ M^{-1}$, previously obtained [52].

The addition of α-cyclodextrin to RuC1C19/DNA (and RuC19C19/DNA) solutions produces the de-condensation of the globular state of the surfactant/polynucleotide complex. This is due to the formation of inclusion complexes of surfactants with the

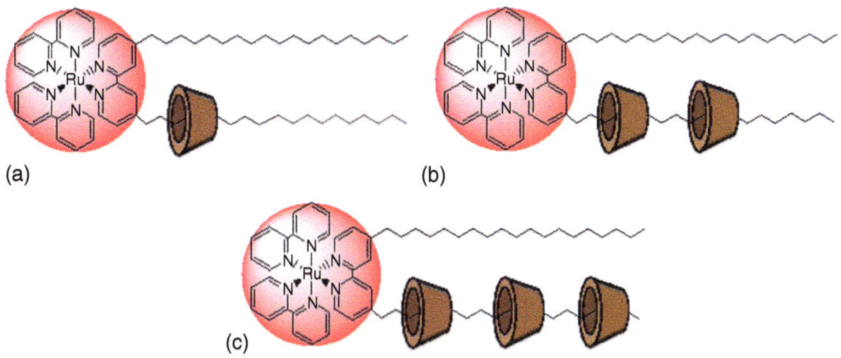

Figure 8.8 Structural representations of RuC19C19/cyclodextrin complexes with different stoichiometries 1 : 1 (a), 1 : 2 (b) and 1 : 3 (c).

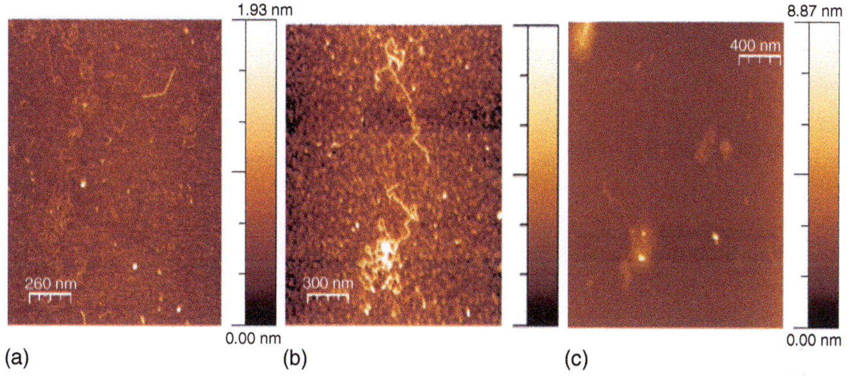

Figure 8.9 AFM images of DNA (a) and RuC1C19/DNA complexes at [S]/[DNA] = 0.5 (b) and 1 (c).

macrocyclic rings exhibiting different stoichiometry 1 : 1, 1 : 2, and 1 : 3 (Figure 8.8), as -observed with Ru-based surfactants with shorter hydrocarbon chains [48]. Comparing the equilibrium constants for the formation of the 1 : 1, 1 : 2, and 1 : 3 surfactant:CD complexes, it is demonstrated that all CD molecules are located in the same hydrophobic tail, as illustrated with the RuC11C11 metallosurfactant in [46]. The importance of the hydrophobic forces in the condensation process of the nucleic acid is demonstrated again. AFM images show the globular structure of the polynucleotide in the presence of the different Ru surfactants studied (Figure 8.9).

Recently, the interactions between the metallosurfactants $[Ru(DMP)_2(CA)Cl](ClO_4)$ and $[Ru(DMP)_2(CA)_2](ClO_4)_2$, where DMP = 2,9-dimethyl-1,10-phenanthroline and CA = cetylamine, and ct-DNA; and the equilibrium binding constants of the surfactant/DNA complexes were estimated from electronic absorption and spectral titration [53]. Hyperchromism, a slight shift of the band, and a competitive binding between the ethidium bromide and the Ru(II) metallosurfactants were observed. According to the authors, these results indicate a binding of intercalative nature between the amphiphilic molecules and the DNA, due to the planar structure of the phenanthroline ligand, and a strong hydrophobic association between the methylene groups of the hydrophobic tails and the polynucleotide. Values of K_b about $10^5\ M^{-1}$ were obtained, which agree with those obtained for other Ru(II) metallosurfactants [54]. These high binding constants are in agreement with an intercalative process, as was confirmed by the authors from viscosity measurements. *In vitro* assays of these Ru(II) metallosurfactants were carried out with human cervical cancer (HeLa) cell lines. Cytotoxicity depended on the dosage used, as well as on the exposure time with the cells. Results showed a high cytotoxicity against HeLa cells at low doses, but a lower toxicity against normal cells. This indicates that these Ru metallosurfactants could be used in anticancer treatments. Higher the hydrophobicity of surfactant, stronger cellular damage was observed. Metabolic mechanism in tumor cells requires a high energy. According to the authors, Ru(II) metallosurfactants possess the ability to diminish the energy status in tumor cells, deactivating cellular functions and, hence, causing cell death. Low concentrations of both metallosurfactants induced apoptosis and necrosis in the cells. The inhibitory activity measured as IC_{50} values was lower for the double- than for the single-chained surfactant due to an easier penetration of the hydrophobic complexes in abnormal cells.

Interest in seeking less toxic and generally less health-damaging metal complexes as a substitute of the well-known anticancer drug cisplatin has promoted the preparation of surfactants with metals (or metal complexes) other than Ru. By mixing diverse surfactants and the metal platinum complex $[PtCl_4]^{2-}$with a stoichiometry 1 : 2 (K_2PtCl_4:surfactant), Sharma et al. prepared different type of metallosurfactants in which the metallic centers act as co-ions in the amphiphilic molecules (Figure 8.10) [54–57]. Besides their characterization, *in vitro* biological studies were carried out. In this regard, cytotoxicity assays of the Pt-based surfactants against MCF-7 and MDA-MB-231 human breast cancer cell lines showed strong anticancer properties for all the metallosurfactants when compared with the pure surfactants, that is, the amphiphilic molecules without the metallic centers. The toxicity increased upon increasing the length of the surfactant tails, due to an increment in their hydrophobic nature. Conformational changes in proteins, caused by the presence of the Pt metallosurfactants, were also observed.

Similar metallosurfactants were also prepared with silver [58] and copper [59]. In the case of the silver metallosurfactants, only the interaction of Ag micelles with DNA was studied. On the contrary, in the case of the copper surfactants, a comparison between the pre- and the post-micellar systems was carried out. Both studies will be discussed in Section 8.3.

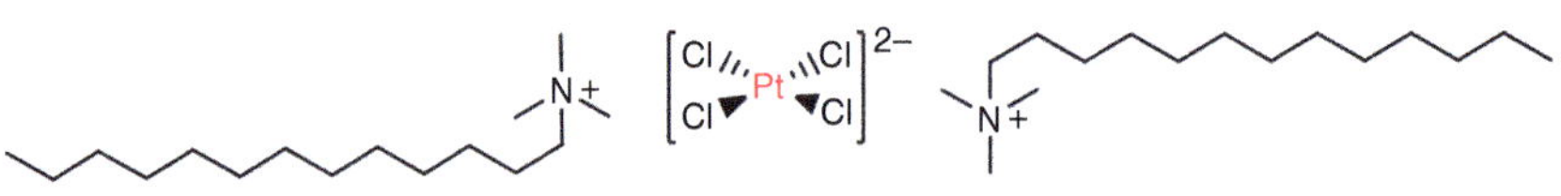

Figure 8.10 Metallosurfactant $[Pt(L)_2Cl_4]$.

8.3 Metallomicelles

The amphiphilic nature of the metallosurfactant monomers favors their aggregation forming assemblies with diverse structures (micelles, liposomes, vesicles, etc.). Their behavior in the transport of genetic material may differ from that of the monomers.

Kaur et al. [58] prepared Ag metallosurfactants with a stoichiometry 1 : 2 (AgB:CTAB), mixing a metal salt and the surfactant CTAB (cetyltrimethylammonium bromide). The surfactant molecules resulting, $[Ag(CTA)_2Br_3]$ cetyltrimethylammonium silver bromide, are self-assembled forming metallomicelles. TEM images showed spherical metallomicelles at a metallosurfactant concentration about 3xCMC. Lower stoichiometries (1 : 0.5) resulted in the formation of nanoparticles. The interaction of the Ag-based metallomicelles with ct-DNA was studied. The hyperchromism effect obtained indicated the complexation of the metallomicelles with the polynucleotide due mainly to electrostatic interactions. The Ag complex was found to be cytotoxic in diverse cancer cell lines (human leukemia HL-60, pancreatic MIA-Pa-Ca-2 and prostate cancer PC-3 cells). Results also demonstrated a high potential of these metallomicelles as antimicrobial agents, showing a stronger effectiveness against Gram-positive bacterial strains than against Gram-negative ones. The cytoplasm of fungal cells was less damaged in comparison to bacterial cells.

The metallosurfactant $[Cu(L)Br_3]$, where LBr is the ligand 1-cetyl-4-aza-1-azoniabiciclo[2.2.2]octane bromide, was synthesized and characterized by Zhiltsova et al. (Figure 8.11) [60]. Here, the metal center is part of the hydrophilic head. It is joined through hydrogen bonds to the bicyclic ring of the ligand. Vesicles were formed at surfactant concentration close to the CMC, and spherical micelles at [surfactant] ≫ CMC. The formation of metallosurfactant/oligonucleotide complexes was confirmed. The binding constant seems to be higher in the presence of the metal in comparison to that obtained with the ligand without the metal center. Besides, the condensation of the genetic material only happens in the presence of the ligand (with or without the metal center), but not when only the $CuBr_2$ salt is used. This result indicates once again the importance of the hydrophobic forces in the nucleic acid condensation. In the opinion of the authors, this metallosurfactant can act as nanovehicle of drugs and genetic material.

The importance of the hydrophobic interactions in the surfactant/DNA binding was also showed with the functionalization of the cationic amphiphilic molecule cetylpyridinium chloride (cpc) with Cu(II), the latter acting as counter ion, with

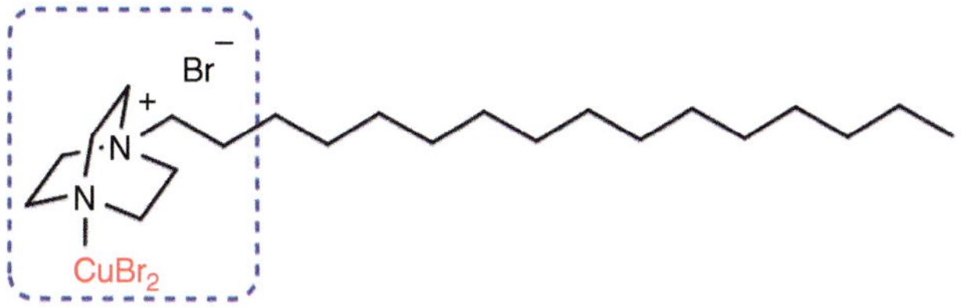

Figure 8.11 Metallosurfactant $[Cu(L)Br_3]$.

two different stoichiometries (1 : 1 and 1 : 2, Cu:cpc and Cu:$(cpc)_2$) [59]. The amphipathic ligand promotes the formation of micelles and a comparison in the binding interaction metallosurfactant/ct-DNA is reported in both the premicellar and the postmicellar region. A strong binding of the monomers (premicellar region) with DNA was observed for both surfactants. According to the authors, the electrostatic interactions between the negative charge of the nucleic acid phosphate groups and the cationic metal-amphiphilic molecule provoke changes in the conformation of the DNA from an extended coil to a globular form. This conformational change seems to be more pronounced in the case of the double-chained surfactant than in the single-chained one. Accordingly, the dimeric copper-based surfactant could act as nanocarrier of DNA in gene transfection processes.

The Ru(II) metallosurfactants prepared by Lebrón et al. in [46], and other analogous amphiphilic molecules, were also used for the formation of metallomicelles to be used as nanocarriers of drugs and genetic material [54]. The metallosurfactants studied were RuC1C11, RuC11C11, RuC1C19, and RuC19C19 (RuC1C19 and RuC19C19 being $[Ru(2,2'\text{-bpy})_2(4\text{-}CH_3\text{-}4'\text{-}(C_{19}H_{39})\text{-}2,2'\text{-bpy})]Cl_2$ and $[Ru(2,2'\text{-bpy})_2(4\text{-}4'\text{-}(C_{19}H_{39})_2\text{-}2,2'\text{-bpy})]Cl_2$, respectively). TEM images revealed the formation of both spherical and elongated micelles depending on the surfactant concentration. The favorable internalization of these metallosurfactants into different cell lines seems to indicate that they could be used as vectors of genetic material. An exhaustive and multidisciplinary study about the binding of the metallomicelles with ct-DNA demonstrated a partial condensation of the polynucleotide with RuC1C11 and a total condensation with the other surfactants. Confocal fluorescence images show the adsorption of the micelles on the DNA surface and its further condensation at a determined micellar concentration. Viability assays resulted in a moderate cytotoxicity at lower surfactant concentration and a high toxicity for higher concentrations. According to these results, the metallomicelles could be used as vectors in gene transfection for low micellar concentrations and as nanocarriers of antibiotics with anticancer and/or antibacterial activity with a determined specificity for LS180, MCF7, and A549 cell lines.

Recently, Nehru et al. have carried out a study about the influence of both the cationic head group size and the tail length on the interaction metallosurfactant/ct-DNA [61]. This type of work contributes to the rational designing of metallodrugs toward biomedical applications [62]. In particular, the interaction of the fluorescent metallosurfactants derived from the $[Ru(bpy)_2Cl_2]^{2+}$ and $[Ru(phen)_2Cl_2]^{2+}$ complexes with ct-DNA was studied. Taking into account the CMC values [63], within the surfactant concentration range used monomers and micelles (or pseudomicelles) were interacting with the polynucleotide. The binding constant values, obtained from absorbance data at different polynucleotide concentrations, were higher for complexes containing phen as ligand than those with bpy. A larger planarity in the ligand (phen > bpy) favors the π-π stacking interactions between DNA base pairs and, hence, the binding constant increases. An increment in the length of the hydrocarbon chains seems to increase the binding constant due to a larger hydrophobic contribution. In agreement with the binding constant values obtained, the authors suggested an interaction of the Ru surfactant with DNA through a moderate intercalation process. Viscosity measurements corroborated this hypothesis.

CD measurements demonstrated that only slight conformational changes in the polynucleotide take place, so they do not seem to be good vectors in transfection processes. This result does not agree to those obtained in the premicellar [46] and postmicellar [54] regions of the metallosurfactants Ru*CiCii* or Ru*CiiCii* ($i = 11$ and 19). This could be due to a difference in the cationic head group size of the metallosurfactant, but the low surfactant concentration used in this work could also be important.

8.4 Metalloliposomes (Metallosomes)

As was previously mentioned, the amphipathic properties of metallosurfactants favor the aggregation of their monomers. Liposomes are spherical structures containing a phospholipid bilayer, non-polar in nature, and an inner aqueous polar region (Figure 8.2). The mixture of different lipids generates the spontaneous formation of a lipid membrane or film that, after a hydration process, forms the liposomes [64, 65]. Given their structure, these aggregates have become one of the best membrane models used to simulate the cell membranes and they can be used as vectors in gene therapy [66–68].

The mixture of metallosurfactants and phospholipids forms vesicles, with structures similar to liposomes, that are known as metallosomes, metalloliposomes, or metallovesicles.

Several metallosomes have been synthesized with biomedical applications [31, 69–74]. Here we are going to discuss only those more related to gene transfection processes.

Cruz-Campa et al. [75] synthesized in 2007 novel supramolecular aggregates with amphiphilic Cu(II) complexes. To the best of author's knowledge, this was the first time that metalloliposomes were used as gene vectors. The complexes obtained were $[Cu(L_{dt})_2](OTF)_2$ and $[Cu(L_{ot})_2](OTF)_2$ (Figure 8.12) where L_{dt} is 1-dodecyl-1,4,7-triazacyclononane, L_{ot} is 1-octadecyl-1,4,7-triazacyclononane and OTF is trifluoromethanesulfonate.

The metalloliposomes were found to bind with the fluorescent pEGFP-N1 plasmid. This plasmid was condensed with both Cu liposomes. TEM images demonstrated the formation of spherical metallo-lipoplexes, that is, spherical vesicles/plasmid complexes. The use of the intercalative SYBR-Green dye proved the encapsulation of the plasmid into the metallovesicle.

It is interesting to note the high transfection efficiency, about 37%, obtained with the complex containing L_{dt} as hydrophobic ligand in the human embryonic kidney 293-T (HEK 293-T) cells. This value is, perhaps, the highest efficiency observed in the presence of metallosurfactants. The efficiency value was much lower for the liposome containing the L_{ot} ligand. Higher transfection efficiency was obtained for the metallosurfactant with a lower CMC value.

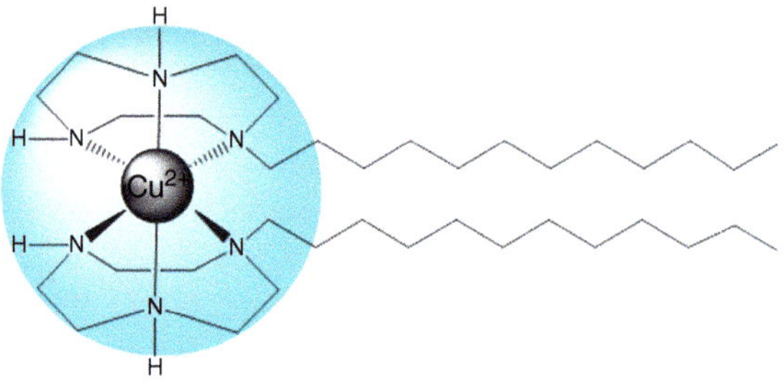

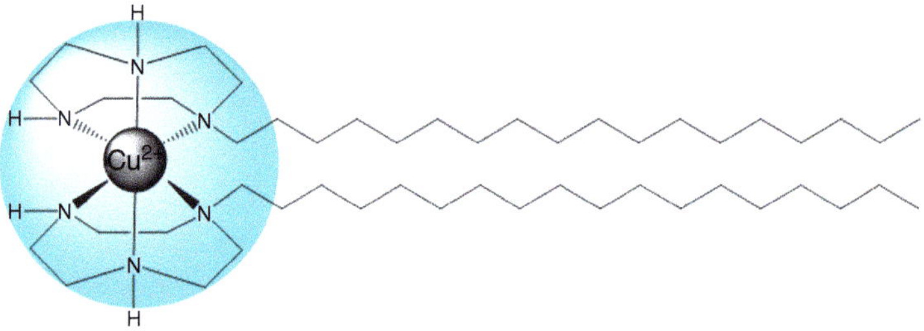

Figure 8.12 Metallosurfactants $[Cu(L_{dt})_2]^{2+}$ and $[Cu(L_{ot})_2]^{2+}$.

The authors also tested the importance of the hydrophobic forces in gene transfection, since no observable transfection was found with the analogous non-amphiphilic copper complex. They also indicated that the redox properties of the metal complex could play an important role in the gene release process.

The self-assembly of the lipophilic Cu(II) complexes ($[Cu(L)_2](OTF)_2$) of 1-alkyl-1,4,7-triazacyclononane (L) (Figure 8.13), with different hydrocarbon tail lengths (alkyl: octyl, decyl, dodecyl, tetradecyl, hexadecyl, and octadecyl), also resulted in the formation of metallosomes able to condense pEGFP-N1 plasmids [76]. Good transfection efficiency values were obtained. As reported in [75], the transfection efficiency in HEK 293-T cells decreases when the length of the hydrocarbon chains of the ligands increases and, hence, when the CMC value of the copper complexes increases. On the other hand, the Zn(II) complex of 1-dodecyl-1,4,7-triazacyclononane gave the same transfection efficiency value as its analogue with Cu(II). Therefore, a change in the metal ion, keeping the alkyl length of the ligand constant, does not influence the *in vitro* transfection process. However, Cu vesicles seem to be more stable than Zn vesicles as DNA vectors in murine models. In fact, liposomes with copper produce more T cells than with zinc at the lymph nodes and spleen samples according to *in vivo* transfection assays carried out with BALB/c strain of mice vaccinated with the pVAX and pBT vectors and the metallolipoplexes.

This group recently prepared a similar amphiphilic Cu(II) compound, the bis-(4-pyridylmethyl hexadecanoate)-(1,4,7-triazacyclononane) copper(II) complex (Figure 8.13) [77]. The metallovesicles (vesicles with structures like liposomes), obtained by the reversed-phase procedure, condensed the pEGFP-N1 DNA plasmid; but showed a low transfection efficiency of about 4% in HEK 293-T cells (the value

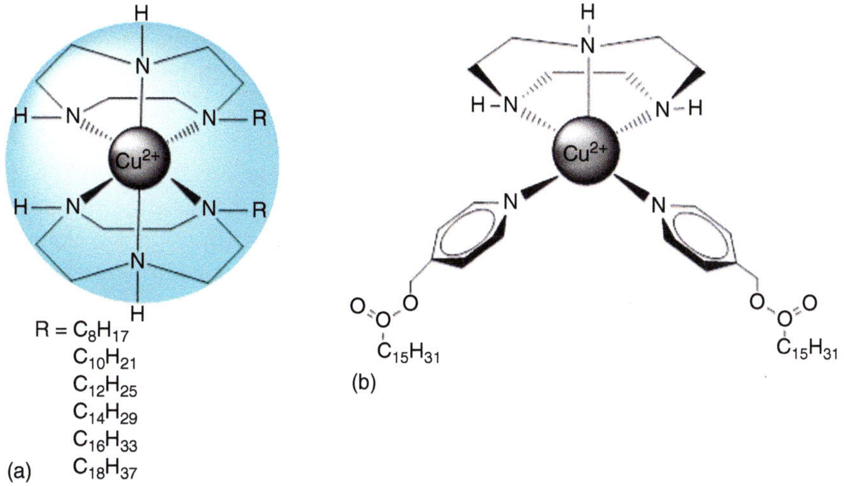

Figure 8.13 Cu-based metallosurfactants.

slightly depended on the method used). No transfection was observed by using the analogous non-amphiphilic Co(II) complex.

Metallosomes containing a mixture of lipids (cholesterol and the double-chained metallosurfactant bis-dodecylaminecopper(II) chloride) provoke slight conformational changes in the ct-DNA double-helix structure [78]. The binding of these aggregates with the polynucleotide induces changes in its CD spectrum. The changes observed, an increase in the ellipticity values of the positive band after the addition of metalloliposomes to the DNA solution, are typical of an intercalative binding. No condensation of the nucleic acid happened. Hence, these metalloliposomes are not good vectors to be used in gene transfection. Changes in the cholesterol concentration modified some properties of the lipid membrane, varying their drug release ability.

A different situation was found in the study of the interaction of unilamellar metalloliposomes containing 1,2-dioleoyl-*sn*-glycero-3-phosphoethanolamine (DOPE) and double-chained Ru metallosurfactants (Figure 8.4) derived from the $[Ru(bpy)_3]^{2+}$ complex with DNA [79]. In this case, a strong binding between the metallosomes and the nucleic acid occurs, resulting in the condensation of the polynucleotide. Equilibrium binding constants are estimated for the metalloliposome/DNA association. Results demonstrated the importance of both electrostatic and hydrophobic forces in such interactions.

Low cytotoxicity values and good internalization results for the Ru-based metalloliposomes were obtained in MCF7 (human breast adenocarcinoma), LS180 (human colonic epithelial adenocarcinoma), HepG2 (human liver carcinoma), A549 (human alveolar basal epithelial adenocarcinoma), and RPE-1 (hTERT-immortalized retinal pigment epithelial) cell lines. GFP (green fluorescent protein) expression of the pEGFP-C1 plasmid in U2OS (human bone osteosarcoma epithelial) cells provided a transfection efficiency value in the range 2.2–2.6%. This value was slightly lower for the aggregates containing the metallosurfactant with a longer hydrocarbon tail. Curiously, this result agrees with those observed with the Cu(II) vesicles previously reported [75–77]. The transfection efficiency decreases when the CMC of the metallosurfactant increases, even if the metal center is different.

Su et al. [80] synthesized different metallosurfactants to be used for biomedical applications, bearing the metal ligand Zn(II)-dipicolylamine as cationic head group. Lipids with different length and nature (saturated and unsaturated linear hydrocarbon chains, steroids, and tocopherol) were used as hydrophobic chains. The self-assembly of these metallosurfactant monomers generated the formation of metallosomes in both the presence and the absence of the neutral lipid DOPE. All the metallosomes prepared interact with DNA, forming lipoplexes. The size of such complexes depended on the +/−charge ratio and a size decrease was observed when the condensation of the nucleic acid (ct-DNA and pGL3 plasmid) took place.

Cytotoxicity assays in the HEK293 cell line showed a high viability for most metallosurfactants. A higher toxicity was found for the surfactants with long unsaturated hydrocarbon chains, while tocopherol and steroid-based lipids gave similar cytotoxicity values. *In vitro* transfection results with HEK293 and HepG2 cell lines also depended on the type of hydrocarbon tails. The best efficiency value was obtained with the linear hydrocarbon chain containing 16 carbon atoms and with unsaturated bonds in the hydrocarbon chain. The presence of tocopherol in the lipids provoked higher transfection efficiency in comparison with steroid-based lipids. The cellular uptake was studied from flow cytometry and the intracellular fate after endocytosis was carried out by confocal laser scanning microscopy. The internalization of the plasmid by endocytosis process was confirmed. Besides, DNA was found to be delivered and released around the cell nuclei. This indicated the nucleic acid scape from the endosome. In any case, the transfection mechanism by these lipids needs a more thorough study.

Recently, Kaur et al. [81] synthesized spherical metallosomes, of multivesicular nature, from cholesterol and double-chained metallosurfactans by mixing the corresponding metallic salt (iron, cobalt, nickel, and copper) and the lipid cetyl pyridinium chloride with 1 : 2 stoichiometry. The metal influenced the type of metallosome prepared: unilamellar vesicles for Fe and multilamellar vesicles for Cu, Ni, and Co. Aggregates with other stoichiometries (1 : 3 and 1 : 4), as well as metallosomes without cholesterol, were also generated to give spherical, multivesicular metallosomes.

The intercalating nature of the metallosomes and a groove binding mode with ct-DNA resulted in a strong interaction of the aggregates with the polynucleotide, more efficient with Cu metallosomes than with those based on Fe, Co, and Ni. Changes in the zeta potential values of the metallosome/DNA complexes were observed depending on the metal ion embedded in the aggregates as counter ion. Conformational and structural changes in DNA were also obtained from gel electrophoresis assays. This confirmed the possibility of using these metallosurfactants in gene therapy processes.

Aggregates with other stoichiometries seemed to give spherical metallosomes, but up till now they have not been used in biomedical applications.

8.5 Future Perspectives

Interesting investigations about the preparation of metallosurfactants, and the interactions of their monomers as well as their aggregates with DNA, have been carried out. Stability, low cytotoxicity, biodegradability, and good transfection efficiency values were obtained for some of them. Results show that the synergistic effect between the properties of the metal ions and those of the amphiphilic tails can lead to enhancing and improving their interest in biomedical applications, especially in anticancer and gene therapies.

Besides, transfection processes occur according to mechanisms still unknown. Both the endocytosis pathways and the mechanism of intracellular internalization processes need to be elucidated. *In vivo* assays and clinical trials need to be performed to demonstrate the advantages that metallosurfactants could provide in biomedicine.

Bearing this in mind, more studies are required on metal surfactants. The design and synthesis of novel gene carriers, but also *in vitro* and *in vivo* assays are essential to clarify all the issues mentioned above. Molecular dynamic studies will be also welcome.

Acknowledgments

This work was funded by the Ayudas a Consolidación de Grupos de la Junta de Andalucía (2019/FQM-206 and 2019/FQM-274) and the European Union (FEDER Funds).

References

1 Tebas, P., Stein, D., Tang, W.W. et al. (2014). Gene editing of CCR5 in autologous CD4 T cells of persons infected with HIV. *The New England Journal of Medicine* 370: 901–910. https://doi.org/10.1056/NEJMoa1300662.

2 Misra, S. (2013). Human gene therapy: a brief overview of the genetic revolution. *The Journal of the Association of Physicians of India* 61: 127–133.

3 Wirth, T., Parker, N., and Ylä-Herttuala, S. (2013). History of gene therapy. *Gene* 525: 162–169. https://doi.org/10.1016/j.gene.2013.03.137.

4 Ribeil, J.-A., Hacein-Bey-Abina, S., Payen, E. et al. (2017). Gene therapy in a patient with sickle cell disease. *The New England Journal of Medicine* 376: 848–855. https://doi.org/10.1056/NEJMoa1609677.

5 Nathwani, A.C., Reiss, U.M., Tuddenham, E.G.D. et al. (2014). Long-term safety and efficacy of factor IX gene therapy in hemophilia B. *The New England Journal of Medicine* 371: 1994–2004. https://doi.org/10.1056/NEJMoa1407309.

6 Griesenbach, U., Pytel, K.M., and Alton, E.W.F.W. (2015). Cystic fibrosis gene therapy in the UK and elsewhere. *Human Gene Therapy* 26: 266–275. https://doi.org/10.1089/hum.2015.027.

7 Gaucer, S., Lwin, S.M., Titeux, M. et al. (2020). EBGene trial: patient preselection outcomes for the European GENEGRAFT *ex vivo* phase I/II gene therapy trial for recessive dystrophic epidermolysis bullosa. *The British Journal of Dermatology* 182: 788–818. https://doi.org/10.1111/bjd.18559.

8 den Hollander, A.I., Black, A., Bennett, J. et al. (2010). Lighting a candle in the dark: advances in genetics and gene therapy of recessive retinal dystrophies. *The Journal of Clinical Investigation* 120: 3042–3053. https://doi.org/10.1172/JCI42258.

9 Rodrigues, G., Shalaev, E., Karami, T. et al. (2018). Pharmaceutical development of AAV-based gene therapy products for the eye. *Pharmaceutical Research* 36: 29–49. https://doi.org/10.1007/s11095-018-2554-7.

10 FoCUS Project Financing of Cures in the US. MIT NEWDIGS Research Brief 2017F211.v011, November 2017.

11 Maldonado, R., Jalil, S., and Wartiovaara, K. (2020). Curative gene therapies for rare diseases. *Journal of Community Genetics* 19: 137–144. https://doi.org/10.1007/s12687-020-00480-6.

12 Potter, H. (2003). Transfection by electroporation. *Current Protocols in Molecular Biology* Chapter 9, Unit 9.3: 1–12. https://doi.org/10.1002/0471142727.mb0903s62.

13 Chow, Y.T., Chen, S., Wang, R. et al. (2016). Single cell transfection through precise microinjection with quantitatively controlled injection volumes. *Scientific Reports* 6 (24127): 1–9. https://doi.org/10.1038/srep24127.

14 Zhou, Y., Lu, Y., Cheng, J. et al. (2020). Highly uniform in-situ cell electrotransfection of adherent cultures using grouped interdigitated electrodes. *Bioelectrochemistry* 132: 107435. https://doi.org/10.1016/j.bioelechem.2019.107435.

15 Warnock, J.N., Daigre, C., and Al-Rubeai, M. (2011). Introduction to viral vectors. *Methods in Molecular Biology* 737: 1–25. https://doi.org/10.1007/978-1-61779-095-9_1.

16 Merten, O.-W. and Gaillet, B. (2016). Viral vectors for gene therapy and gene modification approaches. *Biochemical Engineering Journal* 108: 98–115. https://doi.org/10.1016/j.bej.2015.09.005.

17 Richardson, S.R., Doucet, A.J., Kopera, H.C. et al. (2015). The influence of LINE-1 and SINE retrotransposons on mammalian genomes. *Microbiology Spectrum* 3: 1–26. https://doi.org/10.1128/microbiolspec.MDNA3-0061-2014.

18 Ramamoorth, M. and Narvekar, A. (2015). Non viral vectors in gene therapy – an overview. *Journal of Clinical and Diagnostic Research* 9: GE01–GE06. https://doi.org/10.7860/JCDR/2015/10443.5394.

19 Yin, H., Kanasty, R.L., Eltoukhy, A.A. et al. (2014). Non-viral vectors for gene-based therapy. *Nature Reviews Genetics* 15: 541–555. https://doi.org/10.1038/nrg3763.

20 Xiao, Y., Shi, K., Qu, Y. et al. (2019). Engineering nanoparticles for targeted delivery of nucleic acid therapeutics in tumor. *Molecular Therapy – Methods and Clinical Development* 12: 1–18. https://doi.org/10.1016/j.omtm.2018.09.002.

21 Donkuru, M., Badea, I., Wettig, S. et al. (2010). Advancing nonviral gene delivery: lipid- and surfactant-based nanoparticle design strategies. *Nanomedicine* 5: 1103–1127. https://doi.org/10.2217/nnm.10.80.

22 Uludag, H., Ubeda, A., and Ansari, A. (2019). At the intersection of biomaterials and gene therapy: progress in non-viral delivery of nucleic acids. *Frontiers in Bioengineering and Biotechnology* 7: 131–152. https://doi.org/10.3389/fbioe.2019.00131.

23 Patil, S., Gao, Y.-G., Lin, X. et al. (2019). The development of functional non-viral vectors for gene delivery. *International Journal of Molecular Sciences* 20: 5491–5514. https://doi.org/10.3390/ijms20215491.

24 Humphry-Baker, R., Moroi, Y., Grätzel, M. et al. (1980). Photoinduced processes in copper(II)-crown ether surfactant micelles. *Journal of the American Chemical Society* 102: 3689–3692. https://doi.org/10.1021/ja00531a004.

25 Dogra, V., Kaur, G., Kaur, A. et al. (2018). In vitro assessment of antimicrobial and genotoxic effect of metallosurfactant based nickel hydroxide nanoparticles against *Escherichia coli* and its genomic DNA. *Colloids and Surfaces B: Biointerfaces* 170: 99–108. https://doi.org/10.1016/j.colsurfb.2018.05.069.

26 Kaur, N., Dhairwal, P., Brar, A. et al. (2019). Amphiphilic metallosurfactants as potential scaffolds for facile fabrication of PdO–NiO nanocomposites for environmentally benign synthesis of xanthene derivatives. *Materials Today Chemistry* 14: 100194. https://doi.org/10.1016/j.mtchem.2019.100194.

27 Dogra, V., Kaur, G., Jindal, S. et al. (2019). Bactericidal effects of metallosurfactants based cobalt oxide/hydroxide nanoparticles against *Staphylococcus aureus*. *Science of the Total Environment* 681: 350–364. https://doi.org/10.1016/j.scitotenv.2019.05.078.

28 Sharma, B., Kaur, G., Chaudhay, G.R. et al. (2020). High antimicrobial photodynamic activity of photosensitizer encapsulated dual-functional metallocatanionic vesicles against drug-resistant bacteria *S. aureus*. *Biomaterials Science* 8: 2905–2920. https://doi.org/10.1039/D0BM00323A.

29 Garg, P., Kaur, G., Sharma, B. et al. (2020). Fluorescein–metal hybrid surfactant conjugates as a smart material for antimicrobial photodynamic therapy against *Staphylococcus aureus*. *ACS Applied Bio Materials* 3: 4674–4683. https://doi.org/10.1021/acsabm.0c00586.

30 Ahmad, T., Askari, H., and Ismail, K. (2020). Synthesis, aggregation and adsorption behavior of benzyldimethylhexadecylammonium based double-chained metallosurfactants. *Journal of Molecular Liquids* 299: 112234. https://doi.org/10.1016/j.molliq.2019.112234.

31 Parera, E., Marín-García, M., Pons, R. et al. (2016). Supramolecular arrangement of molybdenum carbonyl metallosurfactants with CO-releasing properties. *Organometallics* 35: 484–493. https://doi.org/10.1021/acs.organomet.5b00917.

32 Graf, N. and Lippard, S.J. (2012). Redox activation of metal-based prodrugs as a strategy for drug delivery. *Advanced Drug Delivery Reviews* 64: 993–1004. https://doi.org/10.1016/j.addr.2012.01.007.

33 Husale, S., Grange, W., Karle, M. et al. (2008). Interaction of cationic surfactants with DNA: a single-molecule study. *Nucleic Acids Research* 36: 1443–1449. https://doi.org/10.1093/nar/gkm1146.

34 Dias, R.S., Magno, L.M., Valente, A.J.M. et al. (2008). Interaction between DNA and cationic surfactants: effect of DNA conformation and surfactant headgroup.

The Journal of Physical Chemistry B 112: 14446–14452. https://doi.org/10.1021/jp8027935.

35 Martínez-Negro, M., Barrán-Berdón, A.L., Aicart-Ramos, C. et al. (2018). Transfection of plasmid DNA by nanocarriers containing a gemini cationic lipid with an aromatic spacer or its monomeric counterpart. *Colloids and Surfaces B: Biointerfaces* 161: 519–527. https://doi.org/10.1016/j.colsurfb.2017.11.024.

36 Grueso, E., Cerrillos, C., Hidalgo, J. et al. (2012). Compaction and decompaction of DNA induced by the cationic surfactant CTAB. *Langmuir* 28: 10968–10979. https://doi.org/10.1021/la302373m.

37 López-López, M., López-Cornejo, P., Martin-Herrera, V.I. et al. (2018). Importance of hydrophobic interactions in the single-chained cationic surfactant-DNA complexation. *Journal of Colloid and Interface Science* 521: 197–205. https://doi.org/10.1016/j.jcis.2018.03.048.

38 Ostos, J., Lebrón, J.A., López-Cornejo, P. et al. (2020). Self-aggregation in aqueous solution of amphiphilic cationic calix[4]arenes. Potential use as vectors and nanocarriers. *Journal of Molecular Liquids* 304: 112724. https://doi.org/10.1016/j.molliq.2020.112724.

39 Sarrión, B., Bernal-Perez, E., Martín-Herrera, V.I. et al. (2016). Binding of 12-s-12 dimeric surfactants to calf thymus DNA: evaluation of the spacer length influence. *Colloids and Surfaces B: Biointerfaces* 144: 311–318. https://doi.org/10.1016/j.colsurfb.2016.04.028.

40 Domínguez-Gutiérrez, D., Surtchev, M., Eiser, E. et al. (2006). Ru(II)-based metallosurfactant forming inverted aggregates. *Nano Letters* 6: 145–147. https://doi.org/10.1021/nl051944v.

41 Griffiths, P.C., Fallis, I.A., James, C. et al. (2010). Structure–property relationships in metallosurfactants. *Soft Matter* 6: 1981–1989. https://doi.org/10.1039/B920143B.

42 Kumar, R.S. and Arunachalam, S. (2008). Synthesis, micellar properties, DNA binding and antimicrobial studies of some surfactant–cobalt(III) complexes. *Biophysical Chemistry* 136: 136–144. https://doi.org/10.1016/j.bpc.2008.05.007.

43 Riyasdeen, A., Senthilkumar, R., Periasamy, V.S. et al. (2014). Antiproliferative and apoptosis-induction studies of a metallosurfactant in human breast cancer cell MCF-7. *RSC Advances* 4: 49953–49959. https://doi.org/10.1039/C4RA06702A.

44 Elmore, S. (2007). Apoptosis: a review of programmed cell death. *Toxicologic Pathology* 35: 495–516. https://doi.org/10.1080/01926230701320337.

45 Clarke, M.J. (2003). Ruthenium metallopharmaceuticals. *Coordination Chemistry Reviews* 236: 209–233.

46 Lebrón, J.A., Ostos, F.J., Moyá, M.L. et al. (2015). Cooperative interaction between metallosurfactants, derived from the $[Ru(2,2'\text{-bpy})_3]^{2+}$ complex, and DNA. *Colloids and Surfaces B: Biointerfaces* 135: 817–824. https://doi.org/10.1016/j.colsurfb.2015.08.052.

47 Campagna, S., Puntoriero, F., Nastasi, F. et al. (2007). Photochemistry and photophysics of coordination compounds: ruthenium. In: *Photochemistry and Photophysics of Coordination Compounds I. Topics in Current Chemistry*, vol. 280 (eds. V. Balzani and S. Campagna), 117–214. Berlin, Heidelberg: Springer https://doi.org/10.1007/128_2007_133.

48 López-López, M., Montilla, F., Olivares, M. et al. (2013). Interaction between monomers of two surfactants derived from the $[Ru(2,2'\text{-}bpy)_3]^{2+}$ complex and α, β and γ-cyclodextrins: formation of [2]- and [3]-pseudorotaxanes. *Dalton Transactions* 42: 6171–6181. https://doi.org/10.1039/C3DT32197E.

49 Ghosh, S., Banik, D., Roy, A. et al. (2014). Spectroscopic investigation of the binding interactions of a membrane potential molecule in various supramolecular confined environments: contrasting behavior of surfactant molecules in relocation or release of the probe between nanocarriers and DNA surface. *Physical Chemistry Chemical Physics* 16: 25024–25038. https://doi.org/10.1039/C4CP03178D.

50 Gabelica, V., De Pauw, E., and Rosu, F. (1999). Interaction between antitumor drugs and a double-stranded oligonucleotide studied by electrospray ionization mass spectrometry. *Journal of Mass Spectrometry* 34: 1328–1337. https://doi.org/10.1002/(SICI)1096-9888(199912)34:12<1328::AID-JMS889>3.0.CO;2-F.

51 Dalglkish, D.G., Peacocke, A.R., Fey, G. et al. (1971). The circular dichroism in the ultraviolet of aminoacridines and ethidium bromide bound to DNA. *Biopolymers* 10: 1853–1863. https://doi.org/10.1002/bip.360101008.

52 Lopes-Costa, T., Sanchez, F., and López-Cornejo, P. (2009). Cooperative and non-cooperative binding of $^*Ru(bpy)_3{}^{2+}$ to DNA and SB4.5G dendrimers. *The Journal of Physical Chemistry B* 113: 9373–9378. https://doi.org/10.1021/jp902110x.

53 Devi, R.S. and Kumaraguru, N. (2020). Interaction of CT-DNA with ruthenium(II) metallosurfactant complexes: synthesis, CMC determination, antitumour and antimicrobial activities. *Asian Journal of Chemistry* 32: 665–677. https://doi.org/10.14233/ajchem.2020.22485.

54 Lebrón, J.A., Ostos, F.J., López-López, M. et al. (2019). Preparation and characterization of metallomicelles of Ru(II). Cytotoxic activity and use as vector. *Colloids and Surfaces B: Biointerfaces* 175: 116–125. https://doi.org/10.1016/j.colsurfb.2018.11.081.

55 Sharma, N.K., Singh, M., and Bhattarai, A. (2016). Hydrophobic study of increasing alkyl chain length of platinum surfactant complexes: synthesis, characterization, micellization, thermodynamics, thermogravimetrics and surface morphology. *RSC Advances* 6: 90607–90623. https://doi.org/10.1039/C6RA20330B.

56 Sharma, N.K. and Singh, M. (2018). New class of platinum based metallosurfactant: synthesis, micellization, surface, thermal modelling and *in vitro* biological properties. *Journal of Molecular Liquids* 268: 55–65. https://doi.org/10.1016/j.molliq.2018.07.041.

57 Sharma, N.K. and Singha, M. (2018). Multifunctional supramolecular ionic metallosurfactants (SMIMSs) for antimicrobial, anticancer and serum albumins binding. *Journal of Molecular Liquids* 263: 463–471. https://doi.org/10.1016/j.molliq.2018.04.138.

58 Kaur, G., Kumar, S., Kant, R. et al. (2016). One-step synthesis of silver metallosurfactant as an efficient antibacterial and anticancer material. *RSC Advances* 6: 57084–57097. https://doi.org/10.1039/c6ra09677h.

59 Bhar, R., Kaur, G., and Mehta, S.K. (2017). Exploring interactions of copper hybrid surfactants with calf thymus-DNA. *Journal of Molecular Liquids* 241: 715–721. https://doi.org/10.1016/j.molliq.2017.06.054.

60 Zhiltsova, E.P., Pashirova, T.N., Ibatullina, M.R. et al. (2018). A new surfactant-copper(II) complex based on 1,4-diazabicyclo[2.2.2]octane amphiphile. Crystal structure determination, self-assembly and functional activity. *Physical Chemistry Chemical Physics* 20: 12688–12699. https://doi.org/10.1039/c8cp01954a.
61 Nehru, S., Veeralakshmi, S., Kalaiselvam, S. et al. (2020). DNA binding, antibacterial, hemolytic and anticancer studies of some fluorescent emissive surfactant-ruthenium(II) complexes. *Journal of Biomolecular Structure and Dynamics* 1–15 https://doi.org/10.1080/07391102.2020.1747547.
62 Nehru, S., Veeralakshmi, S., Kalaiselvam, S. et al. (2020). Protein binding and antioxidant studies of diimine based emissive surfactant–ruthenium(II) complexes. *Journal of Biomolecular Structure and Dynamics* 1–21 https://doi.org/10.1080/07391102.2020.1733664.
63 Nehru, S., Veeralakshmi, S., and Arunachalam, S. (2017). Synthesis, characterisation and self-assembly behaviour of emissive surfactant–ruthenium(II) complexes. *New Journal of Chemistry* 41: 13830–13837. https://doi.org/10.1039/C7NJ02698F.
64 Panahi, Y., Farshbaf, M., Mohammadhosseini, M. et al. (2017). Recent advances on liposomal nanoparticles: synthesis, characterization and biomedical applications. Artificial cells, nanomedicine, and biotechnology. *Artificial Cells, Nanomedicine, and Biotechnology* 45: 788–799. https://doi.org/10.1080/21691401.2017.1282496.
65 Akbarzadeh, A., Rezaei-Sadabady, R., Davaran, S. et al. (2013). Liposome: classification, preparation, and applications. *Nanoscale Research Letters* 8: 102–111. https://doi.org/10.1186/1556-276X-8-102.
66 Lujan, H., Griffin, W.C., Taube, J.H. et al. (2019). Synthesis and characterization of nanometer-sized liposomes for encapsulation and MicroRNA transfer to breast cancer cells. *International Journal of Nanomedicine* 14: 5159–5173. https://doi.org/10.2147/IJN.S203330.
67 Blanken, D., Foschepoth, D., Serrão, A.C. et al. (2020). Genetically controlled membrane synthesis in liposomes. *Nature Communications* 11: 4317. https://doi.org/10.1038/s41467-020-17863-5.
68 Powers, D. and Nosoudi, N. (2019). Liposomes; from synthesis to targeting macrophages. *Biomedical Research* 30: 288–295. https://doi.org/10.35841/biomedicalresearch.30-19-058.
69 Hainfeld, J.F., Furuya, F.R., and Powell, R.D. (1999). Metallosomes. *Journal of Structural Biology* 127: 152–160. https://doi.org/10.1006/jsbi.1999.4145.
70 Osada, K., Cabral, H., Mochida, Y. et al. (2012). Bioactive polymeric metallosomes self-assembled through block copolymer–metal complexation. *Journal of the American Chemical Society* 134: 13172–13175. https://doi.org/10.1021/ja304615y.
71 Garg, P., Kaur, G., Chaudhary, G.R. et al. (2017). Fabrication of metalosomes (metal containing cationic liposomes) using single chain surfactants as a precursor via formation of inorganic organic hybrids. *Physical Chemistry Chemical Physics* 19: 25764–25773. https://doi.org/10.1039/C7CP02908J.

72 Parera, E., Comelles, F., Barnadas, R. et al. (2011). Formation of vesicles with an organometallic amphiphile bilayer by supramolecular arrangement of metal carbonyl metallosurfactants. *Chemical Communications* 47: 4460–4462. https://doi.org/10.1039/C0CC05493C.

73 Marín-García, M., Benseny-Cases, N., Camacho, M. et al. (2017). Low-toxicity metallosomes for biomedical applications by self-assembly of organometallic metallosurfactants and phospholipids. *Chemical Communications* 53: 8455–8458. https://doi.org/10.1039/C7CC04945E.

74 Marín-García, M., Benseny-Cases, N., Camacho, M. et al. (2018). Metallosomes for biomedical applications by mixing molybdenum carbonyl metallosurfactants and phospholipids. *Dalton Transactions* 47: 14293–14303. https://doi.org/10.1039/c8dt01584h.

75 Cruz-Campa, I., Arzola, A., Santiago, L. et al. (2007). A novel class of metal-directed supramolecular DNA-delivery systems. *Chemical Communications*: 2944–2946. https://doi.org/10.1039/b703201c.

76 Arroyo, I.Z., Gomez, C., Alarcón, H. et al. (2018). Alkyl length effects on the DNA transport properties of Cu(II) and Zn(II) metallovesicles: an in vitro and in vivo study. *Journal of Drug Delivery*: 1–11. https://doi.org/10.1155/2018/2851579.

77 Pal, S., Islam, M.T., Moore, J.T. et al. (2017). Self-assembly of a novel Cu(II) coordination complex forms metallo-vesicles that are able to transfect mammalian cells. *New Journal of Chemistry* 41: 11230–11237. https://doi.org/10.1039/C7NJ02161E.

78 Kaur, B., Chaudhary, G.R., and Kaur, G. (2019). Cholesterol-induced physicochemical changes in dodecylamine-based metallosomes: drug entrapping ability and interactions with biological molecules. *Journal of Materials Chemistry B* 7: 3679–3691. https://doi.org/10.1039/C9TB00607A.

79 Lebrón, J.A., Ostos, F.J., López-López, M. et al. (2020). Metallo-liposomes of ruthenium used as promising vectors of genetic material. *Pharmaceutics* 12: 482. https://doi.org/10.3390/pharmaceutics12050482.

80 Su, R.-C., Liu, Q., Yi, W.-J. et al. (2017). Zn(II)-dipicolylamine-based metallo-lipids as novel non-viral gene vectors. *Journal of Biological Inorganic Chemistry* 22: 867–879. https://doi.org/10.1007/s00775-017-1465-0.

81 Kaur, G., Kaur, B., Garg, P. et al. (2020). A study of synthesis, characterization and metalloplex formation ability of cetylpyridinium chloride based metallosomes. *Journal of Molecular Liquids* 300: 112326. https://doi.org/10.1016/j.molliq.2019.112326.

9

Metallosurfactants as Nanoreactors for Nanoparticle Synthesis

Jaspreet Kaur, Rekha Bhar, Khushwinder Kaur, and Surinder K. Mehta

Panjab University, Department of Chemistry and Centre of Advanced studies in Chemistry, Sector-14, Chandigarh 160014, India

9.1 Introduction

Amphiphiles constitute a very special class of soft materials, known to mankind since many centuries in the capacity of soaps, detergents, wetting agents, and emulsifiers. A single molecule of surfactant possesses two different kinds of chemical nature, i.e. a part of the molecule is hydrophobic, while the other is hydrophilic [1]. The competitive and balancing forces between the two lead to many interesting and useful properties. Perhaps the most fundamental and fascinating property of the surfactants/amphiphiles is the self-assembly and ability to get adsorbed at interfaces such as the air/water. Beyond a particular threshold, they tend to assemble in rather intricate structures, with "micelle" being the simplest of them all. At still higher concentrations, liquid crystalline phases are formed. However, conventional surfactants are predominantly naive in terms of chemical reactivity, and hydrophilic/hydrophobic balance of the system tends to impart them their unique properties [2].

"Metallosurfactants" represent new class of materials in which the polar headgroup of the surfactant molecule contains a transition/inner-transition metal center as an integral structural component, while in a few cases it may also be present in the hydrophobic part [3]. Metallosurfactants can be broadly categorized into two subgroups; one with metal ion as an integral part of the head group of surfactants, while the other one contains metal as counter ion along with amphiphilic part. The incorporation of d- and f-block metal ions into amphiphilic structures brings features like bright colors, redox nature, and paramagnetism to the conventional surfactants [4]. In addition to the above properties, the self-assembly of surfactants aids in formation of well-defined metal complex aggregates of nuclearities equal to the aggregation number of the micelles themselves (Figure 9.1) [5]. These surfactants are rarely

*Jaspreet Kaur and Rekha Bhar contributed

Metallosurfactants: From Fundamentals to Catalytic and Biomedical Applications, First Edition. Edited by Surinder K. Mehta and Ravneet Kaur.

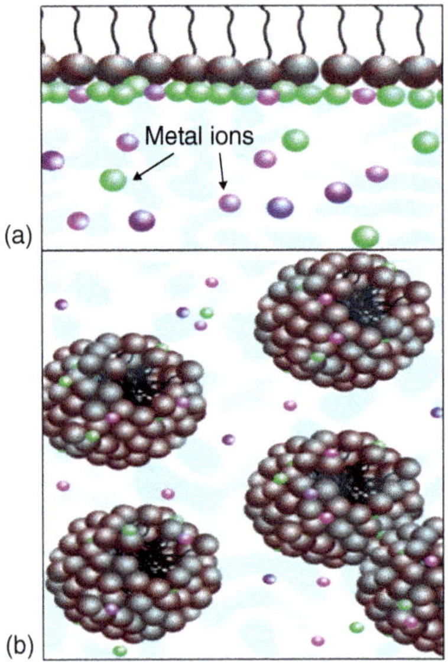

Figure 9.1 Proposed organization of metallo-surfactants, (a) at the air–water interface and (b) in the bulk aqueous solution. Source: Kaur et al. [5].

compared with the traditional surfactants but have established substantial potential applications in multifarious fields such as magnetic resonance imaging, molecular machine for porous materials, thin-film optoelectronics, interfacial photophysics, and homogeneous catalysis. [4].

There is dearth of research in investigating the role of metallosurfactants in the field of nano synthesis. The knowledge is limited on their potential use as nanoscale reactors for the fabrication of nanostructures and their subsequent modification to impart the desired functionality on nanoparticles (NPs) [6]. Nanometal and semiconductor colloids have attracted considerable interest due to their well-defined morphology, size-dependent characteristics, optical and electronic properties, which finally lead to prospective applications in optoelectronics, catalysis, reprography, etc. Clearly, the nanoregime has infiltrated our daily lives, but there will still be a lack of simple synthetic approaches for nanostructure fabrication. Conventional synthetic routes to metallic nanostructures are restricted and are followed by several drawbacks, including cost effectiveness, waste generation, use of harsh chemicals, pH stabilization, higher temperature and pressure requirements, presence of inert atmosphere during the synthesis process, to name a few. Therefore, it is of utmost importance that facile, cheap, and environmentally sustainable methodologies be established for nanostructure synthesis [4].

Metallosurfactants are suitable alternatives because they fulfill the above requirements and deliver a range of advantages over traditional synthetic processes. The distinctive nature of metal surfactant complex to act as both a metal ion source and NPs protector opens up a new arena of synthesis thus eliminating the necessity for any stabilizing or capping agent. Thus, this study focuses on the applicability and modification of metallosurfactants as building blocks for NPs synthesis.

9.2 Metallosurfactants as Reactors for Nanoparticle Synthesis

9.2.1 Colloidal Synthesis

Synthesis procedures involving NPs fabrication must provide the stability, shape, and size control to the generating NPs. The most common problem is the surface passivation or shape mediation of the nanomaterials. However, there are a number of shape controlling/directing agents reported in the literature, *viz*. polymers, surfactants, biomolecules, etc. but the role of ionic surfactants in NP generation is most explored and well understood. In a general synthetic procedure, metal ion (M^{n+}) is firstly reduced to its zero valent state (M^0) by suitable reducing agent. M^0 atoms thus formed tiny nucleating centers, and at this stage surfactants play an active role to control their growth by getting preferentially adsorbed on the surface of growing nanocrystals (Figure 9.2) [7].

Mandeep S. Bakhsi [8] thoroughly investigated the role of ionic surfactants in nanocrystal growth. It was established that selective adsorption of ionic surfactant on [100] plane rather than on [111] plane results in anisotropic growth of nanomaterials. Generally, polar head group of the surfactant gets adsorbed on the charged NPs surface due to electrostatic interactions, which is followed by bilayer formation owing to a hydrophobic stability in aqueous media. However, optimization of surfactant concentration, reaction temperature, pH of the media (ionic strength of the reaction medium), point of surfactant injection, time of the reaction, etc. is still a matter of concern. Another aspect associated with surfactant-assisted NP synthesis is diffusion-controlled motion of surfactant molecules, which is governed by solvents properties like dielectric constant, polarity, and viscosity. Metal-decorated surfactant complexes provide solution to some of the above-raised questions and prove to be one molecule precursor for NPs synthesis. In metallosurfactants, metal ion is closely associated with surfactant moiety; thus, it can provide metal ion as well as shape-directing agent, i.e. surfactant needed for the production of homogeneous NPs.

In this regard, Lisiecki and Pileni [9, 10] presented the synthesis of copper metallic clusters in reverse micellar media of $Cu(AOT)_2$ surfactant (AOT = dioctyl sulfosuccinate). Reverse micelle solution was prepared by dissolving only the

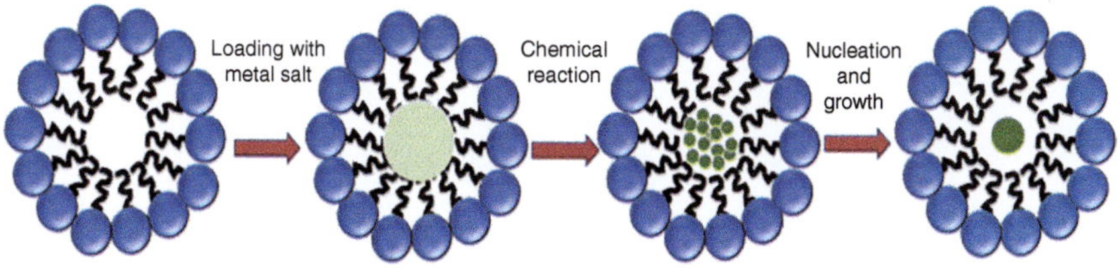

Figure 9.2 Schematic representation of the steps involved in the formation of a micellar structure and its subsequent use as nanoreactor for preparation of a metal NP. Source: Modified from Lombardo et al. [7].

sodium bis(2-ethylhexyl)sulfosuccinate (AOT) surfactant as well as mixture of AOT and $Cu(AOT)_2$ complex in isooctane. Addition of an aqueous solution of reducing agent to the medium produced metallic copper particles surrounded by a copper oxide layer and pure copper oxide particles. Change in the shape of surfactant aggregates modified the shape of metallic clusters as well. It was established that by increasing the water content in the core of aggregates, size of the NPs also increased as indicated by the shift in surface plasmon resonance peak in ultra violet (UV)–vis spectrum. Even the change in solvent can lead to a change in the shape and size of the particles. Interactions between aggregate droplets formed in isooctane and in cyclohexane were different, and produced smaller-sized copper NPs.

In another report, copper NPs were generated from copper dodecyl sulfate with the help of sodium borohydride [10]. Size of NPs originating from surfactant-assisted method strongly depends upon CMC. Below CMC, copper oxide NPs (30 nm) and at/above CMC, copper metallic NPs (280 nm) were obtained (Figure 9.3).

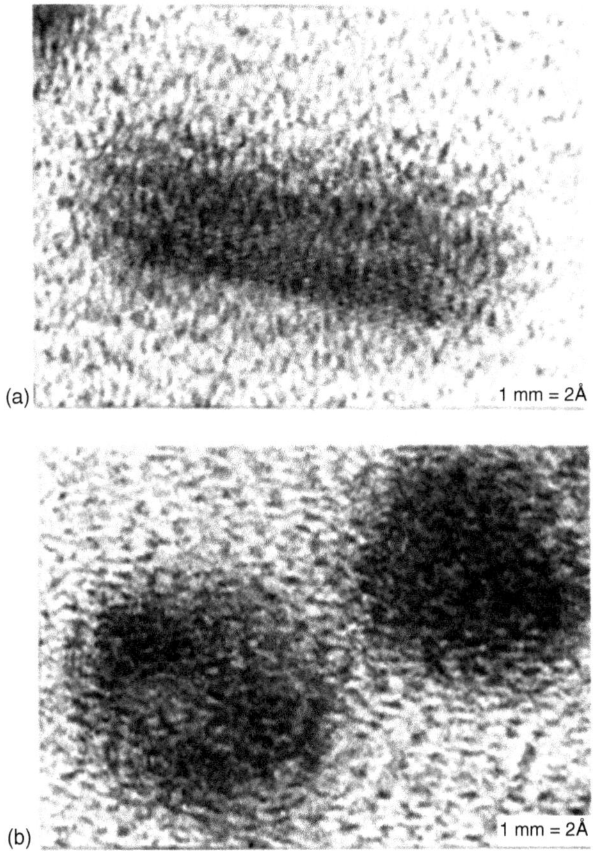

Figure 9.3 High resolution transmission electron microscope (HRTEM) images of various shapes of copper metallic particles; $[Cu(AOT)_2] = 4 \times 10^{-2}$ M, $(N_2H_4) = 8 \times 10^{-2}$ M. (a) Cylindrical particles; (b) Spherical particles (Reproduced at 64% of original size). Source: Lisiecki and Pileni [10].

Then, in 1997, AOT complexed Cd, Mn, and Zn were employed for the fabrication of $Cd_{1-y}Mn_yS$ magnetic semiconductor clusters and $Cd_yZn_{1-y}S$ nanosized particles by Levy et al. [11] and Cizeron and Pileni [12]. $Cd_yZn_{1-y}S$ NPs were obtained immediately after mixing of three micellar solutions, *viz.* AOT-containing sulfide ions (Na_2S), $Zn(AOT)_2$, and $Cd(AOT)_2$. Likewise, $Cd_{1-y}Mn_yS$ nanoclusters were also prepared by coprecipitating the micellar solution of the desired components. Instant formation of NPs on mixing of micellar media attributed to the dynamic properties of reversely aggregated surfactants. The authors concluded that reverse micelles proved to be efficient microreactors for the fabrication of size and shape-controlled NPs.

Further, Tanori et al. [13] produced pure copper metallic as well as Fe–Cu alloy nanosized particles. For the synthesis, 0.1 M solution of $Fe(AOT)_2$ (30%) and $Cu(AOT)_2$ (70%) was prepared in isooctane and mixed with 0.2 M $NaBH_4$. The solution was kept under constant stirring until black precipitates appeared, followed by washing with isooctane and acetone. Particle size/shape was monitored based on the water content present inside the reverse micellar phase. At low water concentration, metal ions were not fully hydrated and thus available for crystal growth, and vice versa when water amount increases. Talking about the shape, as the content of water increases, shape transition takes place from spherical to cylindrical to rod like; however, at very high concentration again spherical NPs were obtained.

AOT surfactant complex with Ag and Zn also yielded Ag and Zn NPs. In connection with this, Bagwe and Khilar [14] studied the inter-micellar exchange rate of the synthesis of Ag-AOT-assisted synthesis of silver chloride NPs. Exchange rate considerably affects the size of the NPs, with NPs being smaller sized at higher exchange rate and bigger at lower exchange rate.

Later on, silver nanodisks were obtained by mixing equal amount of Ag(AOT) and Na(AOT) by Maillard et al. [15], while Quinlan et al. [16] described the synthesis of ZnSe NPs having radii lesser than that of Bohr exciton from reverse micellar solution of $Zn(AOT)_2$ complex and $NaSe_2$. Size of the synthesized NPs was in the range of 3.3–3.5 nm, supporting the fact that reverse micellar approach leads to a narrow size distribution.

Advancing the phenomena, Wang et al. [17] combined the ionic liquids with metallosurfactants to produce nanoceria. Metallosurfactant ionogels, i.e. gelation behavior of hexadecylpyridinium ion $[Cp]^+$ with different counter ions, i.e. Br^-, $[FeCl_3Br]^-$, $[CeCl_3Br]^-$ was checked in various imidazolium ionic liquids. In the presence of ethylammonium nitrate, $[CeCl_3Br]^-$ $[Cp]^+$ yielded ceria (CeO_2) NPs of diverse sizes (Figure 9.4). CeO_2 NPs possessed oxidase and peroxidase activity and proved to be an efficient catalyst for decomposition of H_2O_2. Bubbling of O_2 got enhanced over the passage of time with the decomposition of H_2O_2, and the phenomenon was captured by optical microscopy.

Presence of metal ion along with closely associated surfactant moiety thus produced uniform NPs in terms of size and shape number of other organometallic compounds having amphiphilic character. In this direction, Torigoe and Esumi [18] fabricated gold NPs from water-insoluble ionic mixture of $AuCl_4^-$ with some cationic surfactants, *viz.* dodecyl and hexadecylpyridinium chloride (DPCl, CPCl),

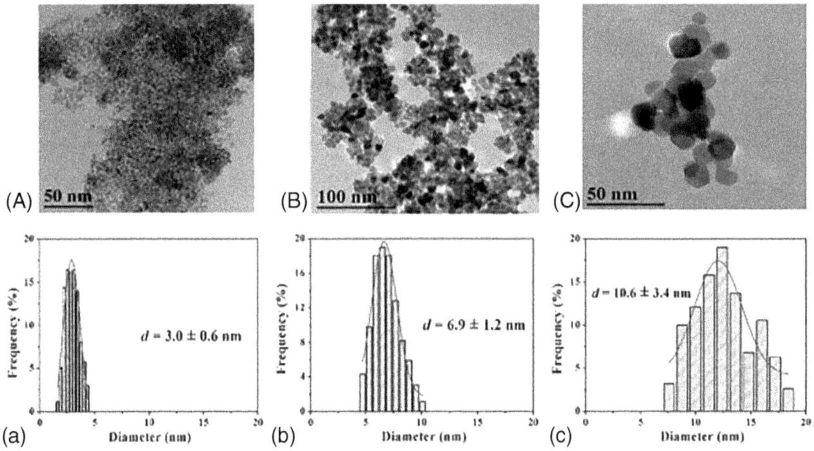

Figure 9.4 TEM images and corresponding histograms of CeO_2 NPs obtained after calcination of HBCe/EAN with (A and a) 12 wt% at 400 °C (B and b) 20 wt% at 500 °C and (C and c) 20 wt% 600 °C. Source: Wang et al. [17].

dodecyl and hexadecyltrimethylammonium chloride surfactants (dodecyltrimethylammonium chloride [DTAC], cetyltrimethylammonium chloride [CTAC]). Net interaction between auric salt, i.e. $HAuCl_4{\cdot}4H_2O$ with the corresponding surfactant resulted in the formation of metallosurfactant of the type $[Surf]^+$ $[AuCl_4]^-$. To prepare the gold colloidal sol, the metallosurfactant solution of particular concentration was irradiated with UV radiation. Pyridinium group was strongly adsorbed onto the surface of NPs; thus, DPCl and CPCl were found to be more effective in controlling shape and size of gold nanocolloids as compared to other two mentioned surfactants. Further, long alkyl chain was found to be more effective in providing protection to the gold sol as indicated by the long-lived transition state of $[CP^+][AuCl_4]^-$ during photolysis, which can only be explained by assuming that CPCl forms a thick adsorption layer around the growing particles; hence, homogeneously dispersed gold NPs were obtained (Figure 9.5).

Similarly, Kameo et al. [19] reported the production of gold fibers by radiating a solution of $AuCl_4$ and alkyltrimethylammoniumchlorides (C_nTAC) with UV light where $n = 10, 12, 14, 16$. The authors concluded that as the length of alkyl chain of the surfactant increases, length of gold nanofibers also increases.

Contrary to this, size of cubooctahedral-shaped Pd NPs decreased with increase in chain length of the surfactant as reported by Veisz and Király [20]. The Pd colloidal particles were synthesized from metallomicelle of the surfactants-type $[C_nTA]^{2-}$ $[PdBr_4]^{2+}$, where the value of n varies from 8 to 16. Reduction of the metallosurfactant solution was carried out by hydrazine. Long-term stability of Pd hydrosol was ensured by the bilayer protection of surfactants around the NPs. Size of the NPs not only depends on the length of the alkyl chain but also varies as a function of surfactant and Pd precursor concentration. A lesser amount of both precursor and surfactant concentration results in smaller-sized particles.

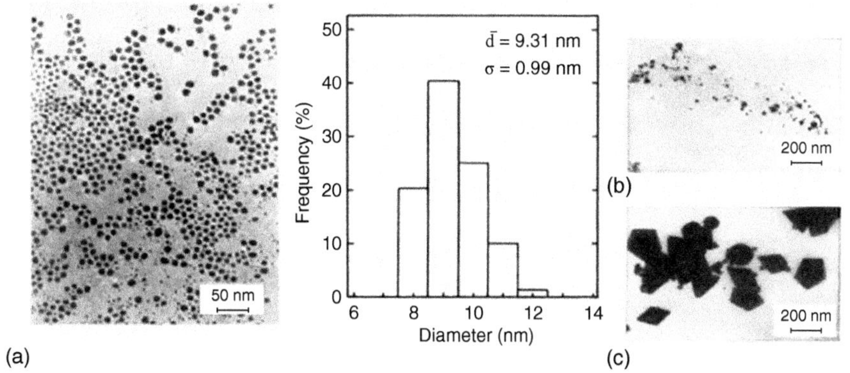

Figure 9.5 TEM of gold particles: (a) CPC1 (20 mmol dm^{-3}), irradiation for 150 minutes; (b) CTAC (20 mmol dm^{-3}), irradiation for 600 minutes; (c) DPC1 (20 mmol dm^{-3}), irradiation for 150 minutes; [$HAuCL_4$] = 0.5 mmol dm^{-3}.

Zero valent Pd NPs were also prepared from palladium complexed with *n*-octylethylenediamine surfactant, i.e. [Pd(oct-en)$_2$]Cl_2. Synthesis was achieved in W/O micro-emulsion of [Pd(oct-en)$_2$]Cl_2/water/$CHCl_3$. Again the sizes of the NPs were found to be sensitive toward the solvent content inside the reverse micellar structure and concentration of $NaBH_4$, the reductant. In 2001, Manna et al. [21] came up with a scheme (Figure 9.6.) describing utilization of *N*-hexadecylethlenediamine silver nitrate complex as precursor for generating silver NPs. Synthesis was carried out in two-phase redox reaction, i.e. *n*-heptane/water mixture. The authors highlighted the importance of metal–hybrid surfactant as an *in situ* phase transfer catalyst required during the two-phase redox reaction. Further, silver–surfactant complex was found to be very useful in developing 2D monolayers and Langmuir–Blodgett films of silver particles having different valences.

Then, in 2005, Sathyamurthy et al. [22] synthesized room-temperature photoluminescent ceria NPs in water in oil microemulsion from cetyltrimethylammonium bromide (CTAB)–Ce complex. For oil phase, *n*-octane was selected, while butanol and aqueous solution acted as co-surfactant in dispersed phase, respectively (Figure 9.7). CTAB was reported as one of the highly coordinating cationic surfactants and formation of reverse micelle provided protection to the growing NPs. A peculiar feature about the photoluminescence of the ceria NPs was that dilution led to the red shifting the emission peak (Figure 9.7c).

In another interesting work, gold nanocrystals were prepared from TPAuFO organometallic compound [23]. Here, TP stands for triphenylphosphine, while FO was perfluorooctanoate. Published work was one of its kind and contained the information about the use of single phase supercritical fluid, i.e. $scCO_2$. Schematic representation of the apparatus employed by the authors for synthesis of gold NPs in $scCO_2$ along with the chemical structure of amphiphilic organoauric(III) complex as shown in the Figure 9.8. Reduction with dimethyl amine borane in above-mentioned method yielded the gold NPs as small as 1 nm (Figure 9.8).

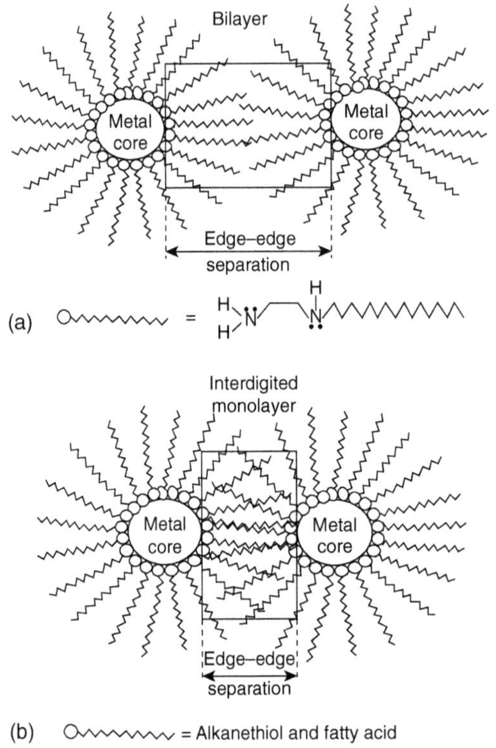

Figure 9.6 Schematic illustration of NPs structure: (a) the normal bilayer between *N*-hexadecyl ethylenediamine-protected silver NPs and (b) the conventional interdigitated monolayer between alkanethiol or fatty acid protected NPs. Source: Manna et al. [21].

Likewise bis-cyclooctatetraene titanium precursor, i.e. $Ti(COT)_2$ organometallic complex was exploited for the production of titania NPs [24]. $Ti(COT)_2$ was refluxed with different surfactants, *viz.* tributylphosphine (TBP), tributylphosphine oxide (TBPO), trioctylphosphine oxide (TOPO) in dichlorobenzene. Color of the solution changed from dark red to pale yellow on addition of dimethyl sulfoxide indicating the formation of titania NPs. A noticeable fact was that color change appeared almost immediately at 120 °C but took two days when the reaction mixture was refluxed at room temperature. Transmission electron microscope (TEM) images confirmed that TiO_2 nanocrystals were internally crystalline, highly monodispersed and 15 nm in size (Figure 9.9a). Nanocrystals were very soluble in the reaction medium and difficult to separate, hence, X-ray diffraction (XRD) pattern of the material could not be recorded, nevertheless selected area electron diffraction (SAED) pattern confirmed the internal phase as well as crystallinity of the TiO_2 nanocrystals (Figure 9.9b). A clear examination of SAED pattern revealed that TiO_2 NPs crystallized dominantly into brookite phase.

Successful synthesis of differently sized 3, 5, 7, 9 nm magnetite NPs was achieved by dissolving iron precursors (Fe^{2+} and Fe^{3+}) in a turbid water in oil emulsion of sodium dodecyl benzenesulfonate in xylene [25]. The reaction was heated up to 90 °C and then addition of hydrazine led to the formation of black precipitates, the iron NPs. The authors reported a very economical and environment-friendly approach for the synthesis of highly crystalline iron NPs. Beauty of the reaction

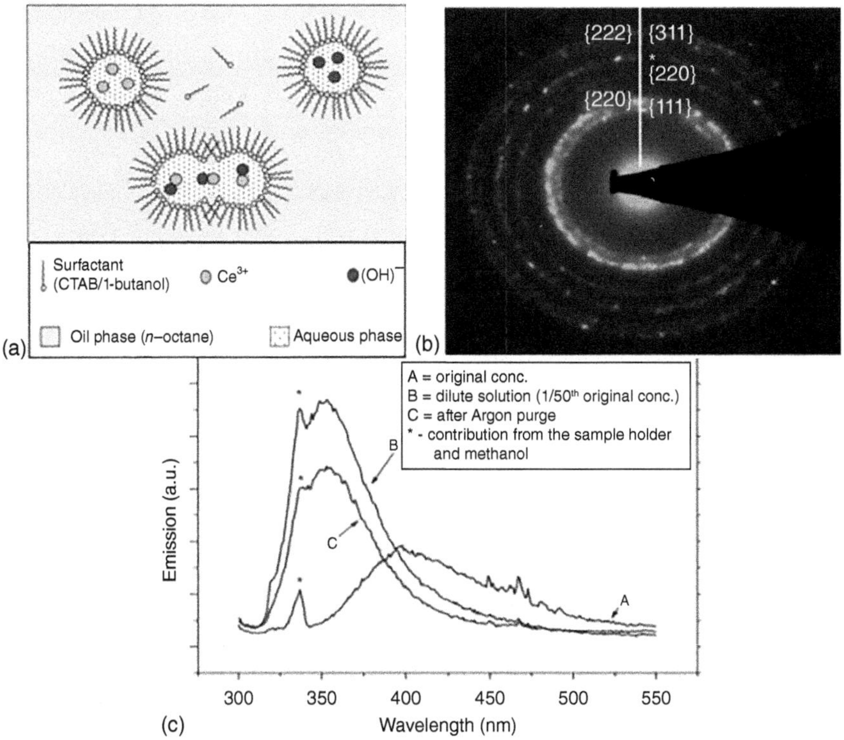

Figure 9.7 (a) Schematic illustration of NPs synthesis using water-in-oil inverse microemulsions, (b) electron diffraction pattern of CeO_2 NPs, and (c) photoluminescence spectra of CeO_2 NPs dispersions in methanol (λ_{ex} = 290 nm). Source: Sathyamurthy et al. [22].

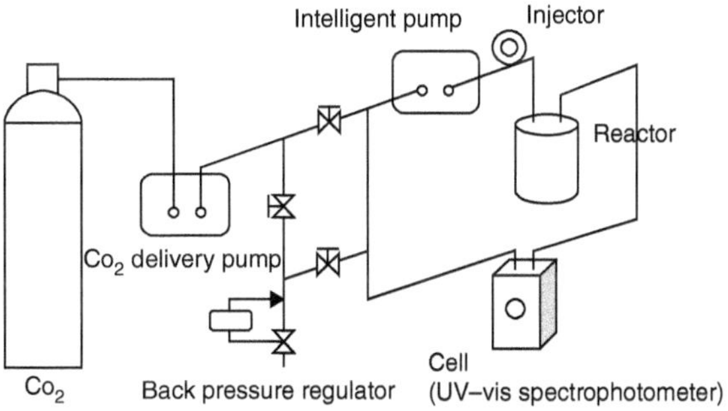

Figure 9.8 Representation of the apparatus employed by the authors for synthesis of gold NPs in $scCO_2$ along with the chemical structure of amphiphilic organoauric(III) complex. Source: Esumi et al. [23].

was that within a single reaction, multi-gram production of NPs was achieved, which could be further scaled up for higher yield. The same methodology suited the fabrication of mixed metal ferrites like cobalt, nickel, zinc ferrite.

A significant contribution into the field of fabrication of highly crystalline and monodispersed NPs was made by Park et al. [26]. Metal oleate complex was synthesized by reacting the $FeCl_3{\cdot}H_2O$ and sodium oleate in a mixture of ethanol, water, and hexane. Then, to obtain iron nanocrystals, iron–oleate organometallic complex

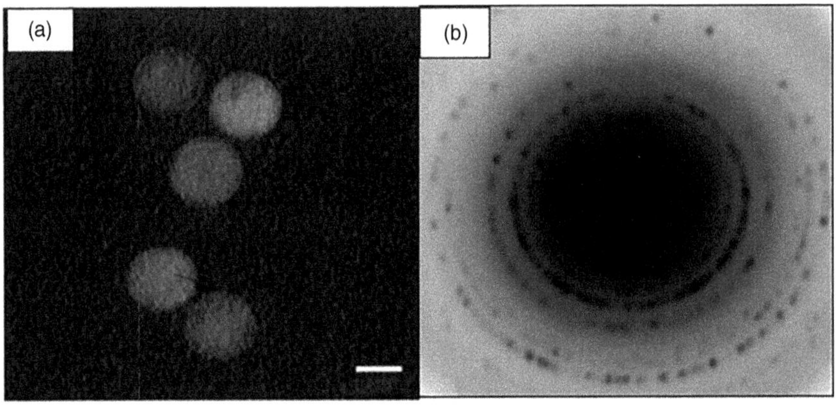

Figure 9.9 (a) TEM image of TiO_2 particles obtained by injecting dimethyl sulfoxide (DMSO) to $Ti(COT)_2$ at 120 °C in the presence of TOPO. (b) Electron diffraction pattern of the TiO_2 NPs. These rings index well to brookite TiO_2. Source: Tang et al. [24].

along with the oleic acid was heated at 320 °C. At this temperature, abrupt reaction took place resulting in appearance of black-colored precipitates in contrast to the initial transparent solution. Similar treatment led to the synthesis of MnO and CoO nanocrystals. It was observed that γ-Fe_2O_3 was the dominating phase in iron oxide NPs, and content of Fe_3O_4 gradually increased as the size of NPs increases. All the presented NPs were highly segregated, crystalline, and uniform in size and shape, affirming the authenticity of the metal–oleate complex as an effective precursor for nanomaterial production.

9.2.2 Thermal Decomposition

Thermal decomposition route is one of the most attractive approaches for NP synthesis. The low decomposition temperature of the metal complexes can be fairly used in the fabrication of metal NPs via thermal treatment. This facile process can be applied to metal surfactant complexes to yield monodisperse and uniform NPs.

Hyeon et al. [27, 28] extensively worked on the synthesis of nanomaterials from surfactant–metal complex. Iron oxide, iron-cobalt mixed ferrite NPs were prepared by thermal pyrolysis of metal oleate complex (Figure 9.10.). Synthetic procedure reported by the scientist has several advantages over the conventional methods for NPs synthesis. Firstly, the method provided a size selection free methodology as monodispersed NPs were obtained, so no special efforts were needed for NPs synthesis of desired size and shape. Secondly, NPs did not show any aggregation when dissolved in different organic media, which further justified the effectiveness of metallosurfactants as two in one precursor for NPs. Lastly, reproducibility of the synthetic procedure was excellent and slight modification of reaction parameters could provide NPs of preferred shape and size.

Metal–oleylamine complex was further exploited in synthesis of semiconducting lead, cadmium, manganese, and zinc sulfide nanomaterial. Using the thermal decomposition approach, Joo et al. [29] synthesized semiconducting metal sulfide

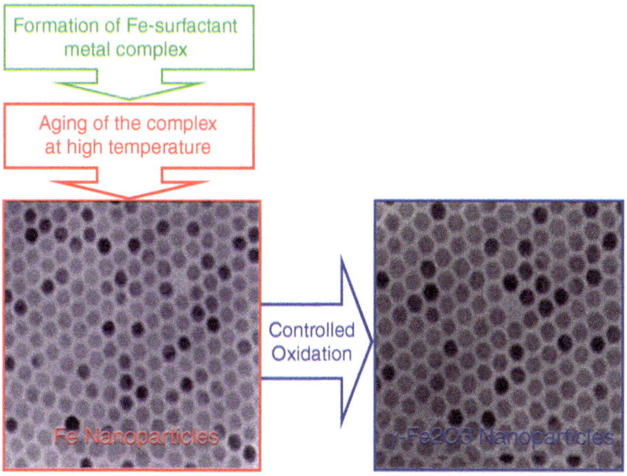

Figure 9.10 Synthetic procedure for obtaining monodisperse g-Fe_2O_3 NPs.Source: Hyeon et al. [27].

nanocrystals of PbS, ZnS, CdS, and MnS with various dimensions and morphology from the thermal reaction of metal chloride and sulfur in oleylamine (Figure 9.11). The shape of the nanocrystals was controlled by changing the metal complex to sulfur ratio to form highly crystalline rods, bipods, tripods, and bullets. The synthetic thermal decomposition procedures developed in this study offer several very important advantageous features for the synthesis of sulfide nanocrystals. Firstly, it uses cheap and commercially available metal chlorides and sulfur with easy scale-up. Secondly, this generalized process can be applied to a wide variety of reactions to generate sulfide nanocrystals with different uniform shapes and sizes.

9.2.3 Biphasic Redox Reaction

Kaur and Mehta [30] synthesized Fe, Co, Cu, Ni, Zn metal nanostructures in heptane/water biphasic media from metallosurfactants and successfully characterized by UV–vis spectra, TEM analysis, Fourier transform infrared spectroscopy (FTIR) spectroscopy, which was further supported by density functional theory (DFT) studies. First indication about the formation of NPs was given by an instant change in color on addition of aqueous solution of $NaBH_4$ in heptane-containing solution of $[M(CH_3COO)_4]$ $[C_{12}H_{25}NH_3]_2$ (Figure 9.12). Their study revealed that capping mechanism or preferences are different for different metallic nanostructures, i.e. Fe, Zn were protected by carboxylate moiety, while Cu, Co, Ni had alkylamine capping. As a result of which Fe, Zn NPs remained dispersed in aqueous layer, but Cu, Co, Ni were dispersed in heptane. Further, shape of the metallomicelle was totally reflected in the shape of NPs. Elliptical metalloaggregate of Zn-surfactant complex generated capsule-like nanostructure, whereas spherical metallomicelle (Fe, Cu, Co, Ni) yielded spherical NPs. Catalytically, the efficiency of NPs in conversion of 4-nitrophenol to 4-aminophenol followed the Co > Ni > Cu > Zn > Fe order. Kaur et al. not only contributed to this field

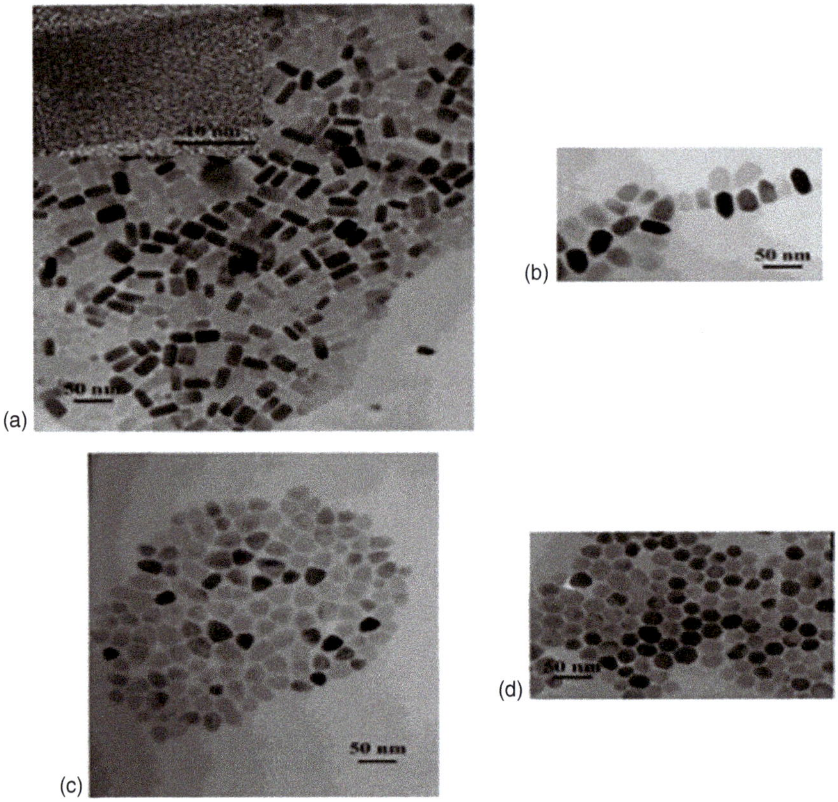

Figure 9.11 TEM images of MnS nanocrystals. (a) MnS nanorods. Inset is an HRTEM image of a single MnS nanorod. (b) Long bullet-shaped MnS nanocrystals. (c) Short bullet-shaped MnS nanocrystals. (d) Hexagon-shaped MnS nanocrystals. Source: Joo et al. [29].

Figure 9.12 Schematic representation of metal NPs fabrication using various metallosurfactants. Source: Kaur and Mehta [30].

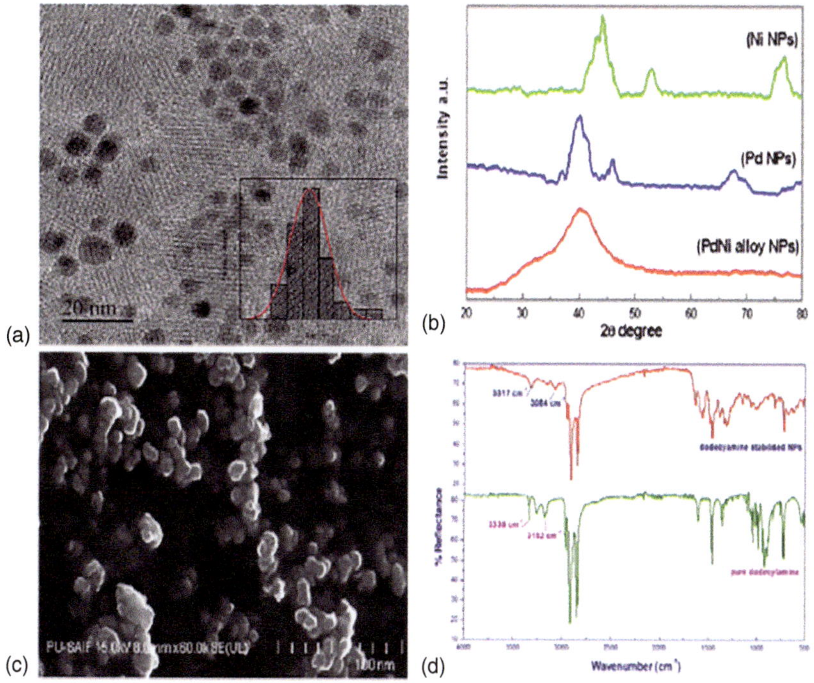

Figure 9.13 (a) TEM image (b) XRD pattern (c) Field emission scanning electron microscope (FESEM) image, and (d) FTIR spectrum of Pd–Ni NPs. Source: Kaur et al. [31].

experimentally but also critically reviewed the role of metallosurfactant in fabrication of various types of nanomaterials.

In 2018, Kaur et al. [31] used metallosurfactants to synthesize Pd–Ni alloy NPs with uniform shape and dimensions. They employed two-phase redox methodology forming mixed reverse metallomicelles fabricated from Pd and Ni metallosurfactants. The metal heads assembled in the cores were subsequently reduced by $NaBH_4$ to form zerovalent metal atoms. The spherical morphology of the bimetallic NPs was confirmed by TEM images with an average size of 5 nm as shown in Figure 9.13.

In another report [32], the authors investigated ligand insertion method to synthesize novel double-chained metallocomplex bisdodecylaminepalladium(II) chloride $[Pd(C_{12}H_{25}NH_2)_2]Cl_2$. This complex was further utilized to form Pd NPs by a two-phase (water/dichloromethane [DCM]) redox reaction (Figure 9.14). $NaBH_4$ was employed as a reducing agent for metal ions to form zerovalent metallic state. The surfactant layer on the NPs prevented the agglomeration of Pd(0) and led to the formation of monodispersed spherical NPs of 3–5 nm as shown in TEM analysis.

Hence, it can be inferred that metallosurfactants play a vital role as reactors for nanoparticle synthesis. By varying the amount of precursors or the reaction conditions, diverse types of metal nanoparticles can be obtained as shown in Table 9.1.

Table 9.1 Some examples of metallosurfactants as nanoreactors.

S. No.	Metal surfactant complex	Synthetic methodology	Morphology	Size	References
1.	[(Oleylamine)AuCl] complex	Slow reduction	Au nanowires	1.8 nm	[33]
2.	Pd-trioctylphosphine complex	Thermal decomposition	Pd nanoparticles	3.5–7 nm	[34]
3.	$K_2[PdCl_4]/C_nTABr$ complex	Hydrazine reduction	Pd cubo-octahedral nanoparticles	1.6–6.8 nm	[20]
4.	Zinc-oleylamine trioctylphosphine oxide (TOPO) complex	Size-selective precipitation	ZnS nanocrystals	11 nm	[29]
5.	Bisdodecylamine Pd(II) Cl and bis-dodecylamine Ni(II) Cl	$NaBH_4$ reduction of reverse micelles	Pd–Ni nanoparticles	5 nm	[31]
6.	CTAB, $HAuCl_4$, and PbS	Ascorbic acid reduction	PbS–Au (Nanostar–nanoparticle) heterodimer	Nanostar 80 nm and nanoparticle 50 nm	[35]
7.	Cobalt 2-ethylhexanoate ($Co(EH)_2$) and Aerosol-OT	Irradiation of reverse micellar system	Co_3O_4 and Bi nanoparticles	Co_3O_4 5 nm Bi 7–10 nm	[6]
8.	$[CeCl_3Br]$ (Hexadecylpyridinium Br, HBCe; $C_{21}H_{38}N^+$ $[CeCl_3Br]$)	Calcination	CeO_2 nanoparticles	3–11 nm	[17]
9.	Bis(1,5-cyclooctadiene) dirhodium(I) dichloride (Rh-COCl)	Microemulsion reduction	Rh nanoparticles	2–8 nm	[36]
10.	Bis-(cyclooctatetraene) titanium	Precipitation	TiO_2 nanoparticles	5.3 nm	[24]
11.	Fe(II) and Fe(III) oleic acid–oleylamine	Thermal decomposition	Fe_3O_4 nano-crystallites	5 nm	[37]
12.	Bis(dodecyl-ethylenediamine) $AgNO_3$	$NaBH_4$ reduction of microemulsion	Ag nanoparticles	—	[38]
13.	Copper(II) bis(2-ethyl hexyl) sulfosuccinate $Cu(AOT)_2$	Hydrazine reduction of microemulsion media	Cu spheres and cylinders	Sphere-5-10 nm and cylinder <7 nm	[39]
13.	Barium bis(2-ethylhexyl) sulfosuccinate $Ba(AOT)_2$ and sodium chromate $(Na_2CrO_4)^-$	Microemulsion media	$BaCrO_4$ linear chains	50–500 nm	[40]

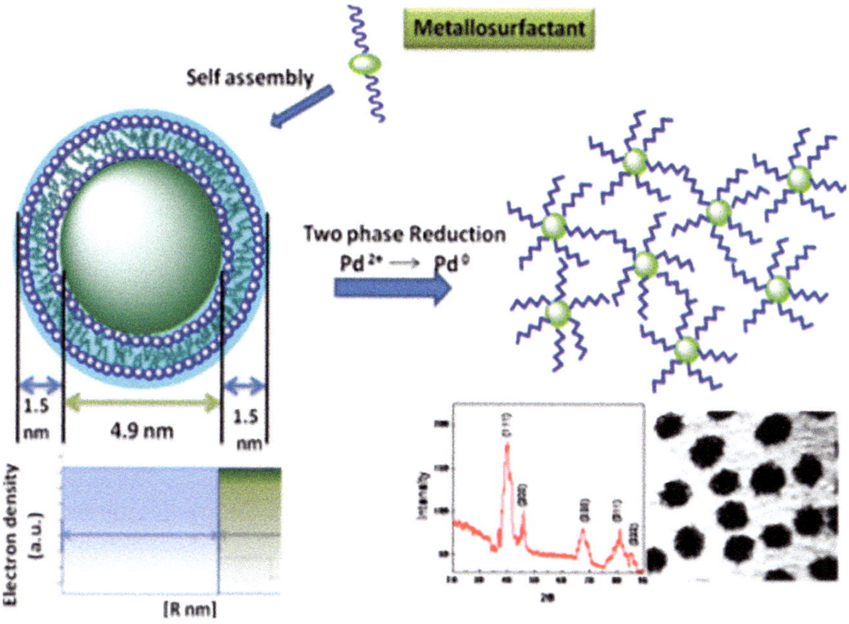

Figure 9.14 Schematic representation of Pd(0) NPs via two-phase redox reaction. Source: Chaudhary et al. [32].

9.3 Future Perspective

By utilizing metal–surfactant complex as a precursor for NPs fabrication, a number of concerns were resolved like homogeneity, reproducibility, flexibility, and environmental issues. However, it is still difficult to produce pure metallic NPs of certain metals like copper as copper is very much prone to the air/solvent oxidation. Thus, a methodology might be undertaken to produce metallic copper NPs, but it may result in copper oxide NPs formation eventually. Kaur et al. [30, 41] provided a remedy to this problem, using metallosurfactants. The method proposed by them delivers extreme stability to the colloidal sol even in the aqueous environment. To achieve the unmatchable stability copper surfactant, $[Cu(CH_3COO)_4][C_{12}H_{25}NH_3]_2$ was mixed with poly acrylic acid (PAA). Then, the solution was subjected to vigorous

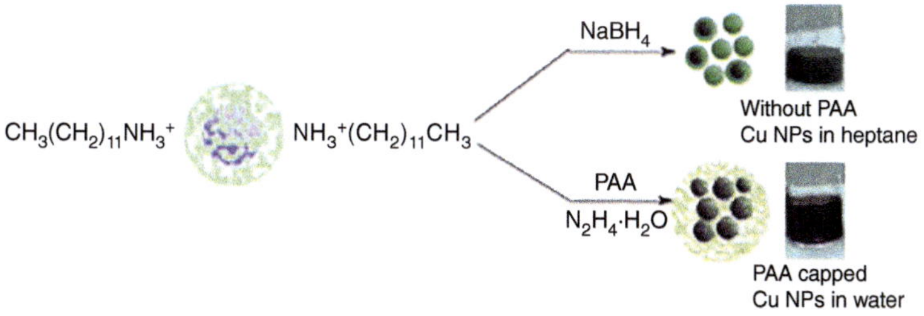

Figure 9.15 Pictorial representation of CuNP synthesis. Source: Kaur et al. [41].

stirring, and reducing agent $NaBH_4$ was added to the reaction mixture (Figure 9.15). Appearance of reddish color indicated the formation of copper colloids in zero valent state. Copper NPs thus formed were found to be stable even after a month. A size control on the size of copper metallic NPs could be attained by just varying the concentration of PAA. Homogeneously dispersed NPs were employed as catalyst for the redox reaction of 4-nitrophenol to 4-aminophenol conversion in aqueous medium. Hence, metallosurfactants have paved a way for new mechanistic studies for nanostructure formation with homogeneity and reproducibility, which will help in overcoming the drawbacks of conventional methods.

9.4 Conclusion

It can be effectively inferred that logical design and modification provide the possibility of regulating the unique characteristics of metallosurfactants. The structural architecture of the metallosurfactant can be precisely tuned to cater the requirements and specifications of the applications. Solid-state metallosurfactants and even self-assembled frameworks obtained in solution form can be used effectively in synthesizing NPs because of their surface dynamism. As a result, the self-assembly mechanism of these metallosurfactants offers a variety of morphologies resulting in a consistent and monitored arrangement of functionalized or derived nanostructures. This approach offers many advantages over traditional NPs fabrication methods such as (i) facile and cost-effective methodology without any extreme temperature/pressure conditions, (ii) controlled structural morphology, and (iii) absence of any templating or capping agent to name a few. Due to its ameliorating properties, this subject area has rightly drawn scientific attention from both organometallic and nanoscience arenas. This can pave a way for new breakthroughs in the field of surfactant and nanochemistry by serving as a connection between two disciplines.

Acknowledgments

The authors acknowledge Panjab University, Chandigarh, for its support. SKM and JK are thankful to Council of Scientific and Industrial Research (CSIR) for fellowship. KK is thankful to Department of Science and Technology (DST) Inspire and RB acknowledge University Grants Commission-Basic Science Research (UGC-BSR) for fellowship.

References

1 Griffiths, P.C., Fallis, I.A., Chuenpratoom, T., and Watanesk, R. (2006). Metallosurfactants: interfaces and micelles. *Advances in Colloid and Interface Science* 122 (1): 107–117.

2 Scarso, A. (2016). Micellar Nanoreactors. In: *Encyclopedia of Inorganic and Bioinorganic Chemistry* (ed. R.A. Scott), 1–16. Chichester: Wiley.

3 Owen, T. and Butler, A. (2011). Metallosurfactants of bioinorganic interest: coordination-induced self assembly. *Coordination Chemistry Reviews* 255 (7): 678–687.

4 Kaur, R. and Mehta, S.K. (2014). Self aggregating metal surfactant complexes: precursors for nanostructures. *Coordination Chemistry Reviews* 262: 37–54.

5 Kaur, R., Gupta, S., Mehta, S.K. et al. (2014). Probing the self-aggregation behavior and counter ion distribution of a copper surfactant complex. *New Journal of Chemistry* 38 (8): 3925–3932.

6 De Oliveira, R.J., Brown, P., Correia, G.B. et al. (2011). Photoreactive surfactants: a facile and clean route to oxide and metal nanoparticles in reverse micelles. *Langmuir* 27 (15): 9277–9284.

7 Lombardo, D., Kiselev, M.A., Magazù, S., and Calandra, P. (2015). Amphiphiles self-assembly: basic concepts and future perspectives of supramolecular approaches. *Advances in Condensed Matter Physics* 2015: 1–22.

8 Bakshi, M.S. (2016). How surfactants control crystal growth of nanomaterials. *Crystal Growth & Design* 16 (2): 1104–1133.

9 Lisiecki, I. and Pileni, M.P. (1995). Copper metallic particles synthesized "in situ" in reverse micelles: influence of various parameters on the size of the particles. *Journal of Physical Chemistry* 99 (14): 5077–5082.

10 Lisiecki, I. and Pileni, M.P. (1993). Synthesis of copper metallic clusters using reverse micelles as microreactors. *Journal of the American Chemical Society* 115 (10): 3887–3896.

11 Levy, L., Feltin, N., Ingert, D., and Pileni, M.P. (1997). Three dimensionally diluted magnetic semiconductor clusters $Cd_{1-y}Mn_yS$ with a range of sizes and compositions: dependence of spectroscopic properties on the synthesis mode. *Journal of Physical Chemistry B* 101 (45): 9153–9160.

12 Cizeron, J. and Pileni, M.P. (1997). Solid solution of $Cd_yZn_{1-y}S$ nanosized particles: photophysical properties. *Journal of Physical Chemistry B* 101 (44): 8887–8891.

13 Tanori, J., Duxin, N., Petit, C. et al. (1995). Synthesis of nanosize metallic and alloyed particles in ordered phases. *Colloid & Polymer Science* 273 (9): 886–892.

14 Bagwe, R.P. and Khilar, K.C. (2000). Effects of intermicellar exchange rate on the formation of silver nanoparticles in reverse microemulsions of AOT. *Langmuir* 16 (3): 905–910.

15 Maillard, M., Giorgio, S., and Pileni, M.P. (2002). Silver nanodisks. *Advanced Materials* 14 (15): 3.

16 Quinlan, F.T., Kuther, J., Tremel, W. et al. (2000). Reverse micelle synthesis and characterization of ZnSe nanoparticles. *Langmuir* 16 (8): 4049–4051.

17 Wang, X., Yang, Q., Cao, Y. et al. (2016). Metallosurfactant ionogels in imidazolium and protic ionic liquids as precursors to synthesize nanoceria as catalase mimetics for the catalytic decomposition of H_2O_2. *Chemistry A European Journal* 22 (49): 17857–17865.

18 Torigoe, K. and Esumi, K. (1992). Preparation of colloidal gold by photoreduction of tetracyanoaurate(I)-cationic surfactant complexes. *Langmuir* 8 (1): 59–63.

19 Kameo, A., Suzuki, A., Torigoe, K., and Esumi, K. (2001). Fiber-like gold particles prepared in cationic micelles by UV irradiation: effect of alkyl chain length of cationic surfactant on particle size. *Journal of Colloid and Interface Science* 241 (1): 289–292.

20 Veisz, B. and Király, Z. (2003). Size-selective synthesis of cubooctahedral palladium particles mediated by metallomicelles. *Langmuir* 19 (11): 4817–4824.

21 Manna, A., Imae, T., Iida, M., and Hisamatsu, N. (2001). Formation of silver nanoparticles from a *N*-hexadecylethylenediamine silver nitrate complex. *Langmuir* 17 (19): 6000–6004.

22 Sathyamurthy, S., Leonard, K.J., Dabestani, R.T., and Paranthaman, M.P. (2005). Reverse micellar synthesis of cerium oxide nanoparticles. *Nanotechnology* 16 (9): 1960–1964.

23 Esumi, K., Sarashina, S., and Yoshimura, T. (2004). Synthesis of gold nanoparticles from an organometallic compound in supercritical carbon dioxide. *Langmuir* 20 (13): 5189–5191.

24 Tang, J., Redl, F., Zhu, Y. et al. (2005). An organometallic synthesis of TiO_2 nanoparticles. *Nano Letters* 5 (3): 543–548.

25 Lee, Y., Lee, J., Bae, C.J. et al. (2005). Large-scale synthesis of uniform and crystalline magnetite nanoparticles using reverse micelles as nanoreactors under reflux conditions. *Advanced Functional Materials* 15 (3): 503–509.

26 Park, J., An, K., Hwang, Y. et al. (2004). Ultra-large-scale syntheses of monodisperse nanocrystals. *Nature Materials* 3 (12): 891–895.

27 Hyeon, T. (2003). Chemical synthesis of magnetic nanoparticles. *Chemical Communications* 8: 927–934.

28 Hyeon, T., Chung, Y., Park, J. et al. (2002). Synthesis of highly crystalline and monodisperse cobalt ferrite nanocrystals. *Journal of Physical Chemistry B* 106 (27): 6831–6833.

29 Joo, J., Na, H.B., Yu, T. et al. (2003). Generalized and facile synthesis of semiconducting metal sulfide nanocrystals. *Journal of the American Chemical Society* 125 (36): 11100–11105.

30 Kaur, R. and Mehta, S.K. (2017). Metallomicelle templated transition metal nanostructures: synthesis, characterization, DFT study and catalytic activity. *Physical Chemistry Chemical Physics* 19 (28): 18372–18382.

31 Kaur, N., Kaur, G., Bhalla, A. et al. (2018). Metallosurfactant based Pd–Ni alloy nanoparticles as a proficient catalyst in the Mizoroki Heck coupling reaction. *Green Chemistry* 20 (7): 1506–1514.

32 Chaudhary, G.R., Singh, P., Kaur, G. et al. (2015). Multifaceted approach for the fabrication of metallomicelles and metallic nanoparticles using solvophobic bisdodecylaminepalladium(II) chloride as precursor. *Inorganic Chemistry* 54 (18): 9002–9012.

33 Lu, X., Yavuz, M.S., Tuan, H.-Y. et al. (2008). Ultrathin gold nanowires can be obtained by reducing polymeric strands of oleylamine–AuCl complexes formed via aurophilic interaction. *Journal of the American Chemical Society* 130 (28): 8900–8901.

34 Kim, S.-W., Park, J., Jang, Y. et al. (2003). Synthesis of monodisperse palladium nanoparticles. *Nano Letters* 3 (9): 1289–1291.

35 Zhao, N., Li, L., Huang, T., and Qi, L. (2010). Controlled synthesis of PbS–Au nanostar–nanoparticle heterodimers and cap-like Au nanoparticles. *Nanoscale* 2 (11): 2418.

36 Sanchez-Dominguez, M., Boutonnet, M., and Solans, C. (2009). A novel approach to metal and metal oxide nanoparticle synthesis: the oil-in-water microemulsion reaction method. *Journal of Nanoparticle Research* 11 (7): 1823–1829.

37 Ye, X.R., Daraio, C., Wang, C. et al. (2006). Room temperature solvent-free synthesis of monodisperse magnetite nanocrystals. *Journal of Nanoscience and Nanotechnology* 6 (3): 852–856.

38 Iida, M. and Harada, M. (2005). EXAFS study for the formation of noble metal nanoparticles prepared from reverse micelles of metallosurfactants. *Photon Factory Activity Report* 22: 183.

39 Pileni, M.P., Tanori, J., and Filankembo, A. (1997). Biomimetic strategies for the control of size, shape and self-organization of nanoparticles. *Colloids and Surfaces A: Physicochemical and Engineering Aspects* 123: 561–573.

40 Li, M., Schnablegger, H., and Mann, S. (1999). Coupled synthesis and self-assembly of nanoparticles to give structures with controlled organization. *Nature* 402 (6760): 393–395.

41 Kaur, R., Giordano, C., Gradzielski, M., and Mehta, S.K. (2014). Synthesis of highly stable, water-dispersible copper nanoparticles as catalysts for nitrobenzene reduction. *Chemistry – An Asian Journal* 9 (1): 189–198.

10

Metallosurfactants and Their Biological Attributes: Anticancer and Antimicrobial Properties

Neha Jindal[1] *and Shivani Uppal*[2]

[1] *DAV College, PG Department of Chemistry, Bathinda 151001, Punjab, India*
[2] *Panjab University, Centre of Advanced Studies in Chemistry, Department of Chemistry, Chandigarh 160014, India*

10.1 Introduction

Surfactants, the surface-active agents, are amphiphilic compounds that render special properties owing to their ability to solubilize in both polar and non-polar solvents. They have the unique ability to gather themselves at a particular concentration to form aggregates. This ability of self-aggregation of the surfactants in the presence of metal into metalloaggregates was first described in the 1980s [1], and since then metallosurfactants have been extensively studied for diverse applications ranging from electron storage systems in photo-redox processes to biological applications [2]. Although surfactants themselves can be used in the development of many drugs, the incorporation of metal ions gives a fascinating blend with desired properties. The surface of metal surfactants leads to change in surface conformation that affects the biological function of the DNA or protein (building blocks of the living systems), which can be exploited in therapeutics. Biologically active metal complexes are of prodigious significance in medical field as a reasonable substitute to organic compounds accompanied by acute side effects. Although the metallosurfactants exhibit numerous captivating properties, their prospects in the medicinal applications have not yet been fully explored and are a relatively new research field.

Medicinal inorganic chemistry is still a juvenile discipline, which consists of escalating pharmacopeia of therapeutic and diagnostic agents having metal ions, an insight into their metabolism and mode of action [3]. It is nearly impractical to study this field without taking into consideration some of the metallocomplexes (or ligands alone, anticipated to bind to endogenous metal ions) that are in different stages of development. The most common example that is cited is of platinum compounds, cisplatin (*cis*-$[Pt(NH_3)_2Cl_2]$), carboplatin, and oxaliplatin, used for the treatment of cancer. Although prescribed on limited basis, Auranofin, is another metal-based therapeutic, which is used for relieving arthritis pain. A technetium radiopharmaceutical, ^{99m}Tc-pertechnetate $[^{99m}TcO_4]^-$, is an indispensable part of

Metallosurfactants: From Fundamentals to Catalytic and Biomedical Applications, First Edition.
Edited by Surinder K. Mehta and Ravneet Kaur.

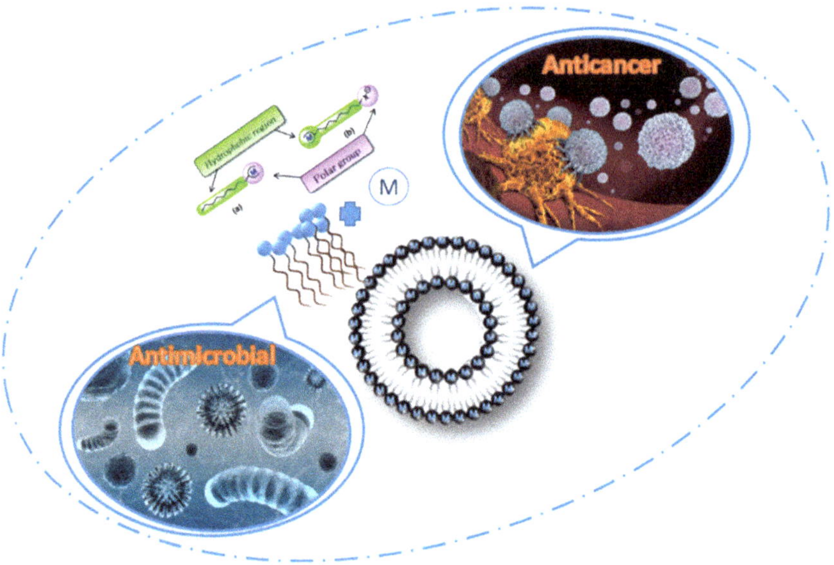

Figure 10.1 Metallosurfactants and their biological attributes: anticancer and antimicrobial properties.

radio diagnosis. Lanthanides and transition metals (Gd, Fe, Mn) as paramagnetic contrast agents for magnetic resonance imaging have opened new horizons by the development of novel complexes having the aptitude to target specific tissues and physiological states [3].

The interactions of the organic amphiphiles with the living organisms and the cell membranes are extensively studied in literature [4, 5]. The metallosurfactants have demonstrated capacity for expansion as contemporary and effectual therapeutic agents. Research is being done to understand the variations in the microenvironment of an amphiphilic metal in comparison to the non-amphiphilic metal complex and its effect on the biological activity. Efforts are concerted to unravel the reasons that lead to the enhancement in the activity of the self-assembly of metallosurfactants as compared to the non-aggregated compounds.

The chapter primarily focusses on two main biological attributes of the metallosurfactants, namely, the antimicrobial efficacy and the potential to combat the deadly cancer (Figure 10.1).

10.2 Antimicrobial Activity

Infections have plagued mankind since the birth, which represent a major concern for global health so far. Hence, the treatment of infections is one of the crucial solicitudes of medical professionals. The rapid emergence of resistant microbes occurring worldwide is making the problem further complicated by endangering the efficacy of antibiotics. Some of the newly developed antibiotics, such as oxazolidinone (linezolid), streptomycin (quinupridine/dalfopristin), and glycolipid (daptomycin), have an activity range limited to Gram-positive pathogens. For scientists, this presents a challenge to design antibiotics with unique properties and more effective action mechanisms.

The mesmerizing world of metallosurfactants plays a key role in the development of new genre of pharmaceutically active agents, which offer a promising approach to combat the increasing microbial resistance. The hydrophobicity of microbial cell membrane is attributed to their unique structure consisting of lipids or proteins arranged in a bilayer making the membrane selectively permeable. This controls all the biological activities occurring in the microbes. Any aspect that manipulates the functioning of membrane proves fatal. Cationic metallosurfactants have been in the limelight recently owing to their ability to self-assemble into a supramolecular structure having sharp polarity gradient at the interface and obvious hydrophobic area in the aqueous solution.

10.2.1 Metallosurfactants Effective against Bacteria

McDonnell and Russell [6] endorsed the fact that cationic surfactants exhibit antimicrobial properties, which are based on their aptitude to disturb the bacterial membrane by cumulative hydrophobic and electrostatic adsorption phenomenon at the membrane water interface. Pathogens primarily have negatively charged membrane, which facilitates their interactions with the positively charged cationic amphiphiles lending them biocidal activity. The amphiphiles adsorb at the interface of water-cell membrane, which leads to disruption of biochemical reactions or metabolic pathway occurring within the cell cytoplasm [7, 8]. The consequent physical disruption is due to the incursion of the hydrophobic chains into the cellular membrane due to their similarity in chemical nature and structure.

Antimicrobial ferrosurfactants (-di[chlorium ferriate] complexes of morpholin-4-yl-alkan-1-one) prepared by Aiad et al. [9] demonstrate good inhibition properties for microorganisms (Gram-positive, Gram-negative bacteria, and some fungal strains as mold and yeast).

The length of the hydrophobic chain of amphiphiles significantly affects the biological activity of the metallosurfactant. As the hydrophobicity increases, biological activity also increases. This fact is endorsed by Adawy and Khowdiary [10] in their work where they have prepared metallocationic surfactants by complexing the quaternary ammonium salt with nickel(II) and copper(II) chlorides. Out of the two metals used, copper exhibits higher activity, which has been attributed to its lower electronegativity and a larger volume [11].

According to Hafiz [12], metallosurfactant complexes tend to dissolve and disintegrate the constituents of biological membranes. The intake capacity of the surfactants is influenced by critical micelle concentration (CMC). It is observed that the metallo amphiphiles having large-sized micelles at lower CMC are more favorable from biological perspective. The complex undertaken was benzyl triphenyl phosphonium chlorometallate having copper or iron ions. The metalloamphiphiles exhibit antibacterial activity against the growth of Gram-positive and Gram-negative bacteria.

The literature shows that cationic surfactants as derivatives of quaternary ammonium salts (+1 charge) have biological activity [13]. However, until recently, there have been few studies on similar water-soluble molecules with a higher charge on the head group. Behm et al. [14] synthesized a series of highly charged large bicyclic caged cobalt complexes with lipophilic substituents. The authors propose that this molecule can affect membranes in many ways. The high charge (+3) and the size of the head group may affect the electron transport system bound to the membrane or inactivate the negative binding site on the enzyme. It is observed that compounds with paraffin tails (C-8) or larger can pierce biofilms. The synthesized amphiphiles were tested for ectoparasites, nematodes, trematodes, and protozoa. They show surprising effects on observed organisms even at millimolar concentrations. Amphiphiles having long single paraffin tails ($>C_{10}$ or equivalent) or bifurcated tails are effective in killing the parasites. On the other hand, molecules having smaller chains/lower lipophilicity are found to be ineffective. Biodistribution studies revealed that unlike organic compounds, the metallo amphiphiles are confined to the gastrointestinal tract, which implies that this would be an effective method for the treatment of gastrointestinal parasites.

10.2.2 Metallosurfactants Effective Against Fungus

Kumar and Arunachalam [15, 16] developed a new type of antibacterial metal amphiphile- *cis*-$[Co(X)_2(C_{14}H_{29}NH_2)Cl]^{2+}$ and *cis*-$[Co(X)_2(C_{11}H_{23}NH_2)]$-type cobalt(III) complex ion containing N,N-donor ligand (X = ethylenediamine (en), bpy, phen or X_2 = trien). The synthetic complex binds to DNA and also inhibits the growth of certain bacterial and fungal species. They were tested for Gram-positive (*Staphylococcus aureus* and *Bacillus subtilis*) and Gram-negative bacteria (*Escherichia coli* and *Pseudomonas aeruginosa*) and *Candida albicans* (pathogenic yeast). The complex encompassing trien ligand has high activity against all tested species, so its performance is better than other ligands.

Organic compounds containing heterocycles are known to be active pharmaceutical agents in most of the biocidal drugs. Derivatives of pyrazole are of great significance because they portray considerable anticancer, anti-HIV, antibacterial [17], monoamine oxidase inhibitory activity [18], insecticidal and herbicidal properties. Considering this, Negm et al. [19] synthesized copper and tin complexes of pyrazolium derivatives (Figure 10.2) having different substituents. The cationic surfactants without metal illustrate reasonable antibacterial (*S. aureus* and *E. coli*) activity, with A = H and B = CH_3 being the most active compounds. However on complexation with metal ions (i.e. $[MCl_2Br_2]^{2-}$ counterions), the activity increased tremendously. It is observed that complexes having a higher positive charge on the nitrogen atoms exhibit lower activity against the tested microorganisms and vice versa, which in turn is dependent on the electropositivity of the metal ions. The amphiphiles were found to be active against the fungus *C. albicans*, but not against *Aspergillus flavus*. The resistance toward the latter fungus was attributed to the rigid outer membrane of the fungus. In this case also, the antifungal activity was found to be enhanced upon

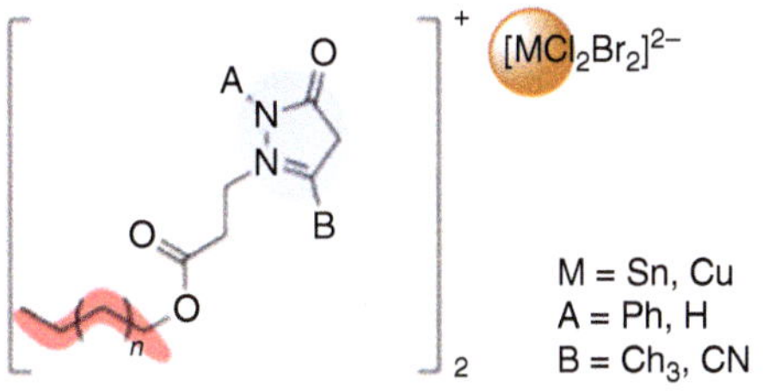

Figure 10.2 Heterocyclic cationic surfactant with tin and copper as counterions (M = Sn, Cu and *n* = 7,9).

complexation with metal ions. In addition to this, it is seen that the resistance of *A. flavus* began to decrease for the metal complexes reflecting their higher potency.

Chandar et al. [20] screened cobalt-containing Schiff bases and their complexes for biological activities. Under identical conditions, they exhibit higher activity than the corresponding free ligands against pathogenic bacterial and fungal species. Chelation decreases the polarity of metal ion, thereby increasing its permeation through the lipid bilayer [21].

Negm et al. [19] also used Schiff base cationic surfactants with the complexes of divalent Co, Cu, and Mn transition metal ions. Different Schiff bases were synthesized using four fatty amines that included dodecyl, tetradecyl, hexadecyl, and octadecyl amine in association with 4-diethyl aminobenzaldehyde. The antibacterial and antifungal properties of the cationic surfactants were then evaluated against several bacterial and fungal strains, which included: *S. Auerus, Staphylococcus typhus, Bacillus subtilis, E. coli, Aspergillus niger, and A. flavus*. The process of adsorption eases penetration of amphiphiles into the cellular membranes. The positive charge on the head group of surfactant molecules stabilizes their adsorption by the interaction with the negative charges on the cellular membranes. Compared with their metal counterparts, it can be seen that the latter has higher biocidal activity than its parent, which is due to the polarity of the Schiff base in the form of cation and the positive charge (N^+) on the nitrogen atom. This is explained based on the concept of overtone [22, 23] and chelation theory [24].

The overtone theory states that the lipid component in the cell membrane aids the channel of fat-soluble substances. According to the chelation theory, the polarity of the metal ion is largely reduced, due to the overlap between the ligand orbitals and the partial sharing of the positive charge of the metal ion with the donor group. This increases the delocalization of Π electrons on the chelating ring, which in turn enhances the lipophilicity of the complex. The lipophilicity of metal complexes augments their capacity to penetrate lipid membranes and blocks metal-binding sites on microbial enzymes. Therefore, the bactericidal activity is higher than that of the parent compound.

Among different metals used, Cu(II) was found to show highest activity, which is in accordance with the previous studies by the same group. High electronegativity (Cu > Co > Mn) and large atomic radius (Cu > Co > Mn) facilitate the absorption of electron density from nitrogen donor atoms, which leads to increased hydrophobicity and bactericidal activity of the corresponding metal complexes. Besides this as discussed earlier, the hydrophobic chain length also has an important effect on the potency of metalloamphiphiles.

Antibacterial and antifungal activities of some nonionic Schiff bases having Cu(II) and Fe(III) have been explored by Negm group [25]. The biocidal action of Schiff's bases is found to be directly proportional to their hydrophobic chain length. As anticipated, the metal chelates showed increased antibacterial and antifungal activity compared to the free ligand. This is attributable to enhanced cell permeability, which in turn is related to delocalization of electrons over Schiff's base that increases the hydrophobicity of alkyl chain. However, the biological activities of the prepared metal complexes diminished on increasing the hydrocarbon chain length, which is due to the adsorption ability of the complexes containing long chains at the interfaces due to which they form micelles. This decreases their biological activities as compared to amphiphiles with shorter hydrocarbon chain lengths. Among the two metals, Cu(II) complexes showed higher activity against bacteria and fungi, which was attributed to the electronic configuration of the tridentate complex. This could further lead to the fact that antibacterial and antifungal activities are related to the molecular structure of the ligands and the captivity of the complex formed [26, 27].

Metal ions increase the biological attributes of already-known molecules was also supported by another study [28]. Ni(II) complexed with anionic Schiff base amphiphiles were explored against *S. aureus, Streptococcus pyogenes* as Gram-positive and *Pseudomonas phaseolicola, Pseudomonas fluorescens* as Gram-negative bacteria. As expected, the biological active compounds become more bacteriostatic and carcinostatic upon chelation with metal ions.

Recently, metal complexes of natural peptides have garnered significant interest as they exhibit an array of antimicrobial properties through structural changes of peptides upon metal binding and redox chemistry.

Antimicrobial peptides (AMP) are usually short chains (less than 50 amino acid residues), consisting of both hydrophobic and hydrophilic domains, and exhibiting cogent cationic properties. They are, thus, often referred as cationic antimicrobial peptides (CAMP) [29, 30]. The net positive charge ensures remarkable collaboration with the negative charge residing on the bacterial membranes depicting substantial specificity and genetic variability. They are known to cause bacterial cell death through either membranolytic and non-membranolytic mechanisms or even through the interaction with intracellular targets, such as DNA, RNA, and proteins. Four main models of (inner) membrane perturbation have been proposed: aggregate, toroidal pore, barrel-stave, and carpet models. However, their metal complexes present different mechanism of actions. They can lead to destruction of bacterial plasma membranes as well as hydrolytic or oxidative cleavage of nucleic acids, which is usually promoted by metal-based aggregation [31].

The widely explored metal AMP motif is the amino terminal Cu(II)- and Ni(II) (ATCUN) having binding sequence with an XXH stretch, an amino terminal, where X is any amino acid.

Another family of potent AMPs are the Piscidins found in the teleosts. Piscidin 1, P1 (FFHHIFRGIVHVGKTIHRLVTG) and piscidin 3, P3 (FIHHIFRGIVHAGRSIGRFLTG), are 22-residue-long, cationic linear peptides, amphipathic and membrane disruptive isoforms [32]. They exhibit broad range of antibacterial activity, although their mechanism of action is quite different, P1 has higher antimicrobial activity than P3, whereas P3 exhibits higher specificity as compared to P1.

For P1, the ATCUN motif is Phe-Phe-His- and for P3 it is Phe-Ile-His and binds Cu^{2+} and Ni^{2+} ions with high affinity through the N-terminal amino group, two amide nitrogens of the backbone, and the δ-nitrogen of the His-3 [33]. Both P1 and P3 localize on bacterial cell wall as well as intracellularly on DNA at sub-inhibitory concentrations. P1 is more hydrophobic and membrane disruptive, whereas P3 is more neutralizing and condensing to isolated DNA, which led to the hypothesis that P3 may be targeting DNA rather than membranes to achieve strong potency.

Histatin-5 (Hst-5) (24 residues) and hepcidin (Hpc-25) (25 residue peptides) also contain the ATCUN motif [34]. Hst-5 is a conformational kinetic peptide, tightly adsorbed on nickel triacetate resin [35], indicating that the peptide has an affinity for divalent metal ions with obvious copper and zinc sites [36]. Hpc-25 is the main iron regulator that binds to the main iron transporter in the human body. Here, the N-terminal is responsible for this interaction, and it also covers ATCUN, thereby enhancing its antimicrobial ability [37]. CAMPs constitute another interesting template for a new generation of antimicrobial agents, which can quickly kill Gram-negative and -positive bacteria.

Bacteriocin (also known as bovine dodecapeptide; RLCRIVVIRVCR) is one of the smallest naturally occurring cationic antimicrobial peptides. Sub5 (a peptide with five substituents) is a short synthetic AMP derived from a substitution library of bacteriocin 2 (a linear variant of bacteriocin). After adding the ATCUN motif to Sub5, the antibacterial activity against various microorganisms was increased by two to three times, and the minimum inhibitory concentration was in the low micromolar range.

10.3 Anticancer Activity

The main cause of cancer onset is genome mutation, which leads to the destruction of basic biological functions, such as cell division, differentiation, angiogenesis, and relocation. For this changing system in cancer development, multi-target treatment is required. One of the most successful and effective cytotoxic agents against cancer is cisplatin. However, owing to its chemical instability, poor water solubility, and low lipophilicity, the clinical application is restricted. Oncologists are currently using combination therapy, which has shown marginal benefit of common therapy due to the development of drug resistance or related toxicity. It is estimated that by 2030, the number of cancer deaths will almost double. Continuous research is being carried on alleviating the side effects by either coupling them with nanocarriers or by incorporation into micellar system. Some of the complexes of cisplatin with polymers are under clinical trials [38].

At present, metal surfactant-based complexes have recognized increasing considerations because of their good medicinal value and fewer side effects. The latest report on a complex having a metal surfactant structure, although not in clinical trials, contains platinum coordinated with the carboxylic acid group of an amphiphilic

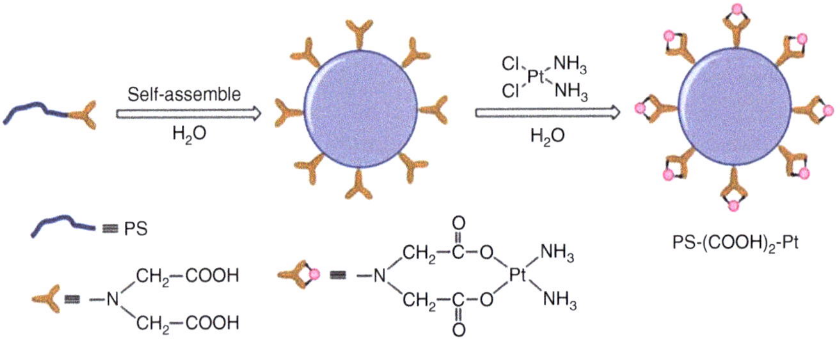

Scheme 10.1 Micelles containing chelated cisplatin for anticancer therapy by Qi et al. Source: Reproduced with permission from Qi et al. [39].

polystyrene oligomer (Scheme 10.1) [39]. By calculating the viability of Sk-Br3 cancer cells after cellular uptake of the complex, the efficacy of the oligomer-based Pt complex was explored. Polymer micelles without cisplatin did not inhibit cell growth at any concentration. However, as expected, cisplatin-coordinated oligomers show effective tumor growth inhibition and reduced systemic toxicity in mice. The efficiency of the metallocomplex is even higher than cisplatin.

The organic–inorganic hybrid, biscetylpyridiniumtetrachloroplatinate (Pt-CPC) prepared by Sharma and Singh [40], was found to exhibit anticancer activity, which was comparable to the control drug adriamycin.

10.3.1 Metallosurfactants Effective Against Breast Cancer

The cytotoxic activity of the cobalt complexes prepared by Kumar and Arunachalam [15, 16] was determined on human breast cancer cells using MTT assay and some specific staining techniques. It was observed that the cells succumbed to apoptosis thus reducing the viability of the cells. It is expected that these complexes might prove beneficial in target-based cancer therapy as the mechanism of cell death appears to be essentially apoptosis, but necrosis also is one of the desired endpoints in cancer therapy.

Photochemistry is an elegant tool, which has been into limelight recently for conversion of non-toxic prodrugs to active cytotoxic species. Light-sensitive ruthenium(II) polypyridyl complexes have been recently proposed as prodrugs for photoactivatable anticancer therapy (PACT), which permits spatial as well as temporal patterns thus alleviating the toxicity of an anticancer drug [41].

Thioether-cholesterol ligand (SRR′) is used to anchor the amphiphilic ruthenium complex $[Ru(tpy)(bpy)(SRR')]^{2+}$ to the pegylated 1,2-dimyristoyl group-*sn*-glycero-3-phosphocholine (DMPC) liposomes in the lipid bilayer [42]. This photoactivatable complex has a photosensitive Ru—S bond, which is stable in the dark, but will be hydrolyzed to aquatic species under blue light irradiation $[Ru(tpy)(bpy)(H_2O)]^{2+}$ (Figure 10.3). The *in situ* up-conversion is promoted by using photosensitizer (PS) and annihilator (A) to cause triplet–triplet up-conversion

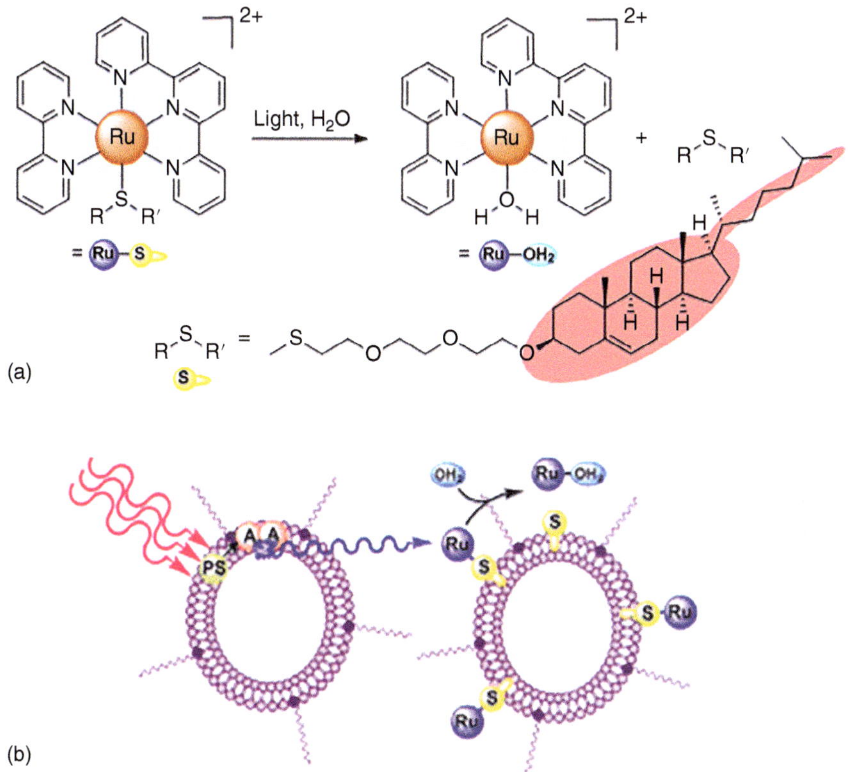

Figure 10.3 (a) Activation of photodissociative amphiphilic Ru(II) complex and (b) implantation of amphiphile into liposomes. Source: Reproduced with permission from Askes et al. [42].

(TTA-UC), which leads to the photoluminescence of Ru—S bond in ruthenium liposome solution. Up-conversion is done with the help of a PDT (photodynamic therapy) laser source, which is used clinically (630 nm). It is suggested that this may provide promising applications in bioimaging and PDT chemotherapy.

Another interesting metal-based amphiphile is a ruthenium-based polypyridyl complex, which induces cell death via different mechanisms depending on its concentration [43]. At lower concentration, i.e. below cac, it acts as a monomer and penetrates the cell membrane, whereas at higher concentration, i.e. above cac, it forms supramolecular aggregates and induces non-apoptotic cell death. They were able to permeate the cell membranes and extract lipids and membrane proteins. Different mechanism of action due to collective behavior above cac was described as "chemical swarming" analogous to biological swarming of bees. Here the amphiphilic nature of the metallodrug leads to unexpectedly high and nonselective cytotoxicity profile against a range of human cancer cell lines. These findings do issue a word of caution for future designing of metallodrugs. The combination of charged metal-based head with a lipophilic tail makes self-aggregation feasible,

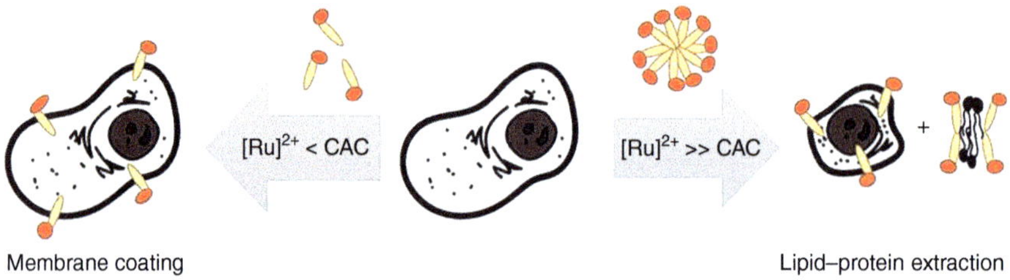

Figure 10.4 Concentration-dependent mechanisms of cell death in amphiphilic Ru(II) complexes. Source: Siewert et al. [43].

which should not be overlooked as it changes the concentration of complex that can penetrate the cell as well as its mode of action (Figure 10.4).

The non-emissive complex [Ru(tpy)(bpy)Cl]Cl complex on coupling to a dodecyl chain via an amide linker on the 4′ position of the terpyridine yields a new amphiphilic ruthenium complex [44]. This leads to stabilization of MLCT level, which is just adequate to switch on red emission. It exhibits self-assembly in aqueous solution, which is dependent of chloride concentration. The emissive amphiphilic complex is found to be more cytotoxic than the reference compound as well as cisplatin. Lipophilic nature of the complex facilitates cellular uptake, first by insertion into the membrane, and then by migration to the peri-nuclear region (Figure 10.5). It is possible to modulate such complexes simply by tuning the functional group attached to the amide like the introduction of cancer-targeting groups could lead to the development of PACT complexes.

In another study, four metallosurfactants derived from the $[Ru(bpy)_3]^{2+}$ complex have been prepared having same polar head group but different hydrophobic tails. *In vitro* studies showed specificity for the LS180, MCF7, and A549 cell lines [45].

In a similar study by another group, amine-based surfactant–ruthenium(II) complexes with different head groups and variable chain lengths have been synthesized to understand the strategy of metal drug design [46]. It is observed that the cell

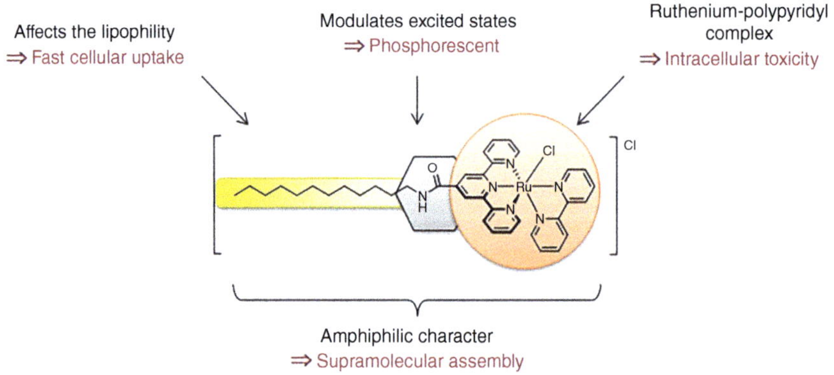

Figure 10.5 Blueprint of an emissive Ru(II) amphiphilic complex depicting red phosphorescence, fast cellular uptake, supramolecular assembly, and intracellular toxicity. Source: Siewert et al. [44].

damage caused by these complexes is related to the size of the head group and the chain length of the alkylamine ligands present in the complex. This further indicates that the desired biological activity can be achieved by adjusting the head and tail base of the metal drug.

10.3.2 Metallosurfactants Effective Against Lung Cancer

Metal containing Schiff base amphiphilic complexes also offer promising anti-cancer properties. Anti-proliferative property of two cobalt(III) Schiff base complexes, *trans*-$[Co(salen)(DA)_2](ClO_4)$ and *trans*-$[Co(salophen)(DA)_2](ClO_4)$ (where salen: *N,N′*-bis(salicylidene)ethylenediamine, salopen: *N,N′*-bis(salicylidene)-1,2-phenylenediamine, DA: dodecylamine) were examined [47]. Both complexes bind to DNA through insertion and hydrophobic modes, and the latter shows better binding affinity than the former. Compared with salen complex, it is found that the salopen-containing complex substantially affects the viability of lung cancer cells through apoptosis. This result is consistent with earlier studies, which revealed a correlation between DNA binding and anti-cancer activity [48, 49]. It has been proposed that the synthetic complexes could potentially be used for cancer chemotherapy and, in general, the inhibition of angiogenesis in lung cancer, particularly.

Some other surfactant–Co(III) complexes, which act as efficient artificial proteases and bind to DNA, also exhibit high cytotoxicity in cancer cells. Co(III) cyclen complexes [34] show high cytotoxicity toward A549 lung cancer cells. Here, the hydrophobic interactions with the proteins rather than micelle formation were assumed to be responsible for cytotoxic activity. Both *cis*-$[Co(trien)(4CNP)(DA)]^{3+}$ (trien = triethylenetetramine, 4CNP = 4-cyanopyridine and DA = dodecylamine) [50] and *cis*-$[Co(L)_2(C_{11}H_{23}NH_2)Cl]^{2+}$ (L = bpy or phen) [51] bind to nucleic acids. They act as excellent antitumor agent even at very low concentration, which is due to the fact that the cobalt complexes increase tumor hypoxia and decrease their energy status. The extent of cellular damage inflicted by these similar complexes also depends on the type of axial ligand. The phenanthroline ligand binds more strongly than the bipyridine ligand and hence exhibits lower IC_{50} value. This approach is expected to have potential application in target-based cancer therapy since the mechanism of cell death appears to be essentially apoptosis. Another surfactant–Co(III) complex that reflects DNA damage in MCF-7 breast cancer cell is the one with tetradecylamine chain, *cis*$[Co(trien)(C_{14}H_{29}NH_2)Cl](ClO_4)_2$ and affects their viability at very low concentrations [52]. The mode of cell death is essentially apoptosis, though necrosis has also been noticed.

10.4 Conclusion and Future Challenges

In an era of increasing resistance to most synthetic drugs, fighting malignancies has become increasingly difficult. Metallosurfactant-based complexes are alternatives to the organic drugs and seem to have the potential to fight a variety of pathogens.

The amphiphilic character of metallosurfactants is a boon as the hydrophobic moiety interacts with the lipophilic part of cell membrane or DNA/proteins and the hydrophilic part having inherent Lewis acidity/redox properties promotes chemical reactivity. They can be tuned according to the requirement. The results of *in vitro* and *in vivo* studies have shown their therapeutic potential as antibacterial and antifungal agents. Although many studies have been reported, the precise mechanism leading to its antimicrobial activity is still not fully understood, and it may vary depending on the type of microbes. They have been also studied in cancer prevention approaches and have gained more interest in recent years. Conferring to numerous works, the metal-surfactant based aggregates have been associated to a low possibility of cancer growth, promoting their protective role. However, several studies have revealed that its fundamental interpretation could lead to confrontation in chemotherapy. Although the exact mechanism is not known, these characteristics make them excellent applicants for preventive chemotherapy against tumor stimulation and development. In future, an improved strategy is essential to review the metallosurfactant-working mechanism in cancer prevention and therapy, though more studies to initiate the safety profile and efficiency of these substances will be critical for imminent human trials. The current facts suggest that they are powerful anticancer agents. Yet, more experiments are needed that illuminate global toxicity and experimental to decide its clinical benefits. We expect that advances in technology will yield new tools and approaches to detect and cure diseases.

References

1 Le Moigne, J. and Simon, J. (1980). A new type of surfactants. The annelides. Characterization of organized metal ion assemblies obtained by cationic complexation at the micelle subsurface. *The Journal of Physical Chemistry* 84 (2): 170–177.

2 Schattschneider, C., Kettenmann, S.D., Hinojosa, S. et al. (2019). Biological activity of amphiphilic metal complexes. *Coordination Chemistry Reviews* 385: 191–207.

3 Mjos, K.D. and Orvig, C. (2014). Metallodrugs in medicinal inorganic chemistry. *Chemical Reviews* 114 (8): 4540–4563.

4 Przestalski, S., Sarapuk, J., Kleszczyńska, H. et al. (2000). Influence of amphiphilic compounds on membranes. *Acta Biochimica Polonica* 47 (3): 627–638.

5 De, S., Malik, S., Ghosh, A. et al. (2015). A review on natural surfactants. *RSC Advances* 5 (81): 65757–65767.

6 McDonnell, G. and Russell, A.D. (1999). Antiseptics and disinfectants: activity, action, and resistance. *Clinical Microbiology Reviews* 12 (1): 147–179.

7 Brunt, K.D. and Hill, H.C. (1987). *Biocides for the oil industry*, vol. 201 (eds. K.D. Brunt and H.C. Hill). New York: Wiley.

8 Hugo, W. and Snow, G. (1981). *Biochemistry of Antibacterial Action*. Springer.

9 Aiad, I., Ahmed, M.H.M., Hessein, A., and Ali, M. (2012). Preparation, surface, and biological activities of some novel metallosurfactants. *Journal of Dispersion Science and Technology* 33 (8): 1144–1153.
10 Adawy, A.I. and Khowdiary, M.M. (2013). Structure and biological behaviors of some metallo cationic surfactants. *Journal of Surfactants and Detergents* 16 (5): 709–715.
11 Badawi, A.M., Negm, N.A., and El-Zahar, S.M. (2004). Biocidal activity of some novel cationic metallomicelles. *Metal Ions in Biology and Medicine* 8: 89.
12 Hafiz, A.A. (2008). Crystal structure of benzyl triphenyl phosphonium chlorometallate: some surface and biological properties of their metallosurfactant derivatives. *Journal of the Iranian Chemical Society* 5 (1): 106–114.
13 Ashman, R.B., Blanden, R.V., Ninham, B.W., and Evans, D.F. (1986). Interaction of amphiphilic aggregates with cells of the immune system. *Immunology Today* 7 (9): 278–283.
14 Behm, C.A., Boreham, P.F.L., Creaser, I.I. et al. (1995). Novel cationic surfactants derived from metal ion cage complexes: potential antiparasitic agents. *Australian Journal of Chemistry* 48 (5): 1009–1030.
15 Kumar, R.S. and Arunachalam, S. (2008). Synthesis, micellar properties, DNA binding and antimicrobial studies of some surfactant–cobalt(III) complexes. *Biophysical Chemistry* 136 (2, 3): 136–144.
16 Kumar, R.S., Arunachalam, S., Periasamy, V.S. et al. (2009). Surfactant–cobalt(III) complexes: synthesis, critical micelle concentration (CMC) determination, DNA binding, antimicrobial and cytotoxicity studies. *Journal of Inorganic Biochemistry* 103 (1): 117–127.
17 Tanitame, A., Oyamada, Y., Ofuji, K. et al. (2004). Synthesis and antibacterial activity of a novel series of potent DNA gyrase inhibitors. Pyrazole derivatives. *Journal of Medicinal Chemistry* 47 (14): 3693–3696.
18 Chimenti, F., Bolasco, A., Manna, F. et al. (2004). Synthesis and selective inhibitory activity of 1-acetyl-3, 5-diphenyl-4, 5-dihydro-(1*H*)-pyrazole derivatives against monoamine oxidase. *Journal of Medicinal Chemistry* 47 (8): 2071–2074.
19 Negm, N.A., Said, M.M., and Morsy, S.M. (2010). Pyrazole derived cationic surfactants and their tin and copper complexes: synthesis, surface activity, antibacterial and antifungal efficacy. *Journal of Surfactants and Detergents* 13 (4): 521–528.
20 Chandar, S.C.N., Santhakumar, K., and Arumugham, M.N. (2009). Metallosurfactant Schiff base cobalt(III) coordination complexes. Synthesis, characterization, determination of CMC values and biological activities. *Transition Metal Chemistry* 34 (8): 841.
21 Jadon, S.C., Gupta, N., and Singh, R.V. (1995). Synthetic and biochemical studies of some hydrazine carbodithioic acid derivatives of dioxomolybdenum (VI). *Indian Journal of Chemistry. Section A: Inorganic, Physical, Theoretical and Analytical* 34: 733–736.
22 Raman, N., Muthuraj, V., Ravichandran, S., and Kulandaisamy, A. (2003). Synthesis, characterisation and electrochemical behaviour of Cu(II), Co(II), Ni(II)

and Zn(II) complexes derived from acetylacetone and *p*-anisidine and their antimicrobial activity. *Journal of Chemical Sciences* 115 (3): 161–167.

23 Cui, Y., Dang, Y., Yang, Y., and Ji, R. (2005). Synthesis of novel oxazolidinone derivatives for antibacterial investigation. *Current Science* 89: 531–534.

24 Tweedy, B.G. (1964). Plant extracts with metal ions as potential antimicrobial agents. *Phytopathology* 55: 910–914.

25 Negm, N.A. and Zaki, M.F. (2008). Structural and biological behaviors of some nonionic Schiff-base amphiphiles and their Cu(II) and Fe(III) metal complexes. *Colloids and Surfaces B: Biointerfaces* 64 (2): 179–183.

26 Parekh, J., Inamdhar, P., Nair, R. et al. (2005). Synthesis and antibacterial activity of some Schiff bases derived from 4-aminobenzoic acid. *Journal of the Serbian Chemical Society* 70 (10): 1155–1162.

27 Vaghasiya, Y.K., Nair, R., Soni, M. et al. (2004). Synthesis, structural determination and antibacterial activity of compounds derived from vanillin and 4-aminoantipyrine. *Journal of the Serbian Chemical Society* 69 (12): 991–998.

28 Abdel-Salam, F.H. (2010). Synthesis, biological study and complexation behavior of some anionic Schiff base amphiphiles. *Journal of Surfactants and Detergents* 13 (4): 423–431.

29 Mookherjee, N., Anderson, M.A., Haagsman, H.P., and Davidson, D.J. (2020). Antimicrobial host defence peptides: functions and clinical potential. *Nature Reviews Drug Discovery* 19: 1–22.

30 Craik, D.J., Fairlie, D.P., Liras, S., and Price, D. (2013). The future of peptide-based drugs. *Chemical Biology & Drug Design* 81 (1): 136–147.

31 Yu, Z. and Cowan, J.A. (2018). Metal complexes promoting catalytic cleavage of nucleic acids – biochemical tools and therapeutics. *Current Opinion in Chemical Biology* 43: 37–42.

32 Hayden, R.M., Goldberg, G.K., Ferguson, B.M. et al. (2015). Complementary effects of host defense peptides piscidin 1 and piscidin 3 on DNA and lipid membranes: biophysical insights into contrasting biological activities. *The Journal of Physical Chemistry B* 119 (49): 15235–15246.

33 Harford, C. and Sarkar, B. (1997). Amino terminal Cu(II)-and Ni(II)-binding (ATCUN) motif of proteins and peptides: metal binding, DNA cleavage, and other properties. *Accounts of Chemical Research* 30 (3): 123–130.

34 Gusman, H., Lendenmann, U., Grogan, J. et al. (2001). Is salivary histatin 5 a metallopeptide? *Biochimica et Biophysica Acta (BBA)-Protein Structure and Molecular Enzymology* 1545 (1, 2): 86–95.

35 Qian, X., Zhou, W., Khaledi, M.G., and Tomer, K.B. (1999). Direct analysis of the products of sequential cleavages of peptides and proteins affinity-bound to immobilized metal ion beads by matrix-assisted laser desorption/ionization mass spectrometry. *Analytical Biochemistry* 274 (2): 174–180.

36 Barry, B.A., Chen, J., Keough, J. et al. (2012). Proton-coupled electron transfer and redox-active tyrosines: structure and function of the tyrosyl radicals in ribonucleotide reductase and photosystem II. *The Journal of Physical Chemistry Letters* 3 (4): 543–554.

37 Abbas, I.M., Vranic, M., Hoffmann, H. et al. (2018). Investigations of the copper peptide hepcidin-25 by LC-MS/MS and NMR. *International Journal of Molecular Sciences* 19 (8): 2271.

38 Oberoi, H.S., Nukolova, N.V., Kabanov, A.V., and Bronich, T.K. (2013). Nanocarriers for delivery of platinum anticancer drugs. *Advanced Drug Delivery Reviews* 65 (13, 14): 1667–1685.

39 Qi, X., Li, N., Gu, H. et al. (2013). Amphiphilic oligomer-based micelles as cisplatin nanocarriers for cancer therapy. *Nanoscale* 5 (19): 8925–8929.

40 Sharma, N.K. and Singh, M. (2018). New class of platinum based metallosurfactant: synthesis, micellization, surface, thermal modelling and in vitro biological properties. *Journal of Molecular Liquids* 268: 55–65.

41 Zayat, L., Calero, C., Alborés, P. et al. (2003). A new strategy for neurochemical photodelivery: metal–ligand heterolytic cleavage. *Journal of the American Chemical Society* 125 (4): 882–883.

42 Askes, S.H.C., Bahreman, A., and Bonnet, S. (2014). Activation of a photodissociative ruthenium complex by triplet–triplet annihilation upconversion in liposomes. *Angewandte Chemie International Edition* 126 (4): 1047–1051.

43 Siewert, B., van Rixel, V.H., van Rooden, E.J. et al. (2016). Chemical swarming: depending on concentration, an amphiphilic ruthenium polypyridyl complex induces cell death via two different mechanisms. *Chemistry A European Journal* 22 (31): 10960–10968.

44 Siewert, B., Langerman, M., Hontani, Y. et al. (2017). Turning on the red phosphorescence of a [Ru(tpy)(bpy)(Cl)]Cl complex by amide substitution: self-aggregation, toxicity, and cellular localization of an emissive ruthenium-based amphiphile. *Chemical Communications* 53 (81): 11126–11129.

45 Lebrón, J.A., Ostos, F.J., López-López, M. et al. (2019). Preparation and characterization of metallomicelles of Ru(II). Cytotoxic activity and use as vector. *Colloids and Surfaces B: Biointerfaces* 175: 116–125.

46 Nehru, S., Veeralakshmi, S., Kalaiselvam, S. et al. (2020). DNA binding, antibacterial, hemolytic and anticancer studies of some fluorescent emissive surfactant-ruthenium(II) complexes. *Journal of Biomolecular Structure and Dynamics* 39: 1–15.

47 Subramanian, A., Yesaiyan, M., Sankaralingam, A. et al. (2019). Biomolecular interaction, anti-cancer and anti-angiogenic properties of cobalt(III) Schiff base complexes. *Scientific Reports (Nature Publisher Group)* 9 (1): 1–14.

48 Zheng, K., Liu, F., Xu, X.M. et al. (2014). Synthesis, structure and molecular docking studies of dicopper(II) complexes bridged by *N*-phenolato-*N′*-[2-(dimethylamino)ethyl]oxamide: the influence of terminal ligands on cytotoxicity and reactivity towards DNA and protein BSA. *New Journal of Chemistry* 38 (7): 2964–2978.

49 Paul, A., Anbu, S., Sharma, G. et al. (2015). Synthesis, DNA binding, cellular DNA lesion and cytotoxicity of a series of new benzimidazole-based Schiff base copper(II) complexes. *Dalton Transactions* 44 (46): 19983–19996.

50 Perera-Bobusch, C., Hormann, J., Weise, C. et al. (2016). Significantly enhanced proteolytic activity of cyclen complexes by monoalkylation. *Dalton Transactions* 45 (26): 10500–10504.

51 Nagaraj, K. and Arunachalam, S. (2014). Synthesis, CMC determination, nucleic acid binding and cytotoxicity of a surfactant–cobalt(III) complex: effect of ionic liquid additive. *New Journal of Chemistry* 38 (1): 366–375.

52 Riyasdeen, A., Senthilkumar, R., Periasamy, V.S. et al. (2014). Antiproliferative and apoptosis-induction studies of a metallosurfactant in human breast cancer cell MCF-7. *RSC Advances* 4 (91): 49953–49959.

11

Metallosurfactants as Carbon Monoxide-Releasing Molecules

Maribel Marín-García[1], *Núria Benseny-Cases*[1], *Mercedes Camacho*[2], *and Ramon Barnadas-Rodríguez*[1]

[1] *Universitat Autònoma de Barcelona, Biophysics Unit/Center for Biophysical Studies, Department of Biochemistry and Molecular Biology, Faculty of Medicine, Avinguda de Can Domènech, 08193, Cerdanyola del Vallès, Spain*
[2] *Institut de Recerca de l'Hospital de la Santa Creu i Sant Pau - Centre CERCA, Genomics of Complex Diseases, Carrer de Sant Quintí, 77-79, 08041, Barcelona, Spain*

11.1 Why CO?

Possibly, the first question that the title of this chapter suggests to a non-specialized reader is why it is interesting to study molecules that are capable to release carbon monoxide (CO), particularly having in mind that CO is a notorious poison.

As it is known, CO is not detected by any of our sensory organs: since it has no color, no taste, no smell, and it does not irritate the skin nor mucosae; when inhaled, it stealthily reaches the blood flowing through the pulmonary capillaries. Here, most of the CO molecules are taken up by the hemoglobin (Hb) contained in the red blood cells, with an affinity more than 200 times higher than that of oxygen, to form carboxyhemoglobin (COHb), while a few other remain dissolved in the aqueous phase of the blood (plasma). In such a silent manner, COHb can cause a dangerous decrease in the supply of O_2 to the tissues (hypoxic effect) and this circumstance gets worse when, distributed through the body by the blood, the CO reaches the tissues where it binds to other proteins that contain heme groups (non-hypoxic effect). If the subject exposed to a high concentration of CO is in an unconscious state (i.e. sleeping), the early symptoms of poisoning (headache, dizziness, fatigue, blurred vision, etc.) are not detected and death is usually the fatal end.

There are several natural sources of CO such as volcanic eruptions, bushfires, and oxidation of methane in troposphere [1, 2]; however, excluding the acute levels induced by catastrophes, the anthropogenic emissions are those that regularly show a deeper impact on human beings. The incomplete carbon combustion of fuels in industry, motor vehicles, etc. can locally increase over 100 times the world average outdoor CO concentrations, which are about 0.04–0.12 ppm (expressed in volume) [3]. Regarding anomalous indoor CO peak values, they are usually caused by defective heating and cooking systems [4, 5] and, for example, an extremely high risk of

Metallosurfactants: From Fundamentals to Catalytic and Biomedical Applications, First Edition.
Edited by Surinder K. Mehta and Ravneet Kaur.

death is reached at 500 ppm, a value that causes about 40% of COHb [3]. Every year this circumstance originates accidental deaths and serious injuries worldwide, but in this century it has also become an easy method to committing painless suicide using charcoal [6, 7].

With the previous data it does not seem to be that way, but the relationship of humans with the artificially produced CO dates long back to before the Industrial Revolution. For instance, the Greek philosopher Aristotle (384–322 BCE) is usually mentioned as the first person who wrote about the effect of CO on humans but, in fact, his observations were slightly vague [8]. In the same way, and although there is some controversy about the scabrous dead by CO of the Roman Empress Fausta [9], there are strong evidences of CO intoxication and death of two Byzantine Emperors [10].

Thus, as can be observed, the participation of CO in human history is absolutely related with its toxicity. This is precisely what happens with an ancient practice that is still maintained: smoking. Tobacco smoke contains CO and induces a chronic hypoxia in heavy smokers because of a permanent increment of blood COHb (till around five times the normal value). Consequently, in an attempt to compensate the diminished O_2 supply, the body produces more red cells (polycythemia) [11–14], a phenomenon that increases blood viscosity and demands a higher effort to the heart.

With such a *curriculum vitae*, it was difficult to believe many decades ago that CO could have any biomedical application; however, little by little, a series of discoveries showed that, like other molecules, its criminal Mr. Hyde personality can be efficiently modulated to a caring Dr. Jekyll [15].

During the XIX century, some studies were published about the presence of gases – among them CO – in the blood of animals. This presence was detected not only due to high external concentrations of the gases, but also in normal conditions [16–18], and a detailed description of the CO physiological effects was described by Claude Bernard (1813–1878) in 1857 [19, 20]. Thus, at the beginning of the XX century, there was an uncertainty about the reason of the existence of CO in the human blood: was its presence just an environmental consequence or was it related to the body metabolism? The clear first answer to this question was exposed by Torgny Sjöstrand (1907–1987) in 1949 [21, 22] and it was rapidly complemented in the subsequent years by him and other authors [23–25]. These pioneering studies showed that, surprisingly, our body produces CO endogenously and that it is a product of the metabolism of the heme group of Hb.

Nowadays we know that the heme group is degraded by an enzyme called heme oxygenase (HO) (EC 1.14.14.18 in mammalians) and that its catalytic action is by far the main source of CO produced by the body (other minor sources are, for example, lipid peroxidation) [26, 27]. As a result, in regular conditions, an overall rate of 0.006 ml of CO per hour and kg of body weight is obtained [3]. At average CO outdoor concentrations (that is, with no environmental peaks), this production causes less than 0.5% of COHb in the blood, while the total value in a healthy human is about 0.5–1.7%, due to the inhalation of exogenous CO [3, 28–30]. At first sight, and considering the vital relevance of the heme group, the existence of a specialized route for its rapid degradation can look unexpected. However, the free heme group is highly

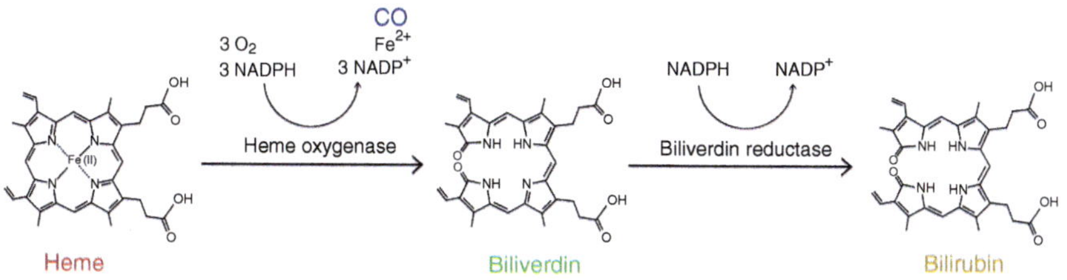

Figure 11.1 Successive steps of the degradation of heme group and release of CO. Heme oxygenase and biliverdin reductase require nicotinamide adenine dinucleotide phosphate (NADPH) as cofactor.

cytotoxic (as is the case of other porphyrin derivatives), and this is why its intracellular and extracellular regulation is fundamental not only in particular situations, such as the erythrocyte turnover [31], hemolytic diseases and hemorrhages [32], but also in normal conditions [33–36]. There are two, or three, isoforms of HO [37–39]: HO-1, whose expression is induced under different stimuli (as a response to oxidative stress), is found at high levels in the spleen, liver, and bone marrow (where the senescent erythrocytes are recycled) [31, 37], and rapidly increases its concentration in other tissues upon transcriptional activation; HO-2, which is continuously synthesized (it is a constitutive enzyme) and is found throughout the body, principally in testes, brain, liver, kidney, and gut [37, 40]; and, possibly, HO-3 with an uncertain function [41–44], which is still not clear if it really is a third isoenzyme or if it stems from HO-2 [32, 45].

Thus, HO has an important role as a cytoprotective agent, which, in mammalians, is accomplished by (i) the cleavage and oxidation of the heme ring (Figure 11.1) to give biliverdin, that next is catalyzed to bilirubin by biliverdin reductase (EC 1.3.1.24); (ii) the release of the iron (as Fe^{2+}) that was coordinated in the heme group; and (iii) the formation of CO. Furthermore, the immediate beneficial function of HO through the free heme degradation is complemented by the biological implications of its reaction products: on one hand, bilirubin exhibits a potent antioxidant activity and it is also implied in immunity processes that, for example, in mice increase transplantation tolerance for diabetes treatment [46, 47]; and, on the other hand, the reactive ferrous cation is rapidly removed from the medium by oxidation to Fe^{3+} and stored into the iron-carrier protein ferritin for its subsequent recycling [37, 48, 49]. Taking into account the above, it should not be a surprise that HO has become a target enzyme for the treatment of several diseases [50–58].

And, what about the CO produced by HO? Is it just a useless and dangerous residual molecule? The astonishing answer is no. CO can be considered as an example of the famous (and partially true) statement of Paracelsus (1493–1541) about what is and what is not a poison: *solely the dose determines that a thing is not a poison* [59]. The famous Renaissantist physician applied his doctrine to the use of inorganic mercury preparations as drugs, but today we know that bioaccumulation is possible and that it can be lethal. This phenomenon explains that the bones of Paracelsus contain about 10 times the normal quantity of mercury, although it is not clear if it was the cause of death (bones that, by the way, have changed their resting place several times) [60]. And what about the endogenous CO? In healthy conditions, it does not accumulate and reaches a non-toxic steady-state concentration in the blood thanks to lung ventilation. But this is not all: surprisingly, it turns out that it is a signaling molecule that regulates diverse biological processes. Thus, it can no longer be considered just as a product of the heme degradation, since, in many occasions, CO is the *desired* molecule, the one that will trigger a given effect at a given place [61–63]. This is the reason why CO is called a gasotransmitter, as it is also the case of NO and H_2S, other endogenously produced gases with signaling activity [64–67] (and perhaps not the unique) [68, 69]. These three gases – which in the liquid body media are molecularly solubilized – easily cross biological membranes, have specific targets (though also share some biological effects), and can influence each other [67, 70]. Even though their complete mechanisms of action are not totally understood, CO activity is mainly related to its binding capacity to heme-containing proteins (for example guanylyl cyclase, some ion channels, cytochrome c oxidase), although its interaction with microbial proteins without heme group has also been reported [71, 72].

Taking all this into account, it is not surprising that CO has a plethora of potential biomedical applications, either by endogenous production (via HO activation/expression) and/or by exogenous administration. Thus, for example, CO shows activity in the cardiovascular system [40, 73–76], has anti-inflammatory properties [37, 77, 78], participates in stem cell regulation [79], alters the perception of pain (nociception) [80–84], facilitates transplantation [85–89], contributes to gastrointestinal protection and healing [90–95], modulates cell metabolism [96, 97], and induces changes in the circadian rhythm [98, 99], just to say some of the effects that are described in literature. At this point, it is clear that CO can play an important role as a drug and, hence, a new question emerges: how can it be delivered?

11.2 How to Deliver CO?

At first glance, the most straightforward way to supply CO to the body is by taking advantage of the respiratory system, that is, by inhalation and, in fact, this procedure is used in some cases. However, this strategy is not as simple as it can seem, since in clinical trials a precise control of the CO dose in the inhaled air is compulsory, and COHb concentration in the blood has to be monitored to avoid any poisoning of the studied subjects. The previous monitorizations can be efficiently accomplished using current technology, but the CO delivery by inhalation has an inherent disadvantage: once the gas reaches red blood cells it is transported all through the body and, consequently, it can affect all the body systems; in other words, a non-targeted delivery is expected. Yet sometimes this is not a handicap: this is the case, for example, of works addressed to assess the effect of CO of cigarette smoke in humans (with inhalations that contain 1200–1500 ppm of CO) [30] and the evaluation of the effectiveness of CO in the treatment of trauma and hemorrhagic shock, in pre-clinical trials conducted with mice and pigs [100]. Even in occasions the systemic distribution of CO has a positive incidence on a given organ, as it is the case of the preventive and therapeutic action of CO in liver injury in mice [101], or the suppression of the pathogenesis of experimental cerebral malaria induced by *Plasmodium berghei*, also in mice [102]. Nevertheless, the local action of inhaled CO is not a general rule. This is the reason why low-dose CO inhalations (100–200 ppm) are common in works intended for the treatment of respiratory diseases [103]: it is expected that small CO doses will not have a systemic effect (or it will be minor), but that should be high enough to act on the first organ that the gas reaches, that is, the lungs. In this way, there are pre-clinical studies that show, for example, the protective role of inhaled CO (and H_2S) on the lungs of mice models of bronchopulmonary dysplasia [104] (a serious lung disease that mostly affects premature newborns) and also many clinical trials for the treatment of chronic obstructive pulmonary disease [105], idiopathic pulmonary fibrosis [106, 107], and acute respiratory distress syndrome [108]. But not all are good news. Some authors have pointed that there is a small correspondence between pre-clinical and clinical studies [103, 109]; furthermore, though some clinical trials have shown some positive results, most of them *only* establish the conditions for a safety low-dose CO inhalation of patients affected by certain lung diseases. However, this is a required step to undertake further studies, and we all know that a trustworthy research needs its time.

There is another approach for CO delivery, which, like inhalation, is related to one of the biological processes involved in respiration: the use of Hb. In fact, it is a Hb derivative, where the protein has been linked to polyethylene glycol (PEG) to avoid extravasation and to maintain several blood properties after transfusion. Strictly speaking, these systems are designed to increase the blood oxygen-carrying capacity, though they are administrated with the PEG–Hb saturated with CO [110, 111]. One of the reasons for that is the protective effect that CO exerts on the protein by decreasing its autooxidation during storage, but it is not the only cause. Besides this, it has been shown that the CO released by PEG–Hb after its transfusion [112] has therapeutic effects related to its known activity on the vascular system [113, 114].

There are plenty of efforts underway to overcome the intrinsic limitations of the two previous CO-delivery strategies (inhalation and PEG–Hb) by using artificial systems. Taking into account the composition/structure of those systems, three different approaches can be considered: (i) metal–carbonyl complexes, that is, molecules that bear CO linked to a transition-metal atom; (ii) alternatively free-metal prodrugs, which generate CO due to a chemical reaction, and; (iii) inorganic and organic nanomaterials/platforms that can absorb and release CO (and/or other gases) or that carry CO precursors. In all these three categories, the delivery of the gas can be triggered by biological factors (pH, proteins, solvent, etc.) or by the use of external stimuli (temperature, magnetic field, X-ray, ultrasounds, etc.) [72, 115–118]. The first two methodologies imply the use of individual molecules, which liberate CO, so these molecules are usually called CO-releasing molecules (CORMs). This term includes

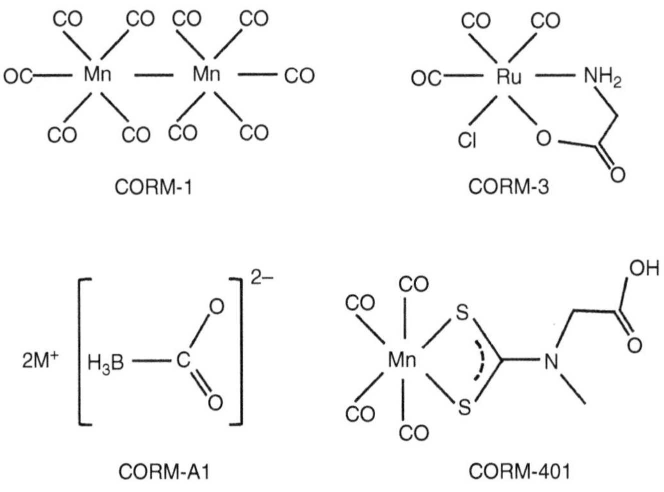

Figure 11.2 Examples of several CORM structures.

several subfamilies such as ET-CORMs or photo-CORMs, which refer to CORMs that release the gas triggered by enzymes or light, respectively [119–121].

The pioneering assays with metal–carbonyl complexes were carried out with *simple* molecules that, in some cases, have low solubility in aqueous media (Figure 11.2) [122, 123]. This handicap for their use in biological systems was rapidly surpassed using molecules with a more complex structure, which confer them with a hydrophilic character. Subsequently, and in a parallel way to what is usually done with other drugs, CORMs based on metal complexes were progressively functionalized with the aim of improving their targeting properties [124, 125]. This strategy leads some authors [126] to consider these molecules constituted by two domains: the *coordination sphere*, composed of the metal and the CO molecules as ligand, and the *drug sphere*, which includes the structure that surrounds the coordination sphere and which determines their pharmacological behavior. However, although it is a clear and efficient way to conceptualize functionalized CORMs, it has to be taken into account that the active principle (that is, the molecule that exhibits therapeutic action) is CO, which is inside the so-called coordination sphere. There are positive pre-clinical results of the use of derivatized and non-derivatized metal-carbonyl based CORMs for the treatment of several diseases (including some that affect the lungs) [127–129], and these molecules are under constant development [108, 117, 119, 130–132].

One handicap of the metal-based CORMs is that, in some cases and depending on the dose, they can be toxic. Their toxicity is related with the degradation products generated by the CO release. CORM degradation of heavy metal complexes can imply processes such as ligand substitution and/or oxidation of the metal atom as is the case, for example, of molybdenum–carbonyl CORMs [72, 115]. However, no general statements can be given and data on human toxicity of molybdenum are limited and with inconsistent findings. Molybdenum nanoparticles are increasingly used in food, industry, and medicine [133], but despite this fact, few toxicology studies are available and it is said to be one of the least toxic metals upon application [134–136]. Their toxic effects (which are dose-dependent, species-specific, and modulated by the particle size) include reactive oxygen species-induced oxidative stress [137], mitochondrial dysfunction [138, 139], membrane integrity disruption [137, 140–142], and DNA damage [138, 143, 144]. In contrast to the above-mentioned studies, there are some others indicating a low and negligible cytotoxicity of molybdenum nanoforms and the possibilities for their use as therapeutic tools [145–149].

As regards the free-metal prodrugs (the second type of artificial systems with CO-releasing capacity), one of their main advantages is the absence of metal transition atoms in their structure, which makes them an interesting alternative to metal–carbonyl CORMs. Most of them are organic molecules that contain a carbonyl group (as carboxylic acid, borocarbonate, borocarbonate ester, borocarbamate, ketone, etc.) that generates CO under certain stimuli or by inter- and intramolecular Diels–Alder reaction [150–152]. This strategy, however, implies a smaller CO-releasing capacity than that of the metal-based CORMs, which usually bear several CO ligands that are able to be released. On the other hand, studies on methylene chloride, a free-metal prodrug with no carbonyl group, showed that it induces the formation of CO when metabolized by mixed-function oxidases in the liver [153, 154], but its toxic doses (oral and inhaled) [155] and its suspected carcinogenesis probably have hampered its use. In any case, there are pre-clinical trials that show the effectiveness of carbonyl-bearing pro-drug CORMs. This is the case of *in vitro* and *in vivo* studies with specific CORMs that have two endogenous triggers (enzyme and pH) and which protect the liver in induced systemic inflammation [156], and that of a family of photo-CORMs, which exhibit anti-inflammatory effects in cell cultures [157].

The last category of systems able to transport and deliver exogenous CO, that is, the inorganic and organic nanomaterials/platforms, comprises a great variety of designs. Some are based in the gas adsorption capacity of porous materials [158], like water-dispersible metal–organic frameworks that form microcrystals, which carry CO and that release the gas due to their biodegradability [159]. The use of different type of porous materials is not restricted to gas adsorption, since CORMs can also be incorporated into these structures without the formation of covalent bonds, due to electrostatic or intermolecular forces (such as hydrogen bonds) [160]. This feature is also the keystone of the entrapment of CORMs into lipid solid nanoparticles [161] and in micelles, where the CO donor, which has poor water solubility, forms the hydrophobic core [162]. Alternatively, in occasions, the strategy is based in the formation of covalent bonds between the carrier platform and the CORMs. Examples of that are, among others, a polymer–protein system that bounds CORM-3 via the protein histidine residues [163], micelle-forming block copolymers, which include a metal-based CORM in the chain [164, 165] and metallodendrimers [166]. Thus, there is a long list of CO-releasing platforms composed of a huge diversity of materials [116, 124, 160, 167, 168], a characteristic that makes possible their functionalization to improve their local CO delivery. As for the case of CORMs (metal-based or metal-free prodrugs), some of the previous nanomaterials have shown positive biological effects in pre-clinical studies [160–162].

11.3 CO-releasing Metallosurfactants

From a conceptual point of view, a metallosurfactant with CO-releasing capacity has a clear design: a transition metal bonded to some CO ligands and to one or more structures that confer an amphipathic character to the molecule. Usually, these structures are hydrocarbon chains that can bear a polar group.

Additionally, the CO ligands must be prone to be released under certain stimuli (acidic or basic pH, temperature, light, enzymes, and/or oxidizing agents), as the previously described CORMs do. With this general design in mind, two particular arrangements are possible: that with the metal embedded into the polar head of the metallosurfactant and the opposite, that is, with the metal located in the hydrophobic domain of the molecule and a polar group (Figure 11.3). As regards the hydrophilic chain, the quantity, length, and saturation grade are important parameters that alongside the volume of the polar head modulates – via the packing factor – the type of the aggregated system that a given surfactant forms [169–171]. This feature is the first difference among most of the regular CORMs and a metallosurfactant CORM (MTS-CORM): due to their intrinsic structure MTS-CORMs are dispersible in aqueous media and give supramolecular aggregates (with no water-solubility problems). It is true that there are a few other CORM systems that form micelles, as it is the case of block copolymers that bear a metal-based CORM [164, 165]. However, there is a substantial difference: these block copolymer chains have molecular weights comprised between 17.000–35.000 Da, while the maximum values for MTS–CORMs are about 1.000 Da, similar to that of the biological membrane-forming phospholipids (about 700–1000 Da). This situation allows MTS–CORMs not only to aggregate when used on their own (usually as micelles if they have one hydrocarbon chain and as vesicles if they have two) [172] but also when mixed with phospholipids, a versatility that allows them to form different type of CO-releasing aggregates (CORAs). For example, the MTS-photo-CORM obtained by Sakla and Jose [173] is a molecule with the polar head constituted by a manganese carbonyl complex and an octadecanoyl chain as hydrophobic

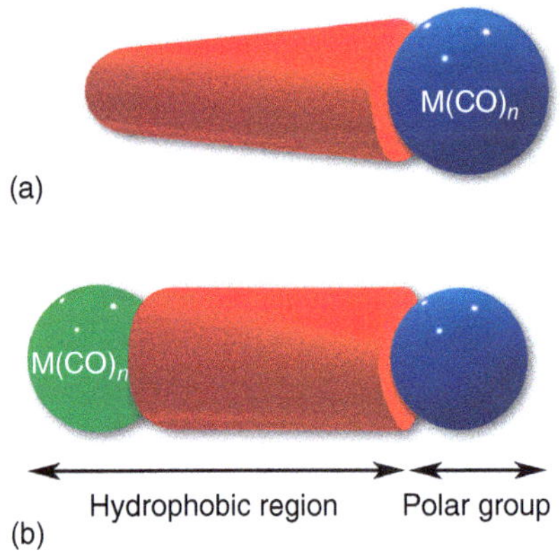

Figure 11.3 Metallosurfactant with the metal–carbonyl complex into the headgroup (a) or into the hydrophobic domain (b).

domain. This inverted cone-shaped MTS (whose packing factor corresponds to a micelle-forming molecule) is efficiently incorporated into liposomes (vesicles) constituted by phospholipid bilayers.

At this point, we have to mention that these vesicular systems (either formed by MTS or MTS/phospholipid mixtures) that contain metal atoms are sometimes called metallosomes. However, this term encompasses a broad spectrum of possibilities. As far as we know, the pioneering paper using this word was published in 1999 by Hainfeld et al., and it was used to describe liposomes that contained some phospholipids covalently attached to gold nanoclusters of 0.8 and 1.4 nm [174]. Thus, in this case, metallosomes describes liposomes that carry groups of metal atoms out of their membrane. This conception of the term fits partially in other works, in which a metal atom is integrated into the surfactant structure (the polar head), which, in turn, is a constitutive part of a vesicle [175–179]. Even more distant to the original use of the word metallosome is the case of vesicular structures obtained by self-assembly of an anticancer molecule and a block copolymer [180]. As can be observed, in spite of the diversity of usages, the most general use of the term metallosome applies to vesicles (with or without phospholipids) that bear molecules that contain a metal atom.

In general, the presence of phospholipids provides some advantages to MTS supramolecular aggregates. Their presence in natural membranes makes phospholipid liposomes a non-toxic drug-delivery system, a role that was glimpsed in 1960s by Alec Bangham, who is considered the father of liposomes [181, 182]. Since then, liposomes that entrap drugs into their bilayers and/or in their inner aqueous medium have been studied and used in drug delivery [183–188] by different routes of administration (oral, topical, parenteral, nasal) [189–192]. These mixed vesicles can be obtained by several methods of preparation (even at industrial scale), which controls their physicochemical characteristics [193–195], can be derivatized (as a targeting strategy) [196–198] and are stable to dilution, a characteristic not always held by aggregates constituted by other surfactants. Even gas-filled liposomes have been described, but they are proposed to imaging applications and do not contain CO [199, 200]. Given this context, the previously described [173] mixed photo-CORAs, which contain the metallosurfactant with the Mn complex in the polar head, are a suitable system for a controlled CO delivery.

However, as it has been previously stated, another approach is possible considering the general structure of MTS–CORMs: a molecule with the metal atom in the non-polar domain. This is the case of molybdenum carbonylic metallosurfactants because of the high hydrophobic character of the Mo—CO bond. These molybdenum-based CORMs (Mo–CORMs) with one or two hydrocarbon chains with a negative group (sulfonate) at the opposite part of the chain, create supramolecular aggregates on their own [201, 202] and also when mixed with phospholipids [203, 204]. The type of structures they form can be regulated by means of the Mo-CORM to phospholipid molar ratio and, like liposomes, can be derivatized; furthermore, the sulfonate group of Mo-CORMs allows to incorporate polymers with amino groups due to the strong interaction between both groups [205].

11.3.1 Synthesis of Molybdenum-Based CORMs

The first step to obtain molybdenum carbonylic CORMs is the synthesis of the ligands, a family of functionalized phosphines with a hydrophobic region and a polar head group that exhibit self-aggregation properties [206]. Both regions will contribute to the desired aggregation properties in aqueous media of the final compounds when coordinated to a metallic center of molybdenum with carbonyl groups. The ligands consist of diphenylphosphines with the third position of the phosphorous atom coordinated to a hydrocarbon chain with a sulfonate group at the end. By modifying the length of the hydrocarbon chain, different aggregation properties are obtained.

The synthesis of the family of ligands, with structure $Ph_2P(CH_2)_nSO_3Na$ where $n = 2$, 6, and 10, is carried out in two steps. For the shorter ligands (L2 ($n = 2$) and L6 ($n = 6$)), the first step allows to obtain a halosulfonated compound (Figure 11.4a). Starting with a commercial precursor, the dihalogenated hydrocarbon compound ($Br(CH_2)_nCl$), a nucleophilic substitution is carried out and one halogen group is changed by a sulfonate group using sodium sulfite (Na_2SO_3). The two halogens of the initial compound are different to avoid the formation of the disulfonate product. In the case of L10 ($n = 10$), it is difficult to find this kind of dihalogenated compound,

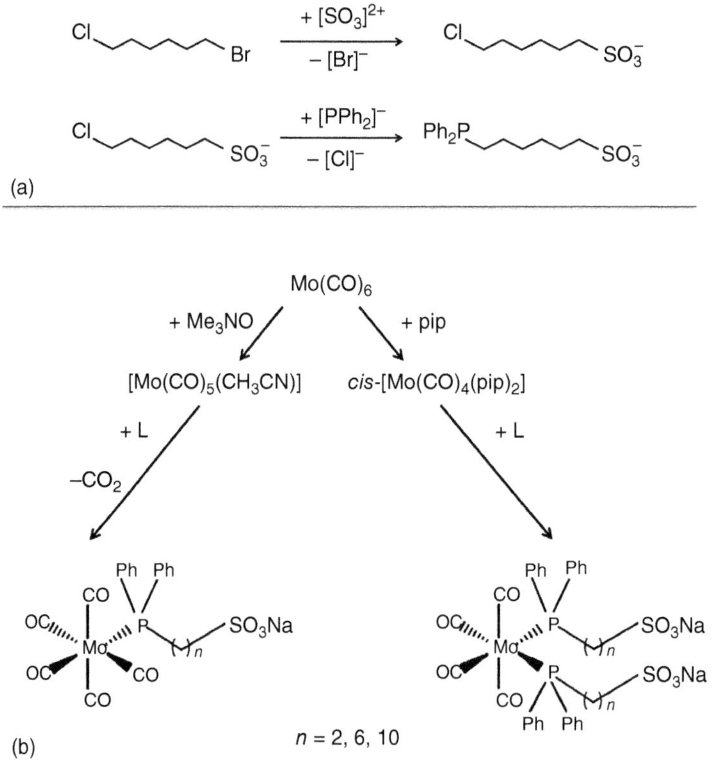

Figure 11.4 (a) Synthesis of the ligand (L) with 6 carbon atoms (L6). (b) Synthesis of the pentacarbonylic and tetracarbonylic Mo-CORMs.

so the precursor is $Cl(CH_2)_{10}Cl$ and extra considerations must be taken to minimize the formation of the disulfonate compound.

The second step allows the introduction of the coordinating diphenylphosphino group to the molecule. It consists of another nucleophilic substitution in which the halogen group is changed by a diphenylphosphine group by means of a reaction in ammoniacal medium. Once these ligands are synthesized and purified, they can be used to obtain two families of metallosurfactants: a family with a metallic center and two ligands coordinated, that is, tetracarbonylic (TCO) compounds $[Mo(CO)_4(L_n)_2]$; and a second family of compounds with only one ligand coordinated, the pentacarbonylic (PCO) compounds $[Mo(CO)_5L_n]$. The synthetic path is different for the two groups of compounds.

The synthesis of molybdenum TCO metallosurfactants (Figure 11.4b) is the same regardless of the length of the chain and consists of a substitution reaction of the two labile piperidine molecules in the compound *cis*-$[Mo(CO)_4(pip)_2]$ by the respective phosphine ligand $Ph_2P(CH_2)_nSO_3Na$ to lead to the TCO compounds *cis*-$[Mo(CO)_4(L_n)_2]$.

Regarding pentacarbonyl derivatives (Figure 11.4b), different synthetic procedures have been reported for their preparation and many of them require high temperatures and long reaction times [207–209]. This approach is possible for the preparation of PCO metallosurfactants [201], but one interesting alternative is the use of trimethylamine oxide dihydrate ($Me_3NO \cdot 2H_2O$) to oxidate one of the carbonyl groups of $[Mo(CO)_6]$ to CO_2 leading to the formation of the complex $[Mo(CO)_5(CH_3CN)]$ [202]. Next, an exchange reaction is carried out between the labile acetonitrile ligand linked to molybdenum by the phosphine ligand $Ph_2P(CH_2)_nSO_3Na$, resulting in the formation of the desired compound, $[Mo(CO)_5L_n]$.

11.3.2 Properties of Mixed CORAs Constituted by Molybdenum-Based Metallosurfactants and Phospholipids

The molybdenum-based MTSs described in 11.3.1 are photo-CORMs and exhibit self-aggregation properties in aqueous medium [201, 202]. However, these aggregates are not stable to dilution, an important property for delivery systems intended for human applications. This is the reason why these Mo-CORMs were mixed with phosphatidylcholine (PC), a natural phospholipid that forms stable bilayers, which can hold other type of molecules. The suspensions were obtained after hydrating and vortexing a thin film of both substances (prepared by rotary evaporation of a $CHCl_3$/methanol solution), which lead to the formation of spontaneous structures at room temperature. The characteristics of these supramolecular systems depend not only on the type of Mo-CORM used, but also on the Mo-CORM to PC ratio (central scheme in Figure 11.5). An example is shown in Figure 11.5a for the case of the tetracarbonyl Mo-CORM with two alkyl chains with 6 carbon atoms each (TCOL6). As can be observed, the size of the aggregates analyzed by dynamic light scattering decreases progressively from 0 to 4 mM concentration of TCOL6 (at a constant PC concentration of 3 mM). At low TCL6 values, the aggregates are mostly large

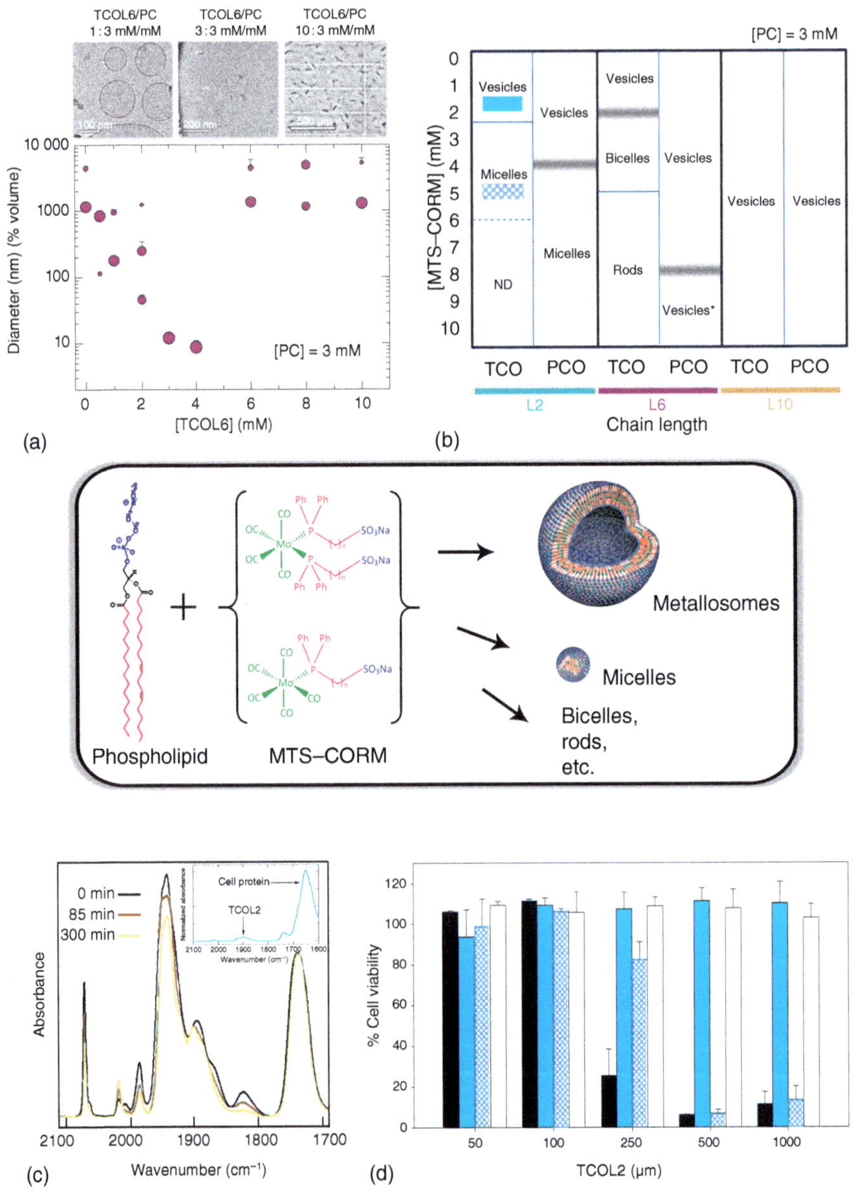

Figure 11.5 Center: scheme of the obtention of Mo-CORMs/phosphatidylcholine (PC) mixed aggregates. (a) Size of different mixed aggregates obtained with tetracarbonyl Mo-CORM containing two carbon chains with 6 carbon atoms (TCOL6) and the corresponding cryo-TEM pictures. (b) Different types of mixed aggregates obtained with PC and tetracarbonyl (TCO) or pentacarbonyl (PCO) Mo-CORMs containing one or two chains with 2 (L2), 6 (L6), and 10 (L10) carbon atoms. ND: not determined. (c) Infrared spectra of PCOL10/PC vesicles after their illumination with UV light. Inset: Infrared spectrum of fibroblasts incubated with TCOL2 vesicles for 12 hours. (d) Fibroblast viability after incubation for 24 hours with free TCOL2 (black), 1 : 3 mol/mol TCOL2/PC vesicles (cyan; see b); 6 : 3 for TCOL2/PC micelles (squared cyan; see b); and PC liposomes (white). Source: Adapted from Marín-García et al. [203, 204].

vesicles, as it was detected by cryo-TEM, while at 3–4 mM of TCOL6 small micelles are formed (see upper pictures). Surprisingly, the system shows a sharp discontinuity at a TCOL6 concentration equal or higher than 6 mM and a very different type of CORA is obtained: they are large rods that can be observed by optical microscopy.

The different type of CORAs that the binary mixtures produce are summarized in Figure 11.5b: vesicles (metallosomes), micelles, bicelles (bilayer fragments), and rods are obtained. The transition from one to another structure can be sharp (continuous horizontal lines) or progressive (coarse faded horizontal lines), and all the systems were stable to dilution except PCOL6 vesicles (indicated with an asterisk) obtained at a Mo-CORM concentration higher than, approximately, 8 mM.

The CO-releasing properties of the molecules upon illumination (with visible or ultraviolet light) were conserved in the aggregates as detected by Fourier transform infrared spectroscopy (FTIR) by measuring the decrease of the CO bands of the Mo-CORMs. It can be observed in Figure 11.5c for the case of PCOL10/PC 30 : 10 mM/mM vesicles, where the main vibrational bands of the Mo-CORM (at 2070 and 1940 cm^{-1}) decrease upon the illumination ($\lambda = 365$ nm) time, while the CO ester band of PC is not altered. FTIR spectroscopy coupled to a microscope and synchrotron light (micro-FTIR) allowed to detect the interaction of the Mo-CORMs with fibroblasts after their incubation with different CORAs. An example is shown in the inset of Figure 11.5c for the case of TCOL2/PC 1 : 3 mM/mM metallosomes after 12 hours of cell incubation. The cell spectrum shows the characteristic CO band of the Mo-CORM located at 1900 cm^{-1} and the intense protein band of the cell at about 1650 cm^{-1}, and it is worth noting that the intensity of the interaction depends on the type of CORA used [204]. This is also the case for the Mo-CORM toxicity in cell cultures, since it is strongly modulated by the type of aggregate used for the incubations, as shown in Figure 11.5d for TCOL2/PC CORAs. Free Mo-CORM and micellar systems are highly toxic, but TCOL2/PC metallosomes exhibit the same behavior than pure PC liposomes in the studied range, that is, no toxicity.

The previously mentioned *in vitro* characterization of the CORA systems constituted by molybdenum-based metallosurfactants and phospholipids is just a first step in their study, but it allows to start thinking in their potential clinical translations. To do that, new problems have to be faced as, for example, an accurate control of the size of the metallosomes, *in vivo* long-term circulation or site-specific targeted delivery. The corresponding solutions can be enlightened by the long experience accumulated in the liposome field. Thus, future works may consider, for example, vesicle size control by extrusion or microfluidics, the incorporation of other lipids (such as cholesterol) to regulate membrane properties (i.e. permeability, resistance to bile salts) and the inclusion of derivatized phospholipids into the CORAs. The last approach should result, for example, in an increase of their *in vivo* circulation time if PEG-phospholipids are used, or in a local delivery if these added phospholipids are derivatized, by instance, with a molecule that binds to a target cellular receptor.

11.4 Conclusions

The relatively new finding that CO is involved in several endogenous (and non-toxic) biological processes as gasotransmitter has spurred the research on its use as drug intended, for example, for the treatment of cardiovalscular diseases, pain regulation, or organ transplantation. The inherent difficulties for its administration are being overcome by the use of molecules that bear bonded CO and that release the gas under the action of biological or external stimuli. In this way, there are a lot of possible candidates that can become a useful CORM/system. One of these actors are metallosurfactants that, for the moment, exhibit two general structures attending to the location of the metal atom (and the CO groups) into the molecule, that is, in the polar domain in the hydrophobic region. Rational design of metallosurfactant (for example, by the number and length of hydrocarbon chains, charges groups, etc.) modulates their characteristics and allows obtaining different aggregated systems with CO-releasing properties. This approach shows some specificities in comparison with the use of other CORMs. For example: metallosurfactants constitute their own drug-delivery system by self-aggregation as micelles, vesicles, or other aggregated systems; they also form several types of mixed systems with phospholipids; and these mixed systems can be modified by the inclusion of other molecules to adjust their *in vivo* behavior. Thus, CO-releasing metallosurfactants have enough merits to be considered as promising molecules, but there is still a long way to go and lots of efforts and creativity (and research funding) are necessary to finally get a commercial drug.

Acknowledgments

We want to thank Prof. Joan Suades for his constant support, wisdom, and generosity before and after his recent (and enviable) retirement, and to Prof. Gregory Gregoriadis and to the Library Services of The Open University for their kind collaboration on identifying references [182] of this chapter; their help was invaluable because, nowadays, those stimulating TV programs were only a faint memory of a teenager's days.

References

1 McConnell, J., McElroy, M., and Wofsy, S. (1971). Natural sources of atmospheric CO. *Nature* 233: 187–188.
2 Weinstock, B. and Niki, H. (1972). Carbon monoxide balance in nature. *Science* 176 (4032): 290–292.
3 U.S. Department of Health and Human Services (2012). *Toxicological profile for carbon monoxide*. Atlanta: Agency for Toxic Substances and Disease Registry.
4 World Health Organization (2010). *WHO Guidelines for Indoor Air Quality: Selected Pollutants*. Copenhagen: WHO Regional Office for Europe.

5 Dubrey, S.W., Chehab, O., and Ghonim, S. (2015). Carbon monoxide poisoning: an ancient and frequent cause of accidental death. *British Journal of Hospital Medicine (London, England)* 76 (3): 159–162.

6 Chiew, A.L. and Buckley, N.A. (2014). Carbon monoxide poisoning in the 21st century. *Critical Care* 18: 221–229.

7 Liu, K.Y., Beautrais, A., Caine, E. et al. (2007). Charcoal burning suicides in Hong Kong and urban Taiwan: an illustration of the impact of a novel suicide method on overall regional rates. *Journal of Epidemiology and Community Health* 61 (3): 248–253.

8 … i.e. as human beings get headaches from, and are often asphyxiated by, the fumes of charcoal, so the lower animals perish from the strong fumes of brimstone and bituminous substancesAristotle (384–322 BC) (1910). Parva naturalia: De sensu et sensili, Chapter V. In: *The works of Aristotle* (eds. J.A. Smith and W.D. Ross). Oxford: Clarendon Press.

9 It is sometimes mentioned that the Roman Empress (Flavia Maxima) Fausta (289–326 AD) was killed by her husband Constantine the Great by means of the incomplete combustion of fire in a bathhouse, where she was locked. However, there is controversy about this (and also about the *reasons* of the murder, which include, among others, an incestuous relationship and abortion).Woods, D. (1998). On the death of the empress Fausta. *Greece and Rome* 45 (1): 70–86.

10 Conversely to the previous case of Fausta (ref. 9), there are strong evidences of accidental CO intoxication in one case, and death in another, of the Byzantine Emperors Julian (331–363 AD) and Jovian (334–364 BC), respectively.Lascaratos, J.G. and Marketos, S.G. (1988). The carbon monoxide poisoning of two byzantine emperors. *Clinical Toxicology* 36 (1 & 2): 103–107.

11 Macdonald, G., Kondor, N., Vandad Yousefi, V. et al. (2004). Reduction of carboxyhaemoglobin levels in the venous blood of cigarette smokers following the administration of carbogen. *Radiotherapy and Oncology* 73 (3): 367–371.

12 Malenica, M., Prnjavorac, B., Bego, T. et al. (2017). Effect of cigarette smoking on haematological parameters in healthy population. *Medical Archives* 71 (2): 132–136.

13 Tadmor, T., Mishchenko, E., Polliack, A. et al. (2011). Hookah (narghile) smoking: a new emerging cause of secondary polycythemia. *American Journal of Hematology* 86 (8): 719–720.

14 AlQahtany, F.S., Algahtani, F.H., Alshebly, M.M. et al. (2020). Association between cigarette & shisha smoking and the severity of polycythemia: a cross sectional study. *Saudi Journal of Biological Sciences* 27 (1): 460–464.

15 To our knowledge, this ambivalent character and its application to substances was first used by S.Tucker, H. (1941). A chemical Dr. Jekyll and Mr Hyde. *Britain Journal of Inebriety* 38 (4): 141–199.

16 Magnus, G. (1837). Ueber die im Blute enthaltenen Gase, Sauerstoff, Stickstoff und Kohlensäure. *Annalen Der Physik Und Chemie* 116 (4): 583–606.

17 Straus, I. (1873). *Les gaz du sang*. Paris: P. Asselin.

18 Grehant, N. (1894). *Les gaz du sang*. Paris: Gauthier-Villars.

19 Sternbach, G.L., Varon, J., and Bernard, C. (2003). The resuscitation greats. Claude Bernard: on the origin of carbon monoxide poisoning. *Resuscitation* 58 (2): 127–130.

20 Bernard, C. (1857). *Leçons sur les effets des substaces toxiques et medicamenteuses*. Paris: J-B Bailliere et Fils.

21 Sjöstrand, T. (1949). Endogenous formation of carbon monoxide in man under normal and pathological conditions. *Scandinavian Journal of Clinical and Laboratory Investigation* 1 (3): 201–214.

22 Sjöstrand, T. (1949). Endogenous formation of carbon monoxide in man. *Nature* 164 (4170): 580–581.

23 Sjöstrand, T. (1951). The *in vitro* formation and disposal of carbon monoxide in blood. *Nature* 168 (4278): 729–730.

24 Coburn, R.F., Blakemore, W.S., and Forster, R.E. (1963). Endogenous carbon monoxide production in man. *The Journal of Clinical Investigation* 42 (7): 1172–1178.

25 Coburn, R.F., Williams, W.J., White, P. et al. (1967). The production of carbon monoxide from hemoglobin in vivo. *The Journal of Clinical Investigation* 46 (3): 346–356.

26 Wolff, D.G. and Bidlack, W.R. (1976). The formation of carbon monoxide during peroxidation of microsomal lipids. *Biochemical and Biophysical Research Communications* 73 (4): 850–857.

27 Archakov, A., Karuzina, I., Petushkova, N. et al. (2002). Production of carbon monoxide by cytochrome P450 during iron-dependent lipid peroxidation. *Toxicology In Vitro* 16 (1): 1–10.

28 Stewart, R.D., Baretta, E.D., Platte, L.R. et al. (1974). Carboxyhemoglobin levels in american blood donors. *JAMA* 229 (9): 1187–1195.

29 Coburn, R.F., Forster, R.E., and Kane, P.B. (1965). Considerations of the physiological variables that determine the blood carboxyhemoglobin concentration in man. *The Journal of Clinical Investigation* 44 (11): 1899–1910.

30 Zevin, S., Saunders, S., Gourlay, S.G. et al. (2001). Cardiovascular effects of carbon monoxide and cigarette smoking. *Journal of the American College of Cardiology* 38 (6): 1633–1638.

31 Knutson, M. and Wessling-Resnick, M. (2003). Iron metabolism in the reticuloendothelial system. *Critical Reviews in Biochemistry and Molecular Biology* 38 (1): 61–88.

32 Li, Q.Q., Li, L.J., Wang, X.Y. et al. (2018). Research Progress in understanding the relationship between heme oxygenase-1 and intracerebral hemorrhage. *Frontiers in Neurology* 9: 682.

33 Yanatori, I., Richardson, D.R., Toyokuni, S. et al. (2019). How iron is handled in the course of heme catabolism: integration of heme oxygenase with intracellular iron transport mechanisms mediated by poly (rC)-binding protein-2. *Archives of Biochemistry and Biophysics* 672: 108071.

34 Pek, R.H., Yuan, X., Rietzschel, N. et al. (2019). Hemozoin produced by mammals confers heme tolerance. *eLife* 8: e49503.

35 Chiabrando, D., Vinchi, F., Fiorito, V. et al. (2014). Heme in pathophysiology: a matter of scavenging, metabolism and trafficking across cell membranes. *Frontiers in Pharmacology* 5: 61.

36 Kumar, S. and Bandyopadhyay, U. (2005). Free heme toxicity and its detoxification systems in human. *Toxicology Letters* 157 (3): 175–188.

37 Ryter, S.W., Alam, J., and Choi, A.M.K. (2006). Heme oxygenase-1/carbon monoxide: from basic science to therapeutic applications. *Physiological Reviews* 86 (2): 583–650.

38 Kochert, B.A., Fleischhacker, A.S., Wales, T.E. et al. (2019). Dynamic and structural differences between heme oxygenase-1 and -2 are due to differences in their C-terminal regions. *The Journal of Biological Chemistry* 294 (20): 8259–8272.

39 Namba, F., Go, H., Murphy, J.A. et al. (2014). Expression level and subcellular localization of heme oxygenase-1 modulates its cytoprotective properties in response to lung injury: a mouse model. *PLoS One* 9 (3): e90936.

40 Leffler Ch.W., Parfenova, H., and Jaggar, J. (2011). Carbon monoxide as an endogenous vascular modulator. *American Journal of Physiology. Heart and Circulatory Physiology* 301 (1): H1–H11.

41 McCoubrey, W.K., Huang, T.J., and Maines, M.D. (1997). Isolation and characterization of a cDNA from the rat brain that encodes hemoprotein heme oxygenase-3. *European Journal of Biochemistry* 247 (2): 725–732.

42 Hayashi, S., Omata, Y., Sakamoto, H. et al. (2004). Characterization of rat heme oxygenase-3 gene. Implication of processed pseudogenes derived from heme oxygenase-2 gene. *Gene* 336 (2): 241–250.

43 Elbirt, K.K. and Bonkovsky, H.L. (1999). Heme oxygenase: recent advances in understanding its regulation and role. *Proceedings of the Association of American Physicians* 111 (5): 438–447.

44 Wilks, A. (2002). Heme oxygenase: evolution, structure, and mechanism. *Antioxidants & Redox Signaling* 4 (4): 603–614.

45 Dennery, P.A. (2014). Signaling function of heme oxygenase proteins. *Antioxidants & Redox Signaling* 20 (11): 1743–1753.

46 Wang, H., Lee, S.S., Dell'Agnello, C. et al. (2006). Bilirubin can induce tolerance to islet allografts. *Endocrinology* 147 (2): 762–768.

47 Rocuts, F., Zhang, X., Yan, J. et al. (2010). Bilirubin promotes de novo generation of T regulatory cells. *Cell Transplantation* 19 (4): 443–451.

48 Gáll, T., Balla, G., and Balla, J. (2019). Heme, heme oxygenase, and endoplasmic reticulum stress-a new insight into the pathophysiology of vascular diseases. *International Journal of Molecular Sciences* 20 (15): 3675.

49 Lanceta, L., Mattingly, J.M., Li Ch. et al. (2015). How heme oxygenase-1 prevents heme-induced cell death. *PLoS One* 10 (8): e0134144.

50 Peterson, S.J. and Frishman, W.H. (2009). Targeting heme oxygenase. *Cardiology in Review* 17 (3): 99–111.

51 Chiang, S.K., Chen, S.E., and Chang, L.C. (2019). A dual role of heme oxygenase-1 in cancer cells. *International Journal of Molecular Sciences* 20 (1): 39.

52 Takahashi, T., Morita, K., Akagi, R. et al. (2004). Heme oxygenase-1: a novel therapeutic target in oxidative tissue injuries. *Current Medicinal Chemistry* 11 (12): 1545–1561.

53 Intagliata, S., Salerno, L., Ciaffaglione, V. et al. (2019). Heme oxygenase-2 (HO-2) as a therapeutic target: activators and inhibidors. *European Journal of Medicinal Chemistry* 183: 111703.

54 Salerno, L., Pittalà, V., Romeo, G. et al. (2015). Novel imidazole derivatives as heme oxygenase-1 (HO-1) and heme oxygenase-2 (HO-2) inhibitors and their cytotoxic activity in human-derived cancer cell lines. *European Journal of Medicinal Chemistry* 96: 162–172.

55 Salerno, L., Floresta, G., Ciaffaglione, V. et al. (2019). Progress in the development of selective heme oxygenase-1 inhibitors and their potential therapeutic application. *European Journal of Medicinal Chemistry* 167: 439–453.

56 Corona, D., Ekser, B., Gioco, R. et al. (2020). Heme-oxygenase and kidney transplantation: a potential for target therapy? *Biomolecules* 10 (6): 840.

57 Chau, L.Y. (2015). Heme oxygenase-1: emerging target of cancer therapy. *Journal of Biomedical Science* 22 (1): 22.

58 Motterlini, R. and Foresti, R. (2014). Heme oxygenase-1 as a target for drug discovery. *Antioxidants & Redox Signaling* 20 (11): 1810–1826.

59 Grandjean, P. (2016). Paracelsus revisited: the dose concept in a complex world. *Basic & Clinical Pharmacology & Toxicology* 119 (2): 126–132.

60 Moran, B.T. (2019). *Paracelsus: An Alchemical Life*. London: Reaktion Books.

61 Maines, M. (1993). Carbon monoxide: an emerging regulator of cGMP in the brain. *Molecular and Cellular Neurosciences* 4 (5): 389–397.

62 Maines, M.D. (1997). The heme oxygenase system: a regulator of second messenger gases. *Annual Review of Pharmacology and Toxicology* 37: 517–554.

63 Wang, R. (2004). The evolution of gasotransmitter. In: *Signal Transduction and the Gasotransmitters: NO, CO, H_2S in Biology and Medicine*. Totowa: Humana Press Inc.

64 Verma, A., Hirsch, D.J., Glatt, C.E. et al. (1993). Carbon monoxide: a putative neural messenger. *Science* 259 (5093): 381–384.

65 Wang, R. (2002). Two's company, three's a crowd: can H_2S be the third endogenous gaseous transmitter? *The FASEB Journal* 16 (13): 1792–1798.

66 de Cabo, R. and Diaz-Ruiz, A. (2019). A central role for the gasotransmitter H_2S in aging. *Cell Metabolism* 31 (1): 10–12.

67 Farrugia, G. and Szurszewski, J.H. (2014). Carbon monoxide, hydrogen sulfide, and nitric oxide as signaling molecules in the gastrointestinal tract. *Gastroenterology* 147 (2): 303–313.

68 Wang, R. (2014). Gasotransmitters: growing pains and joys. *Trends in Biochemical Sciences* 39 (5): 227–232.

69 Kuganesan, M., Samra, K., Evans, E. et al. (2019). Selenium and hydrogen selenide: essential micronutrient and the fourth gasotransmitter? *Intensive Care Medicine Experimental* 7 (1): 71.

70 Głowacka, U., Brzozowski, T., and Magierowski, M. (2020). Synergisms, discrepancies and interactions between hydrogen sulfide and carbon monoxide in the gastrointestinal and digestive system physiology, pathophysiology and pharmacology. *Biomolecules* 10 (3): 445.

71 Motterlini, R. and Foresti, R. (2017). Biological signaling by carbon monoxide and carbon monoxide-releasing molècules. *American Journal of Physiology. Cell Physiology* 312 (3): C302–C313.

72 Heinemann, S.H., Hoshi, T., Westerhausen, M. et al. (2014). Carbon monoxide-physiology, detection and controlled release. *Chemical Communications* 50 (28): 3644–3660.

73 Jing, M., Bina, S., Verma, A. et al. (1996). Effects of halothane and isoflurane on carbon monoxide-induced relaxations in the rat aorta. *Anesthesiology* 85 (2): 347–354.

74 Kim, H.H. and Choi, S. (2018). Therapeutic aspects of carbon monoxide in cardiovascular disease. *International Journal of Molecular Sciences* 19 (8): 2381.

75 Lee, S.R., Nilius, B., and Han, J. (2018). Gaseous signaling molecules in cardiovascular function: from mechanisms to clinical translation. *Reviews of Physiology, Biochemistry and Pharmacology* 174: 81–156.

76 Mistry, R.K. and Brewer, A.C. (2017). Redox regulation of gasotransmission in the vascular system: a focus on angiogenesis. *Free Radical Biology & Medicine* 108: 500–516.

77 Yan, H., Du, J., Zhu, S. et al. (2019). Emerging delivery strategies of carbon monoxide for therapeutic applications: from CO gas to CO releasing nanomaterials. *Small* 15 (49): e1904382.

78 Motterlini, R. (2007). Carbon monoxide-releasing molecules (CO-RMs): vasodilatory, anti-ischaemic and anti-inflammatory activities. *Biochemical Society Transactions* 35 (5): 1142–1146.

79 Wang, B., Huang, C., Chen, L. et al. (2020). The emerging roles of the gaseous signaling molecules NO, H_2S, and CO in the regulation of stem cells. *ACS Biomaterials Science & Engineering* 6 (2): 798–812.

80 Fan, W., Huang, F., Wu, Z. et al. (2011). Carbon monoxide: a gas that modulates nociception. *Journal of Neuroscience Research* 89 (6): 802–807.

81 Pereira de Ávila, M.A., Giusti-Paiva, A., and de Oliveira Nascimento, C.G. (2014). The peripheral antinociceptive effect induced by the heme oxygenase/carbon monoxide pathway is associated with ATP-sensitive K^+ channels. *European Journal of Pharmacology* 726: 41–48.

82 Negrete, R., Hervera, A., Leánez, S. et al. (2014). Treatment with a carbon monoxide-releasing molecule inhibits chronic inflammatory pain in mice: nitric oxide contribution. *Psychopharmacology* 231 (5): 853–861.

83 Wang, H. and Sun, X. (2017). Carbon monoxide-releasing molecule-2 inhibits connexin 43-hemichannel activity in spinal cord astrocytes to attenuate neuropathic pain. *Journal of Molecular Neuroscience* 63 (1): 58–69.

84 Pol, O. (2021). The role of carbon monoxide, heme oxygenase 1, and the Nrf2 transcription factor in the modulation of chronic pain and their interactions with opioids and cannabinoids. *Medicinal Research Reviews* 41 (1): 136–155.

85 Ozaki, K.S., Kimura, S., and Murase, N. (2012). Use of carbon monoxide in minimizing ischemia/reperfusion injury in transplantation. *Transplantation Reviews* 26 (2): 125–139.

86 Nakao, A., Choi, A.M.K., and Murase, N. (2006). Protective effect of carbon monoxide in transplantation. *Journal of Cellular and Molecular Medicine* 10 (3): 650–671.

87 Lee, S.S., Gao, W., Mazzola, S. et al. (2007). Heme oxygenase-1, carbon monoxide, and bilirubin induce tolerance in recipients toward islet allografts by modulating T regulatory cells. *The FASEB Journal* 21 (13): 3450–3457.

88 Steiger, C., Wollborn, J., Gutmann, M. et al. (2015). Controlled therapeutic gas delivery systems for quality-improved transplants. *European Journal of Pharmaceutics and Biopharmaceutics* 97 (Pt A): 96–106.

89 Bhattacharjee, R.N., Richard-Mohamed, M., Sun, Q. et al. (2018). CORM-401 reduces ischemia reperfusion injury in an ex vivo renal porcine model of the donation after circulatory death. *Transplantation* 102 (7): 1066–1074.

90 Babu, D., Motterlini, R., and Lefebvre, R.A. (2015). CO and CO-releasing molecules (CO-RMs) in acute gastrointestinal inflammation. *British Journal of Pharmacology* 172 (6): 1557–1573.

91 Gibbons, S.J., Verhulst, P.-J., Bharucha, A. et al. (2013). Review article: carbon monoxide in gastrointestinal physiology and its potential in therapeutics. *Alimentary Pharmacology & Therapeutics* 38 (7): 689–702.

92 Takagi, T., Uchiyama, K., and Naito, Y. (2015). The therapeutic potential of carbon monoxide for inflammatory bowel disease. *Digestion* 91 (1): 13–18.

93 Naito, Y., Takagi, T., Uchiyama, K. et al. (2016). Multiple targets of carbon monoxide gas in the intestinal inflammation. *Archives of Biochemistry and Biophysics* 595: 147–152.

94 Steiger, C., Uchiyama, K., Takagi, T. et al. (2019). Prevention of colitis by controlled oral drug delivery of carbon monoxide. *Journal of Controlled Release* 239: 128–136.

95 Uddin Md.J., Jeong, S., Zheng, M. et al. (2013). Carbon monoxide attenuates dextran sulfate sodium-induced colitis via inhibition of GSK-3β signaling. *Oxidative Medicine and Cellular Longevity* 2013: 210563.

96 Figueiredo-Pereira, C., Dias-Pedroso, D., Soares, N.L. et al. (2020). CO-mediated cytoprotection is dependent on cell metabolism modulation. *Redox Biology* 32: 101470.

97 Kim, H.J., Joe, Y., Surh, Y.J. et al. (2018). Metabolic signaling functions of the heme oxygenase/CO system in metabolic diseases. *Cellular & Molecular Immunology* 15 (12): 1085–1087.

98 Minegishi, S., Sagami, I., Negi, S. et al. (2018). Circadian clock disruption by selective removal of endogenous carbon monoxide. *Scientific Reports* 8 (1): 11996.

99 Oren, D.A., Sit, D.K., Goudarzi, S.H. et al. (2020). Carbon monoxide: a critical physiological regulator sensitive to light. *Translational Psychiatry* 10 (1): 87.

100 Gomez, H., Kautza, B., Escobar, D. et al. (2015). Inhaled carbon monoxide protects against the development of shock and mitochondrial injury following hemorrhage and resuscitation. *PLoS One* 10 (9): e0135032.

101 Chen, Y., Park, H.-J., Park, J. et al. (2019). Carbon monoxide ameliorates acetaminophen-induced liver injury by increasing hepatic HO-1 and Parkin expression. *The FASEB Journal* 33 (12): 13905–13919.

102 Pamplona, A., Ferreira, A., Balla, J. et al. (2007). Heme oxygenase-1 and carbon monoxide suppress the pathogenesis of experimental cerebral malaria. *Nature Medicine* 13 (6): 703–710.

103 Ryter, S.W., Ma, K.C., and Choi, A.M.K. (2017). Carbon monoxide in lung cell physiology and disease. *American Journal of Physiology. Cell Physiology* 314 (2): C211–C227.

104 Lin, H. and Wang, X. (2020). The effects of gasotransmitters on bronchopulmonary dysplasia. *European Journal of Pharmacology* 873: 172983.

105 Bathoorn, E., Slebos, D.-J., Postma, D.S. et al. (2007). Anti-inflammatory effects of inhaled carbon monoxide in patients with COPD: a pilot study. *The European Respiratory Journal* 30 (6): 1131–1137.

106 Rosas, I.O., Goldberg, H.J., Collard, H.R. et al. (2018). A phase II clinical trial of low-dose inhaled carbon monoxide in idiopathic pulmonary fibrosis. *Chest* 153 (1): 94–104.

107 Casanova, N., Zhou, T., Gonzalez-Garay, M.L. et al. (2019). Low dose carbon monoxide exposure in idiopathic pulmonary fibrosis produces a CO signature comprised of oxidative phosphorylation genes. *Scientific Reports* 9 (1): 14802.

108 Fredenburgh, L.E., Perrella, M.A., Barragan-Bradford, D. et al. (2018). A phase I trial of low-dose inhaled carbon monoxide in sepsis-induced ARDS. *JCI Insight* 3 (23): e124039.

109 Steiger, C., Hermann, C., and Meinel, L. (2017). Localized delivery of carbon monoxide. *European Journal of Pharmaceutics and Biopharmaceutics* 118: 3–12.

110 Khan, F., Singh, K., and Friedman, M.T. (2020). Artificial blood: the history and current perspectives of blood substitutes. *Discover* 8 (1): e104.

111 Misra, H., Kazo, H., and Newmark, J.A. (2014). Toxicology and safety determination for a novel therapeutic dual carbon monoxide and oxygen delivery agent. *Journal of Clinical Toxicology* 4: 4.

112 Abuchowski, A. (2016). PEGylated bovine carboxyhemoglobin (SANGUINATE™): results of clinical safety testing and use in patients. *Advances in Experimental Medicine and Biology* 876: 461–467.

113 Klaus, J.A., Kibler, K.K., Abuchowski, A. et al. (2010). Early treatment of transient focal cerebral ischemia with bovine PEGylated carboxy hemoglobin transfusion. *Artificial Cells, Blood Substitutes, and Immobilization Biotechnology* 38 (5): 223–229.

114 Abuchowski, A. (2017). SANGUINATE (PEGylated carboxyhemoglobin bovine): mechanism of action and clinical update. *Artificial Organs* 41 (4): 346–350.

115 Schatzschneider, U. (2015). Novel lead structures and activation mechanisms for CO-releasing molecules (CORMs). *British Journal of Pharmacology* 172 (6): 1638–1650.

116 Yu, L., Hu, P., and Chen, Y. (2018). Gas-generating nanoplatforms: material chemistry, multifunctionality, and gas therapy. *Advanced Materials* 30 (49): e1801964.

117 Santos-Silva, T., Mukhopadhyay, A., Seixas, J.D. et al. (2011). Towards improved therapeutic CORMs: understanding the reactivity of CORM-3 with proteins. *Current Medicinal Chemistry* 18 (22): 3361–3366.

118 Aki, T., Unuma, K., Noritake, K. et al. (2018). Interaction of carbon monoxide-releasing ruthenium carbonyl CORM-3 with plasma fibronectin. *Toxicology In Vitro* 50: 201–209.

119 Stamellou, E., Storz, D., Botov, S. et al. (2014). Different design of enzyme-triggered CO-releasing molecules (ET-CORMs) reveals quantitative differences in biological activities in terms of toxicity and inflammation. *Redox Biology* 2: 739–748.

120 Ling, K., Men, F., Wang, W.C. et al. (2018). Carbon monoxide and its controlled release: therapeutic application, detection, and development of carbon monoxide releasing molecules (CORMs). *Journal of Medicinal Chemistry* 61 (7): 2611–2635.

121 Divya, D., Nagarajaprakash, R., Vidhyapriya, P. et al. (2019). Single-pot self-assembly of heteroleptic Mn(I)-based aminoquinonato-bridged ester/amide-functionalized dinuclear metallastirrups: potential anticancer and visible-light-triggered CORMs. *ACS Omega* 4 (7): 12790–12802.

122 Motterlini, R., Clark, J.E., Foresti, R. et al. (2002). Carbon monoxide-releasing molecules: characterization of biochemical and vascular activities. *Circulation Research* 90 (2): E17–E24.

123 Rattan, S., Al Haj, R., and De Godoy, M.A.F. (2004). Mechanism of internal anal sphincter relaxation by CORM-1, authentic CO, and NANC nerve stimulation. *American Journal of Physiology-Gastrointestinal and Liver Physiology* 287 (3): G605–G611. (The molecule used in this work is CORM-2).

124 Kautz, A.C., Kunz, P.C., and Janiak, C. (2016). CO-releasing molecule (CORM) conjugate systems. *Dalton Transactions* 45 (45): 18045–18063.

125 Ollivier, A., Foresti, R., El Ali, Z. et al. (2019). Design and biological evaluation of manganese- and ruthenium-based hybrid CO-RMs (HYCOs). *ChemMedChem* 14 (18): 1684–1691.

126 Romão, C.C., Blättler, W.A., Seixas, J.D. et al. (2012). Developing drug molecules for therapy with carbon monoxide. *Chemical Society Reviews* 41 (9): 3571–3583.

127 Shao, L., Liu, C., Wang, S. et al. (2018). The impact of exogenous CO releasing molecule CORM-2 on inflammation and signaling of orthotopic lung cancer. *Oncology Letters* 16 (3): 3223–3230.

128 Shao, L., Gu, Y.-Y., Jiang, C.-H. et al. (2018). Carbon monoxide releasing molecule-2 suppresses proliferation, migration, invasion, and promotes apoptosis in non-small cell lung cancer Calu-3 cells. *European Review for Medical and Pharmacological Sciences* 22 (7): 1948–1957.

129 Lee, C.-W., Chi, M.-C., Hsu, L.-F. et al. (2019). Carbon monoxide releasing molecule-2 protects against particulate matter-induced lung inflammation by inhibiting TLR2 and 4/ROS/NLRP3 inflammasome activation. *Molecular Immunology* 112: 163–174.

130 Motterlini, R. and Otterbein, L.E. (2010). The therapeutic potential of carbon monoxide. *Nature Reviews Drug Discovery* 9 (9): 728–743.

131 García-Gallego, S. and Bernardes, G.J.L. (2014). Carbon-monoxide-releasing molecules for the delivery of therapeutic CO in vivo. *Angewandte Chemie International Edition* 53 (37): 9712–9721.

132 Adach, W. and Olas, B. (2019). Carbon monoxide and its donors - their implications for medicine. *Future Medicinal Chemistry* 11 (1): 61–73.

133 Krishnamoorthy, K., Premanathan, M., Veerapandian, M. et al. (2014). Nanostructured molybdenum oxide-based antibacterial paint: effective growth inhibition of various pathogenic bacteria. *Nanotechnology* 25: 315101.

134 Ding, D., Guo, W., Guo, C. et al. (2017). MoO_{3-x} quantum dots for photoacoustic imaging guided photothermal/photodynamic cancer treatment. *Nanoscale* 9 (5): 2020–2029.

135 Liu, Q., Sun, C., He, Q. et al. (2015). Ultrathin carbon layer coated MoO_2 nanoparticles for high-performance near-infrared photothermal cancer therapy. *Chemical Communications* 51 (49): 10054–10057.

136 Song, G., Shen, J., Jiang, F. et al. (2014). Hydrophilic molybdenum oxide nanomaterials with controlled morphology and strong plasmonic absorption for photothermal ablation of cancer cells. *ACS Applied Materials & Interfaces* 6 (6): 3915–3922.

137 Liu, S., Shen, Z., Wu, B. et al. (2017). Cytotoxicity and efflux pump inhibition induced by molybdenum disulfide and boron nitride nanomaterials with sheetlike structure. *Environmental Science & Technology* 51 (18): 10834–10842.

138 Siddiqui, M.A., Saquib, Q., Ahamed, M. et al. (2015). Molybdenum nanoparticles-induced cytotoxicity, oxidative stress, G2/M arrest, and DNA damage in mouse skin fibroblast cells (L929). *Colloids and Surfaces. B, Biointerfaces* 125: 73–81.

139 Hussain, S.M., Hess, K.L., Gearhart, J.M. et al. (2005). In vitro toxicity of nanoparticles in BRL 3A rat liver cells. *Toxicology In Vitro* 19 (7): 975–983.

140 Braydich-Stolle, L., Hussain, S., Schlager, J.J. et al. (2005). In vitro cytotoxicity of nanoparticles in mammalian germline stem cells. *Toxicological Sciences* 88 (2): 412–419.

141 Chng, E.L., Sofer, Z., and Pumera, M. (2014). MoS_2 exhibits stronger toxicity with increased exfoliation. *Nanoscale* 6 (23): 14412–14418.

142 Wu, R., Ou, X., Tian, R. et al. (2018). Membrane destruction and phospholipids extraction by two-dimensional MoS_2 nanosheets. *Nanoscale* 10 (43): 20162–20170.

143 Fazio, E., Speciale, A., Spadaro, S. et al. (2018). Evaluation of biological response induced by molybdenum oxide nanocolloids on in vitro cultured NIH/3T3 fibroblast cells by micro-Raman spectroscopy. *Colloids and Surfaces. B, Biointerfaces* 170: 233–241.

144 Anh Tran, T., Krishnamoorthy, K., Song, Y.W. et al. (2014). Toxicity of nano molybdenum trioxide toward invasive breast cancer cells. *ACS Applied Materials & Interfaces* 6 (4): 2980–2986.

145 Akhtar, M.J., Ahamed, M., Alhadlaq, H.A. et al. (2015). Antioxidative and cytoprotective response elicited by molybdenum nanoparticles in human cells. *Journal of Colloid and Interface Science* 457: 370–377.

146 Wu, H., Yang, R., Song, B. et al. (2011). Biocompatible inorganic fullerene-like molybdenum disulfide nanoparticles produced by pulsed laser ablation in water. *ACS Nano* 5 (2): 1276–1281.

147 Shah, P., Narayanan, T.N., Li, C.Z. et al. (2015). Probing the biocompatibility of MoS_2 nanosheets by cytotoxicity assay and electrical impedance spectroscopy. *Nanotechnology* 26 (31): 315102.

148 Appel, J.H., Li, D.O., Podlevsky, J.D. et al. (2016). Low cytotoxicity and genotoxicity of two-dimensional MoS_2 and WS_2. *ACS Biomaterials Science & Engineering* 2 (3): 361–367.

149 Pardo, M., Shuster-Meiseles, T., Levin-Zaidman, S. et al. (2014). Low cytotoxicity of inorganic nanotubes and fullerene-like nanostructures in human bronchial epithelial cells: relation to inflammatory gene induction and antioxidant response. *Environmental Science & Technology* 48 (6): 3457–3466.

150 Abeyrathna, N., Washington, K., Bashur, C. et al. (2017). Nonmetallic carbon monoxide releasing molecules (CORMs). *Organic & Biomolecular Chemistry* 15 (41): 8692–8699.

151 Soboleva, T., Esquer, H.J., Benninghoff, A.D. et al. (2017). Sense and release: a thiol-responsive flavonol-based photonically driven carbon monoxide-releasing molecule that operates via a multiple-input AND logic gate. *Journal of the American Chemical Society* 139 (28): 9435–9438.

152 Ji, X. and Wang, B. (2017). Toward carbon monoxide based therapeutics: carbon monoxide in a pill. *Pharmaceutical Patent Analyst* 6 (4): 171–177.

153 Chauveau, C., Bouchet, D., Roussel, J.-C. et al. (2002). Gene transfer of heme oxygenase-1 and carbon monoxide delivery inhibit chronic rejection. *American Journal of Transplantation* 2 (7): 581–592.

154 Kim, S.J., Jung, Y.S., Yoon, M.Y. et al. (2007). Comparative effects of dimethylsulfoxide on metabolism and toxicity of carbon tetrachloride and dichloromethane. *Journal of Applied Toxicology* 27 (1): 25–31.

155 Olson, K.R. (ed.) (2004). *Poisoning & Drug Overdose*. New York: Lange Medical Books/McGraw-Hill.

156 Ji, X., Pan, Z., Li, C. et al. (2019). Esterase-sensitive and pH-controlled carbon monoxide prodrugs for treating systemic inflammation. *Journal of Medicinal Chemistry* 62 (6): 3163–3168.

157 Soboleva, T., Simons, C.R., Arcidiacono, A. et al. (2019). Extracellular vs intracellular delivery of CO: does it matter for a stable, diffusible gasotransmitter? *Journal of Medicinal Chemistry* 62 (21): 9990–9995.

158 Indranil Chakraborty, I. and Mascharak, P.K. (2016). Mesoporous silica materials and nanoparticles as carriers for controlled and site-specific delivery of gaseous signaling molecules. *Microporous Mesoporous Materials* 234: 409–419.

159 Ma, M., Noei, H., Mienert, B. et al. (2013). Iron metal-organic frameworks MIL-88B and NH2-MIL-88B for the loading and delivery of the gasotransmitter carbon monoxide. *Chemistry* 19 (21): 6785–6790.

160 Zhou, Y., Yu, W., Cao, J. et al. (2020). Harnessing carbon monoxide-releasing platforms for cancer therapy. *Biomaterials* 255: 120193.

161 Joshi, H.P., Kim, S.B., Kim, S. et al. (2019). Nanocarrier-mediated delivery of CORM-2 enhances anti-allodynic and anti-hyperalgesic effects of CORM-2. *Molecular Neurobiology* 56 (8): 5539–5554.

162 Yin, H., Fang, J., Liao, L. et al. (2014). Styrene-maleic acid copolymer-encapsulated CORM2, a water-soluble carbon monoxide (CO) donor with a constant CO-releasing property, exhibits therapeutic potential for inflammatory bowel disease. *Journal of Controlled Release* 187: 14–21.

163 Nguyen, D., Oliver, S., Adnan, N.N.M. et al. (2016). Polymer–protein hybrid scaffolds as carriers for CORM-3: platforms for the delivery of carbon monoxide (CO). *RSC Advances* 6: 92975.

164 Hasegawa, U., van der Vlies, A.J., Simeoni, E. et al. (2010). Carbon monoxide-releasing micelles for immunotherapy. *Journal of the American Chemical Society* 132 (51): 18273–18280.

165 Nguyen, D., Adnan, N.N.M., Oliver, S. et al. (2016). The interaction of CORM-2 with block copolymers containing poly(4-vinylpyridine): macromolecular scaffolds for carbon monoxide delivery in biological systems. *Macromolecular Rapid Communications* 37 (9): 739–744.

166 Govender, P., Pai, S., Schatzschneider, U. et al. (2013). Next generation photoCORMs: polynuclear tricarbonylmanganese(I)-functionalized polypyridyl metallodendrimers. *Inorganic Chemistry* 52 (9): 5470–5478.

167 Nguyen, D. and Boyer, C. (2015). Macromolecular and inorganic nanomaterials scaffolds for carbon monoxide delivery: recent developments and future trends. *ACS Biomaterials Science & Engineering* 1 (10): 895–913.

168 Faizan, M., Muhammad, N., Niazi, K.U.K. et al. (2019). CO-releasing materials: an emphasis on therapeutic implications, as release and subsequent cytotoxicity are the part of therapy. *Materials* 12 (10): 1643.

169 Nagarajan, R. (2002). Molecular packing parameter and surfactant self-assembly: the neglected role of the surfactant tail. *Langmuir* 18 (1): 31–38.

170 Sharma, N.K., Singh, M., and Bhattarai, A. (2016). Hydrophobic study of increasing alkyl chain length of platinum surfactant complexes: synthesis, characterization, micellization, thermodynamics, thermogravimetrics and surface morphology. *RSC Advances* 6: 90607.

171 Has, C. and Pan, S. (2020). Vesicle formation mechanisms: an overview. *J Liposome Res. Mar* 3: 1–22.

172 Griffiths, P.C., Fallis, I.A., and James, C. (2010). Structure–property relationships in metallosurfactants. *Soft Matter* 6: 1981–1989.

173 Sakla, R. and Jose, D.A. (2018). Vesicles functionalized with a co-releasing molecule for light-induced CO delivery. *ACS Applied Materials & Interfaces* 10 (16): 14214–14220.

174 Hainfeld, J.F., Furuya, F.R., and Powell, R.D. (1999). Metallosomes. *Journal of Structural Biology* 127 (2): 152–160.

175 Wang, J., Wang, A.Z., Lv, P. et al. (2018). Advancing the pharmaceutical potential of bioinorganic hybrid lipid-based assemblies. *Advancement of Science* 5 (9): 1800564.

176 Kaur, B., Chaudhary, G.R., and Kaur, G. (2019). Cholesterol-induced physicochemical changes in dodecylamine-based metallosomes: drug entrapping ability and interactions with biological molecules. *Journal of Materials Chemistry B* 7: 3679–3691.

177 Kaur, G., Kaur, B., Garg, P. et al. (2020). A study of synthesis, characterization and metalloplex formation ability of cetylpyridinium chloride based metallosomes. *Journal of Molecular Liquids* 300: 112326.

178 Garg, P., Kaur, B., Kaur, G. et al. (2020). A study of the spectral behaviour of eosin dye in three states of metallosurfactants: monomeric, micelles and metallosomes. *Colloids and Surfaces A: Physicochemical and Engineering Aspects* https://doi.org/10.1016/j.colsurfa.2020.125697.

179 Kaur, B., Kaur, G., Chaudhary, G.R. et al. (2020). An investigation of morphological, microscopic dynamics, fluidity, and physicochemical variations in Cu-decorated metallosomes with cholesterol. *Journal of Molecular Liquids* 318: 114034.

180 Osada, K., Cabral, H., Mochida, Y. et al. (2012). Bioactive polymeric metallosomes self-assembled through block copolymer-metal complexation. *Journal of the American Chemical Society* 134 (32): 13172–13175.

181 Bangham, A.D. (1963). Physical structure and behavior of lipids and lipid enzymes. *Advances in Lipid Research* 1: 65–104.

182 https://archive.org/details/bagsoflife and www.open.ac.uk/library/digital-archive/program/video:00525_1107 give a historical perspective of liposomes and membranes.

183 Gao, W., Hu, C.-M.J., Fang, R.H. et al. (2013). Liposome-like nanostructures for drug delivery. *Journal of Materials Chemistry B* 1 (48) https://doi.org/10.1039/C3TB21238F.

184 Vanić, Z., Barnert, S., Süss, R. et al. (2012). Fusogenic activity of PEGylated pH-sensitive liposomes. *Journal of Liposome Research* 22 (2): 148–157.

185 Filipczak, N., Pan, J., Yalamarty, S.S.K. et al. (2020). Recent advancements in liposome technology. *Advanced Drug Delivery Reviews* 156: 4–22. S0169-409X(20)30068-5.

186 Al Sawaftah, N.M. and Husseini, G.A. (2020). Ultrasound-mediated drug delivery in cancer therapy: a review. *Journal of Nanoscience and Nanotechnology* 20 (12): 7211–7230.

187 Montanari, J., Maidana, C., Esteva, M.I. et al. (2010). Sunlight triggered photodynamic ultradeformable liposomes against *Leishmania braziliensis* are also leishmanicidal in the dark. *Journal of Controlled Release* 147 (3): 368–376.

188 Kapoor, B., Gupta, R., Singh, S.K. et al. (2018). Prodrugs, phospholipids and vesicular delivery-an effective triumvirate of pharmacosomes. *Advances in Colloid and Interface Science* 253: 35–65.

189 Cai, S., Yang, Q., Bagby, T.R. et al. (2011). Lymphatic drug delivery using engineered liposomes and solid lipid nanoparticles. *Advanced Drug Delivery Reviews* 63 (10–11): 901–908.

190 Slütter, B., Bal, S.M., Ding, Z. et al. (2011). Adjuvant effect of cationic liposomes and CpG depends on administration route. *Journal of Controlled Release* 154 (2): 123–130.

191 van den Hoven, J.M., Van Tomme, S.R., Metselaar, J.M. et al. (2011). Liposomal drug formulations in the treatment of rheumatoid arthritis. *Molecular Pharmaceutics* 8 (4): 1002–1015.

192 Romero, E.L. and Morilla, M.J. (2011). Topical and mucosal liposomes for vaccine delivery. *Wiley Interdisciplinary Reviews. Nanomedicine and Nanobiotechnology* 3 (4): 356–375.

193 Shah, S., Dhawan, V., Holm, R. et al. (2020). Liposomes: advancements and innovation in the manufacturing process. *Advanced Drug Delivery Reviews* 154-155: 102–122. S0169-409X(20)30078-8.

194 Has, C. and Sunthar, P. (2020). A comprehensive review on recent preparation techniques of liposomes. *Journal of Liposome Research* 30 (4): 336–365.

195 Barnadas-Rodríguez, R. and Sabés-Xamaní, M. (2003). Liposomes prepared by high-pressure homogenizers. *Methods in Enzymology* 367: 28–46.

196 Gabizon, A.A., Shmeeda, H., and Zalipsky, S. (2006). Pros and cons of the liposome platform in cancer drug targeting. *Journal of Liposome Research* 16 (3): 175–183.

197 Accardo, A. and Morelli, G. (2015). Review peptide-targeted liposomes for selective drug delivery: advantages and problematic issues. *Biopolymers* 104 (5): 462–479.

198 Manjappa, A.S., Chaudhari, K.R., Venkataraju, M.P. et al. (2011). Antibody derivatization and conjugation strategies: application in preparation of stealth immunoliposome to target chemotherapeutics to tumor. *Journal of Controlled Release* 150 (1): 2–22.

199 Tian, J., Yang, F., Cui, H. et al. (2015). A novel approach to making the gas-filled liposome real: based on the interaction of lipid with free nanobubble within the solution. *ACS Applied Materials & Interfaces* 7 (48): 26579–26584.

200 Vangala, A., Morris, R., Bencsik, M. et al. (2007). Preparation and characterization of gas-filled liposomes: can they improve oil recovery? *Journal of Liposome Research* 17 (3–4): 263–272.

201 Parera, E., Comelles, F., Barnadas, R. et al. (2011). Formation of vesicles with an organometallic amphiphile bilayer by supramolecular arrangement of metal carbonyl metallosurfactants. *Chemical Communications* 47 (15): 4460–4462.

202 Parera, E., Marín-García, M., Pons, R. et al. (2016). Supramolecular arrangement of molybdenum carbonyl metallosurfactants with CO-releasing properties. *Organometallics* 35 (4): 484–493.

203 Marín-García, M., Benseny-Cases, N., Camacho, M. et al. (2017). Low-toxicity metallosomes for biomedical applications by self-assembly of organometallic metallosurfactants and phospholipids. *Chemical Communications* 53 (60): 8455–8458.

204 Marín-García, M., Benseny-Cases, N., Camacho, M. et al. (2018). Metallosomes for biomedical applications by mixing molybdenum carbonyl metallosurfactants and phospholipids. *Dalton Transactions* 47 (40): 14293–14303.

205 Barnadas-Rodríguez, R. (2013). Effect and mechanism of association of 8-hydroxy-1,3,6-pyrenetrisulfonic acid to chitosan: physicochemical properties of the complex. *Macromolecular Chemistry and Physics* 213: 99–106.

206 Parera, E., Comelles, F., Barnadas, R. et al. (2010). New surfactant phosphine ligands and platinum(II) metallosurfactants. Influence of metal coordination on the critical micelle concentration and aggregation properties. *Langmuir* 26 (2): 743–751.

207 Larpent, C. and Patin, H. (1987). Hydrosoluble transition-metal coordination compounds of triphenylphosphine m-trisulfonate. *Applied Organometallic Chemistry* 1: 529–534.

208 Ardon, M., Hogarth, G., and Oscroft, D.T.W. (2004). Organometallic chemistry in a conventional microwave oven: the facile synthesis of group 6 carbonyl complexes. *Journal of Organometallic Chemistry* 689: 2429–2435.

209 van Assema, S.G.A., de Kanter, F.J.J., Schakel, M. et al. (2006). Decomplexation of phosphirane and phosphirene complexes. *Organometallics* 25 (22): 5286–5291.

12

Supramolecular Metal-Modified Nanocontainers Based on Amphiphilic and Hybrid Matrix: Self-Assembling Behavior and Practical Applications

Elena P. Zhiltsova[1], *Marina R. Ibatullina*[1], *Ruslan R. Kashapov*[1], *Nadezda E. Kashapova*[1], *Albina Y. Ziganshina*[1,2], *Lucia Y. Zakharova*[1], *and Oleg G. Sinyashin*[1]

[1] *A.E. Arbuzov Institute of Organic and Physical Chemistry, FRC Kazan Scientific Center of RAS, 8 Arbuzov str., Kazan 420088, Russia*
[2] *A.M. Butlerov Institute of Chemistry, Kazan Federal University, Department of Organic and Medical Chemistry, 29 Kremlevsky str., Kazan 420008, Russia*

12.1 Introduction

Recently, significant progress has been achieved in the development of drug delivery systems, which makes it possible to increase the safety and efficacy of pharmaceutics, minimize side effects, and improve the quality of treatment of dangerous diseases [1–3]. For these purposes, different families of nanocarriers have been fabricated, including soft supramolecular nanocontainers [4, 5], macrocyclic [6, 7], and hybrid nanostructured materials [8, 9]. Self-assembling supramolecular systems attract special attention due to their tunable morphology and gradient of properties in nanoscale dimension, which provide the tool for multifactor control of the binding/release behavior of encapsulated guests. Beneficial features of nanocontainers based on amphiphilic compounds are their biocompatibility, membrane tropic properties, low toxicity, and binding ability toward both hydrophilic and lipophilic drugs [4, 5]. Importantly, these self-assembling nanocontainers are capable of noncovalent modification that provides simple and versatile way for the fabrication of novel generations of nanocarriers with improved characteristics. Thus, noncovalent strategy involving cationic surfactants was used to engineer the nanocarriers with targeted properties [10], antimicrobial activity [11], vehicles for gene delivery [12, 13], and the nanoformulations capable to overcome the blood–brain barrier [14]. Hybrid nanoplatforms, especially metal-modified nanomaterials [8, 9, 15], are employed to enlarge potential of their practical applications due to fine tailoring the structural characteristics of nanocontainers and involving additional interactions with both guest molecules and bioenvironment, thereby, improving loading capacity, targeting properties, selectivity, and therapeutic activity. Noteworthy, the

Metallosurfactants: From Fundamentals to Catalytic and Biomedical Applications, First Edition. Edited by Surinder K. Mehta and Ravneet Kaur.

introduction of metal opens novel perspectives for technological application of hybrid nanocontainers in the fields of diagnostics, catalysis, protective coating, imaging, etc. These issues are addressed in the chapter.

12.2 Structure, Properties, and Biomedical Application of Metallosurfactants

Metal-containing surfactants ensure their visibility through the expanded range of applications. This is caused by the presence of not only the properties of ordinary surfactants, among which - the spatially oriented adsorption at the phase boundary, the ability to self-assemble with the formation of supramolecular structures of different morphology, and the solubilization of poorly soluble substances are of particular importance. One of the most unique properties of metallosurfactant is of course their enhanced redox activity and magnetic properties [16].

The transition metals, such as copper [17], nickel [18], cobalt [19–21], zinc [22], iron [23], and lanthanides [23], are usually used as a metal part of metallosurfactants. The hydrophobically modified nitrogen-, oxygen- (capable of acting as electron donors to metals), sulfur-, phosphorus-containing compounds of linear (amines, amides, amine acetates, sulfates) [24, 25] and cyclic (pyridine-, imidazole-, triazole-, phenanthroline-based surfactants) [26] structures are used as ligands (Table 12.1). Moreover, the variation of metallosurfactant structure is carried out according to the number of amphiphilic ligands at the metal atom [32], the number of long-chain substituents in ligands [20], and the length of hydrophobic chains [33]. An important feature of metallosurfactants is also the location of the metal in the complex. The metal may be a part of the head group [26], the hydrophobic part of surfactant [28], or act as a counterion of various structures [29] (Table 12.1).

Depending on the synthesis conditions, metallosurfactants can be divided into three types. The synthesis of metallosurfactants of the first type is carried out either by one- (Scheme 12.1) [34] or two-step (Scheme 12.2) [21] procedures and, as a rule, in organic solvents. Molybdenum carbonyl metallosurfactants $[Mo(CO)_5L]$ ($L = Ph_2P(CH_2)_nSO_3Na$, $n = 2, 6, 10$) of the second type were synthesized in methanol/THF from $Mo(CO)_6$ through the intermediate $[Mo(CO)_5(CH_3CN)]$ by decarboxylation with Me_3NO and subsequent reaction with functionalized phosphine. Metallosurfactant $[Mo(CO)_4L_2]$ of the same type was obtained via the intermediate $[Mo(CO)_4(pip)_2]$ complex and its subsequent reaction with a ligand [28]. The synthesis of complexes of the third type can proceed in one (by mixing the ligand and the corresponding salt [29]) or two stages (Scheme 12.3) [16] in ethanol [16, 29], water [30, 31], or dichloromethane [23].

Many morphological structures formed in the metallosurfactant-based systems have been discussed including the aggregates of micellar [19], vesicular [34], liposomal (metallosomes) [35], and lamellar structures [30]. Metal complexes can be part of microemulsions [36], form Langmuir–Blodgett films [37], and also act as a hybrid additive to highly porous silicon-based materials [17]. The possibility of the transition of morphological structures into each other is also noted. Using the

Table 12.1 Different types of metallosurfactant structures.

Metallosurfactants	Metal (M)	References
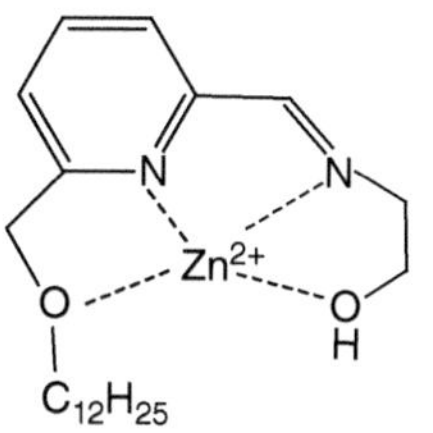	Zn(II)	[26]
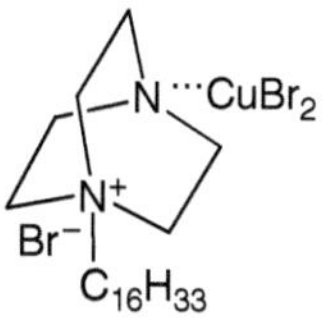	Cu(II)	[27]
$[Mo(CO)_5L]$, $[Mo(CO)_4L_2]$, $L = Ph_2P(CH_2)_nSO_3Na$ ($n = 2, 6, 10$)	Mo(VI)	[28]
$[C_{16}H_{33}(CH_3)_3N^+]_2[MCl_4]^{2-}$	Fe(II), Co(II), Ni(II), Cu(II)	[29]
$[CP]^+$ $[MCl_3]^-$, CP – cetylpyridinium	Co(II)	[19]
$[M(CH_3COO)_4]^{2-}$ $[C_{12}H_{25}NH_3^+]_2$	Cu(II), Ni(II)	[16]
$[C_{14}H_{29}(C_6H_{13})_3P]_2^+$ $[MCl_4]^{2-}$ $[C_{14}H_{29}(C_6H_{13})_3P]^+$ $[MCl_4]^-$ $[C_{14}H_{29}(C_6H_{13})_3P]_3^+$ $[MCl_6]^{3-}$	Co(II), Mn(II) Fe(III) Gd(III)	[23]
$M(DS)_3$ mH_2O, DS – dodecyl sulfate	Al(III), Cr(III), Gd(III), La(III)	[30, 31]
$M(DS)_n$, DS – dodecyl sulfate	Ni(II), Co(II), Cu(II), Mg(II), Mn(II)	[25]
$M(TS)_n$, TS – tetradecyl sulfate	Cu(II)	[31]

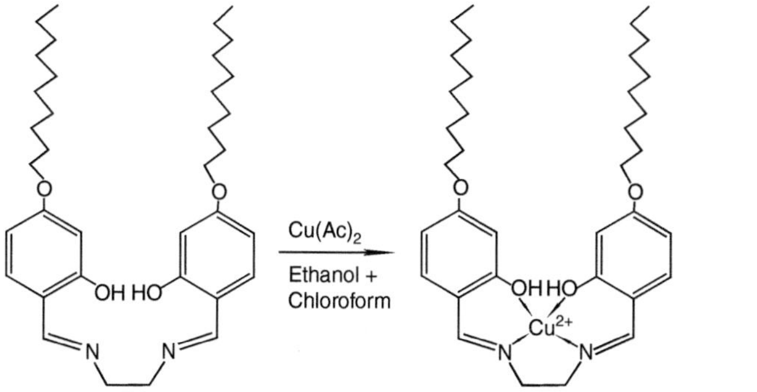

Scheme 12.1 One-stage synthesis of metallosurfactant with Cu(II) in the head group. Source: Liang et al. [34].

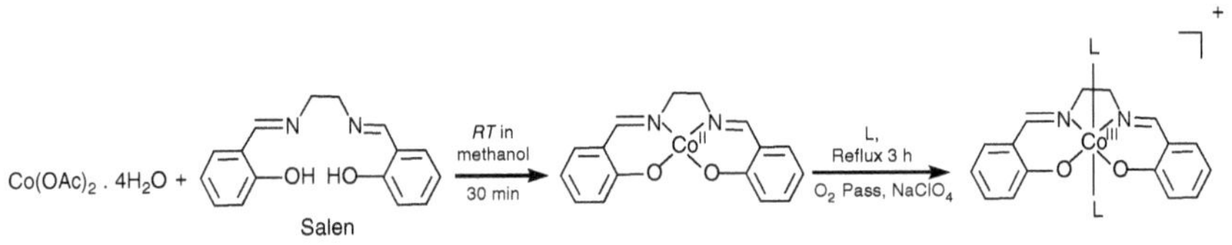

Scheme 12.2 Two-stage synthesis of metallosurfactant with Co(III) in the head group (L = dodecylamine). Source: Ambika et al. [21].

$$C_{12}H_{25}NH_2 + CH_3COOH \rightarrow C_{12}H_{25}NH_3^+\ CH_3COO^-$$

$$M(CH_3COO)_2 + 2C_{12}H_{25}NH_3^+\ CH_3COO^- \rightarrow [M(CH_3COO)_4]^{2-}[C_{12}H_{25}NH_3^+]_2$$

where M: Cu, Ni.

Scheme 12.3 Two-stage synthesis of metallosurfactants with Cu(II), Ni(II) in the counterion. Source: Mehta et al. [16].

cryo-transmission electron microscopy (TEM) method, the authors of [38] showed that metallomicelles can coexist with vesicles or easily transform into them. Such transitions are possible due to the concentration change of the metal complex in solution [38], as well as the presence of a metal center in the metallosurfactant. The latter is discussed in the review by Owen and Butler [39], where the authors describe the surface-active ligands that undergo a change in morphology due to coordination of metal ion (coordination-induced micelle-to-vesicle transition). Amphiphilic siderophores were produced by bacteria and different synthetic amphiphilic ligands (carboxylate surfactants, octadecylamine, imidazolium surfactants, 2,2′-bipyridine surfactants). The mechanism of the metal-induced phase transition was considered in terms of coordination chemistry, nature of ligands, and changes in the molecular geometry as a result of metal coordination.

The conventional functional properties of metallosurfactants are their ability to aggregate in solutions and solubilize important but poorly soluble substances (dyes, drugs) in the micellar phase. This determines the possibility of their use as extractants, nanocarriers, and catalysts. The aggregation parameters (critical micelle concentration [CMC], aggregation number, and micelle shape) substantially depend on the basic structural units of metallosurfactants (head group, metal, and length of hydrocarbon tail) [40]. Herewith, there is a classical decrease of CMC with an increase in the length of the hydrocarbon tail and a non-classical increase of CMC with an increase in the number of methylene units in the cycle of the ligand head group. The latter may be caused by a higher charge on the head group, a steric factor on the micellar surface, and a reduced shielding of the hydrophobic micelle core [40]. The influence of metal in a surfactant molecule and its nature on micelle-forming properties also has its own peculiarities. The introduction of a metal cation can either increase the CMC of the azo-macrocyclic ligand containing Tb(III), Yb(III), and Y(III) [40] or have the opposite beneficial effect in the case of Cu(II)-, Co(II)-, Ni(II)-, and La(III)-containing monoalkylated 1-hexadecyl-1,4-diazabicyclo[2.2.2]octanes (D-16) [27, 32]. The Krafft temperature also depends on the chemical structure of metallosurfactant and the nature of metal included in its composition [41].

Mukherjee et al. studied the thermodynamics of micelle formation of complexes between alkyltrimethylammonium bromides (ATABs), sodium molybdate, and sodium tungstate [42]. The standard free energy of micellization value ($\Delta G^{\circ}{}_{m}$) of both metallosurfactants was close, but higher than their precursor surfactants, ATABs, in absolute value, i.e. the micelle formation of the metallosurfactants was more spontaneous. An increase in the hydrophobicity of the complexes also favored the formation of aggregates. The values of the standard entropy change of micelle formation ($\Delta S^{\circ}{}_{m}$) for all metallosurfactants were positive, significant in magnitude, and higher than $\Delta S^{\circ}{}_{m}$ of the corresponding ATABs. The much smaller values of absolute enthalpy of micellization ($\Delta H^{\circ}{}_{m}$) of complexes, compared with $\Delta G^{\circ}{}_{m}$, indicate that the process is entropy controlled.

Zhiltsova, Ibatullina, et al. studied the solubilizing effect of D-n-based metallosurfactant (n = 14–18) with respect to the water-insoluble dye Orange OT [27, 32] and poorly soluble drugs with a broad spectrum of activities (quercetin [43], nitrofurantoin [44], griseofulvin [45]) and identified the dependence of their functional properties on the nature of metal. D-16 in a complex with copper(II) and lanthanum(III) ions has an increased solubilizing ability. The solubilization capacity of these complexes is up to 2.5 times greater than the capacity of D-16 and significantly higher (up to 6.5 times) than the capacity of cetyltrimethylammonium bromide [32]. The high solubilization ability of metallomicellar systems was also observed toward the drugs used. The increase in the solubility of quercetin reaches 270 times, and the solubility of griseofulvin increases by 1 order of magnitude [43, 44]. It is also known that the metal complexes exhibit good antimicrobial (antibacterial, antifungal) and anticancer activities [46]. At the same time, D-16-based metallosurfactants simultaneously exhibit antibacterial and antifungal activities, which are higher than those of the test compounds in a number of strains [47, 48].

An important feature of metallosurfactants expanding the field of biomedical applications, is also their ability to form complexes with biomolecules. The latter can be proteins (bovine serum albumin, BSA) [29], nucleic acid (DNA) [49–51], or its decamer (oligonucleotide) [27]. From a biomedical point of view, apart from being used as biologically active substances with antimicrobial, antifungal, anticancer activities [20, 47, 48, 50, 51], metallosurfactants are also utilized as nanocontainers for useful substances (including drugs and genetic material) in targeted delivery. Gd^{3+}, $^{64}Cu^{2+}$, $^{111}In^{3+}$-containing metallosurfactants in micellar, polymer–colloidal, or multifunctional liposomal systems can be used for diagnostic purposes in magnetic resonance, optical, and single-photon emission tomography [52]. All this testifies to the high application potentials of metallosurfactants and the prospects for the development of research in the field of biology and medicine.

12.3 Amphiphilic System Based on Metallosurfactants and Macrocycles

The preorganization of binding elements afforded by macrocyclic scaffolds is useful for the development of interesting ligands. The structural modifications of a macrocycle can result in local conformational changes that adapt easily to the geometric requirements of metal cations. The interaction between macrocycles and metals provides not only high values of the binding constants, but can also initiate the aggregation of macrocycles without alkyl tails.

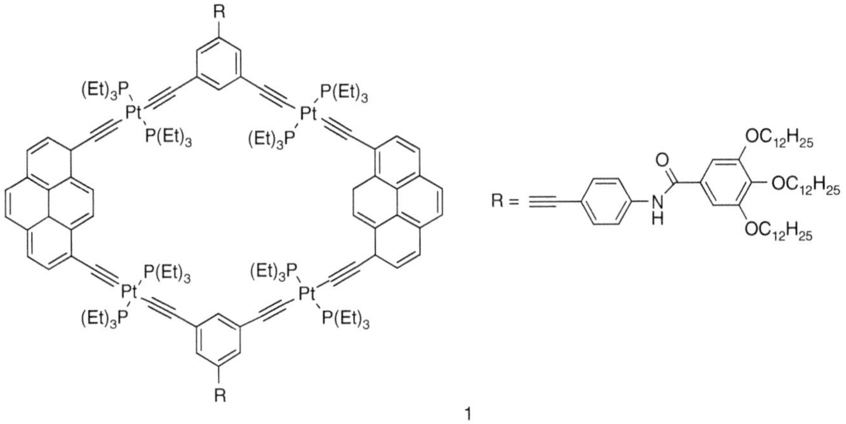

1

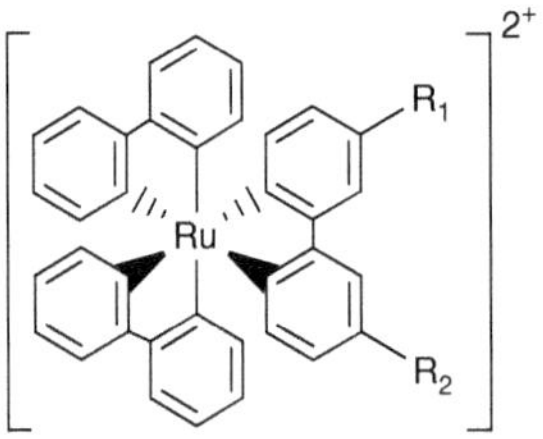

2a $R_1 = C_{11}H_{21}$, $R_2 = C_{11}H_{21}$
2b $R_1 = C_{11}H_{21}$, $R_2 = CH_3$
2c $R_1 = C_{13}H_{27}$, $R_2 = CH_3$
2d $R_1 = C_{13}H_{27}$, $R_2 = C_{13}H_{27}$

Tashiro et al. described the cation-induced aggregation of hexaaza-cyclophanes without long-chain alkyl groups [53]. The coordination between Pd^{2+} and macrocycle initiates the formation of submicrospherical aggregates upon complexation with macrocyclic ligands consisting of three phenylenediamine and *p*-phenylene fragments. The formed trinuclear palladium complexes are considered as amphiphiles, consisting of three hydrophilic Pd^{2+}–*o*-phenylenediamine fragments and six hydrophobic phenylene moieties. These complexes are stable in weak acid medium, while they are slowly colored in basic or neutral aqueous solutions probably due to the oxidation of Pd^{2+}–*o*-phenylenediamine fragments as a result of deprotonation of N–H groups. Nevertheless, for greater stability, the complexation of metal is more favorable with an alkylated macrocycle than with non-alkylated one. The amphiphilic platinum–acetylide metallocycle **1** with clearly defined shape and size was obtained in ref. [54]. The obtained metallocycle based on platinum acetylide proved to be stable to air and humidity at room temperature. The study of its supramolecular aggregation showed that the hydrophobic interaction between alkyl chains and the hydrogen bond between amides are the driving forces for the formation of ordered aggregates.

Lebrón J.A. et al. synthesized two Ru(II)-based metallosurfactants bearing three 2,2′-bipyridine ligands, one of which was alkylated with an undecyl residue and methyl group **2b**, and the other with two undecyl groups **2a** [55]. It was shown the formation of micellar aggregates between the cationic head groups of the metallosurfactant and the anionic phosphate residues of the polynucleotide results in condensation of the polynucleotide at a certain concentration of metallosurfactants. In ref. [56], natural cyclodextrins (CDs) interact with both investigated metallosurfactants **2a**, **2c** forming inclusion complexes, where the undecyl surfactant tails enter the macrocycle cavity. The addition of macrocyclic molecules [55] complicates the self-assembly of metallosurfactants, therefore, preventing the interaction of metallosurfactant with polynucleotide.

The study on mixed systems between metallosurfactants and macrocycle, particularly α-CD, is presented in ref. [57]. At the first stage, the dynamic metallo-supramolecular amphiphiles were obtained in aqueous solution by the interaction of cationic Pt(II) complex of 2,6-bis(*N*-triethyleneglycolbenzimidazol-2-yl)-pyridine **3** with anionic surfactants sodium dodecyl sulfonate (SDS), and sodium dodecanoate (SD). Using spectrophotometric, spectrofluorimetric, and dynamic light scattering (DLS) methods, it was found that the addition of SDS and SD into the aqueous solutions of cationic Pt(II) complex leads to the formation of ordered aggregates with the diameters of 307 and 246 nm, respectively.

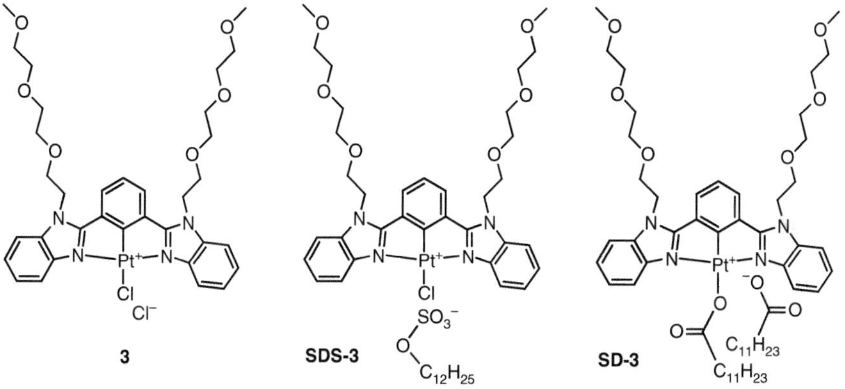

3 **SDS-3** **SD-3**

TEM image proved the formation of rod-like aggregates for SDS-cationic Pt(II) complex and sheet-like aggregates for SD-cationic Pt(II) complex that was explained by the larger hydrophobic size of SD-Pt(II) complex in comparison with SDS-Pt(II) complex. The driving forces of aggregation were the electrostatic, hydrophobic, Pt/Pt, and π–π stacking interactions. At the second stage of the work, the obtained aggregates were titrated with α-CD. As the inclusion complexes with α-CD were formed, the hydrophobic interactions of metallosupramolecular amphiphiles destroyed and the aggregates of SDS and SD with cationic Pt(II) complex dissociated into small particles of 3.2 and 1.4 nm, respectively. Moreover, during the host–guest recognition of α-CD, the signals of the metal–metal-to-ligand-charge transfer (MMLCT) absorption and ^{3}MMLCT emission bands returned back to the original levels, which made it possible to regulate the photophysical properties of the system.

Another study on joint aggregation between metallosurfactant and CD is described by Iza et al. [58]. Using a set of physico-chemical methods, the aggregation properties of chiral metallosurfactant bis(2,2′-bipyridine)(4,4′-ditridecyl-2,2′-bipyridine)ruthenium(II) dichloride **2d** were investigated. Initially, the formation of metallosurfactant aggregates was detected by UV–vis and fluorescence spectroscopies. Then, to determine the shape, the internal structure, and the size of aggregates, the neutron-scattering curves were recorded and DLS experiments were performed. The obtained data indicated the prolate ellipsoid structure of the micelles with the aggregation number of 14. As in ref. [57], the addition of CDs (β-CD and γ-CD) lead to the destruction of micelles due to complexation of macrocycle with metallosurfactant. The formation of inclusion complexes was also confirmed by ^{1}H NMR spectroscopy on the basis of chemical shifts of protons in macrocycle. The stoichiometry of the complexes was determined by Job's method. The spatial structure of the metallosurfactant with two long tails favors the formation of complexes with β-CD and γ-CD in 1 : 3 stoichiometry, where one CD molecule gets through one of the alkyl tails and two CD molecules get through the other tail, close to the polar head group. Finally, it was shown that the complex with β-CD was about 2 orders of magnitude more stable than the complex with γ-CD.

Some amphiphilic metal complexes of synthetic macrocycles attract special attention of researchers as prototypes of active centers in metalloproteins that perform catalytic functions in biological systems. For example, the different micellar and vesicular aggregates based on complexes of octadecanoyl derivative of cyclen with zinc **4a**, **4b**, **5a**, **5b**, **5c** can be used to accelerate DNA cleavage [59]. In addition, zinc complexes of cyclen with long-chain substituents **4c**, **4d** were also investigated in the cleavage of *p*-nitrophenylphosphate in a micellar solution with Triton X-100 [60]. The authors of ref. [61] reported an increase in the cleavage activity of BSA and myoglobin by amphiphilic complexes of cyclen with Cu(II) and Co(III) **6a**–**6f** due to hydrophobic interactions with cleaved proteins, and the cleavage activity increases with increasing chain length. The alkyl chain of these metal complexes is a kind of "anchor" on the hydrophobic domain of proteins, which allows the macrocyclic head to be located next to the peptide bond to be hydrolyzed. On the other hand, in the structure of enzymatic model not only the active center, but also the microenvironment around is considered. Micelles are often used to simulate the microenvironment of enzymes due to the similar identity of the structures of micelles and globular proteins, as well as the similarity of the substrate-binding ability of micelles and enzymes. Based on the foregoing, complexes of aza-macrocyclic compound bearing carboxyl group with lanthanides (lanthanum **7a** and cerium **7b**) can be used as a center of catalytic activity, and a metallomicellar system consisting of lanthanide(III) complex and micelles of zwitterionic surfactant, *n*-lauroylsarcosine, was developed as a mimetic phosphate hydrolase model for the hydrolytic cleavage of bis(4-nitrophenyl)phosphate [62]. The rate of phosphate hydrolysis, catalyzed by a macrocyclic complex in a metallomicellar system, is approximately 50 times higher than in an aqueous solution under the same conditions. In a weakly alkaline medium, lauroylsarcosine is in the form of an anion; therefore, the complex easily dissolves in the Stern layer of the surfactant micelles due to electrostatic attraction between the positively charged complex and the anionic micelle. The latter increases the local concentration and the frequency of collisions of reaction molecules due to the solubilizing effect of micelles and thus promotes the catalytic hydrolysis of phosphate.

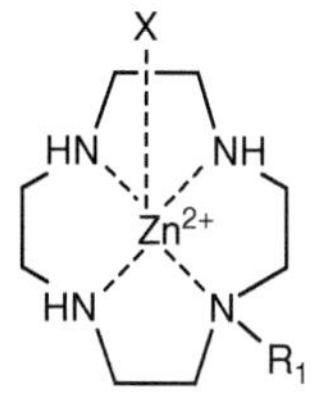

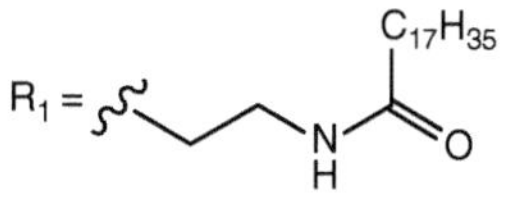

; $X = 2ClO_4^-$ **4a**

$R_1 = H$; $X = 2ClO_4^-$ **4b**

$R_1 = C_{16}H_{33}$; $X = OH_2$ **4c**

$R_1 = C_{16}H_{33}$; $X = OH^-$ **4d**

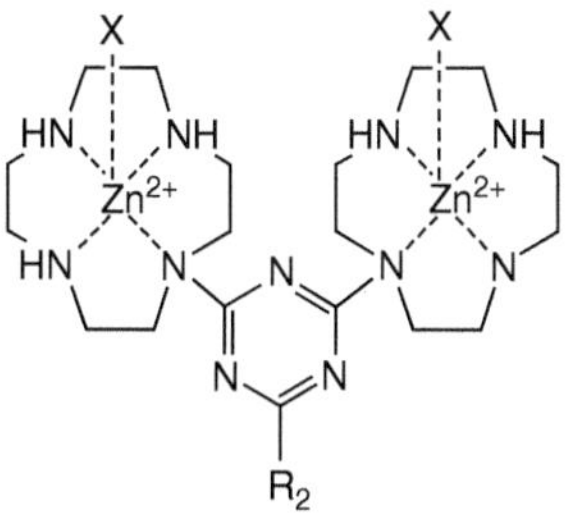

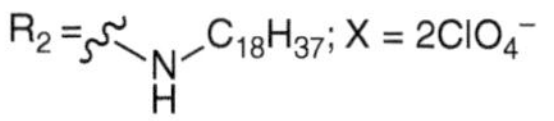

$X = 2ClO_4^-$ **5a**

$R_2 = OMe$; $X = 2ClO_4^-$ **5b**

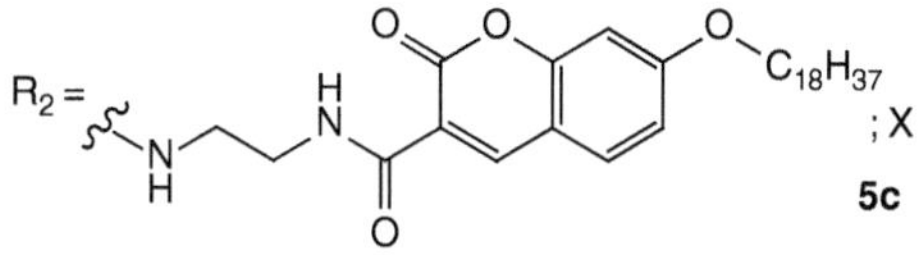

; $X = 2ClO_4^-$ **5c**

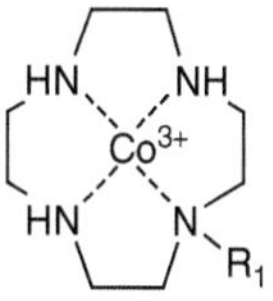

$R_1 = C_{10}H_{21}$ **6a**

$R_1 = C_{12}H_{25}$ **6b**

$R_1 = C_{16}H_{33}$ **6c**

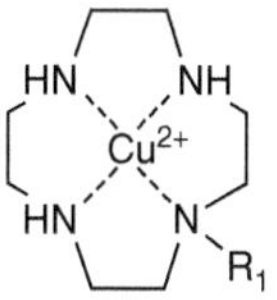

$R_1 = C_{10}H_{21}$ **6d**

$R_1 = C_{12}H_{25}$ **6e**

$R_1 = C_{16}H_{33}$ **6f**

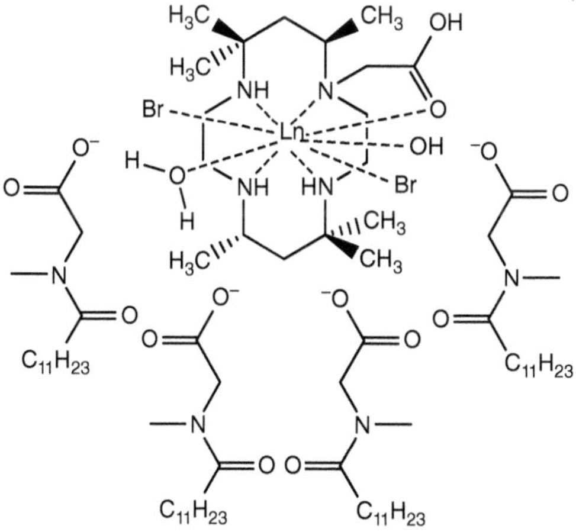

$Ln = La^{3+}$ **7a**

$Ln = Ce^{3+}$ **7b**

The chemical modification of macrocyclic ligands by functionalization with alkyl substituents makes it possible to obtain active amphiphilic functional objects compatible with lipid matrices of colloidal carriers [63]. As noted above, using the reaction of hydrolysis of a model compound, bis-*p*-(nitrophenyl)phosphate, the catalytic activity of synthetic amphiphilic receptor (alkylated bis-cyclic zinc complex **4a**) in micellar and vesicular solutions was investigated [59]. The course of the reaction was monitored by an increase in the absorption intensity of one of the hydrolysis products, nitrophenolate, in the region of 400 nm. It was shown that the catalytic activity of the complex increases significantly even in a micellar solution as compared to a molecular solution of non-alkylated analog. The highest efficiency of catalytic hydrolysis with the participation of an amphiphilic macrocyclic metal complex was achieved in a vesicular system based on 1,2-distearoyl-*sn*-glycero-3-phosphocholine. In continuation of the above work, the mixed Langmuir monolayers of this lipid and the amphiphilic receptor **5a** were obtained [64]. Immobilization of the receptor in the lipid matrix leads to an increase in the rate of decomposition of bis-(*p*-nitrophenyl) phosphate in the monolayer by almost an order of magnitude.

There are few works devoted to investigation of mixed assemblies between metallosurfactant and macrocycles. Roldán-Carmona et al. in ref. [65] described the formation of iridium metallosurfactant–porphyrin film. The latter was formed by Langmuir–Schaefer technique, which allowed authors to combine two luminescent species, amphiphilic monocationic Ir(III) metallosurfactant **8a** and tetraanionic porphyrin **8b**, together. Using macrocycle molecules as counterions with the ratio, when the monolayer is neutral, promotes homogeneity and distribution of metallosurfactant molecules along the air–water interface. The compression does not lead to the release of porphyrin molecules into the aqueous phase, they remain distributed on the surface. Subsequent transfer of metallosurfactant–porphyrin monolayer from the air–water interface to quartz substrates has shown that Langmuir–Schaefer technique is more favorable than Langmuir–Blodgett one. The transfer ratios with the first mentioned technique were close to 1. Further comparison of the intermolecular energy transfer between iridium metallosurfactant and porphyrin molecules in solution and solid state has shown the effectiveness of energy transfer process in the obtained Langmuir–Schaefer films. According to the luminescence spectrum of this film, even when the metallosurfactant and porphyrin molecules were separated by a lipid interlayer, the emission of the iridium metallosurfactant was totally quenched, meaning energy transfer process to the porphyrin molecules. The subsequent separation through the additional implementation of stearic acid interlayer to the above film led to the emission of both metallosurfactant and porphyrin, meaning no energy transfer.

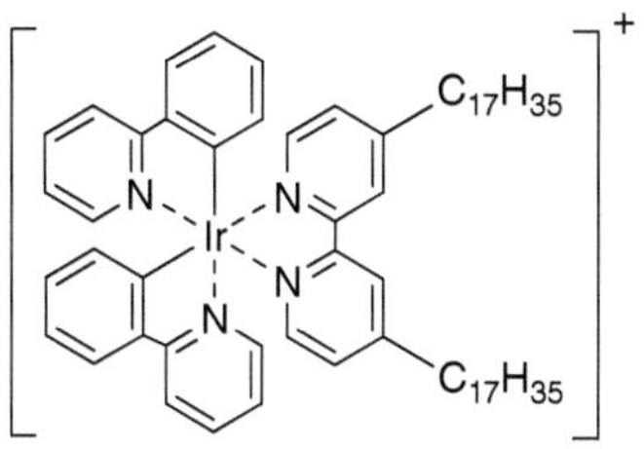

8a

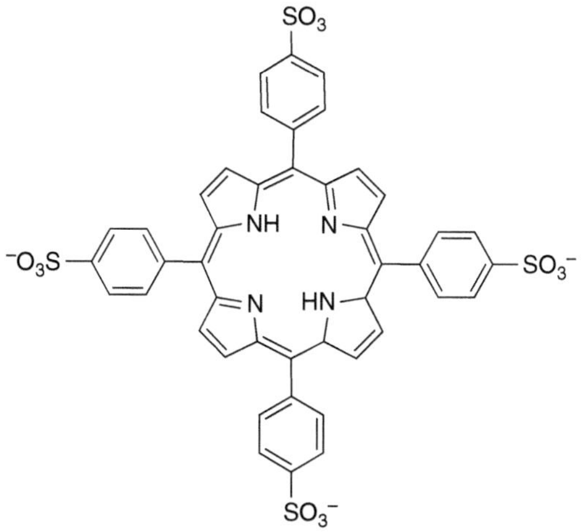

8b

Supramolecular aggregation has also been investigated for platinum complexes in aqueous media. De Cola et al. synthesized 1,3-di(2-pyridine)-benzene complexes of Pt(II) bearing hydrophilic chains of oligo(ethylene glycol) to increase solubility in water. The obtained amphiphilic molecules show a tendency to aggregate into phosphorescent hydrogels through the host–guest interactions between CDs and oligo(ethylene glycol) tail of organoplatinum complexes(II) [66]. In ref. [67], an amphiphilic cationic terpyridyl complex of Pt(II) containing a redox unit of ferrocene aggregated in water with the formation of vesicular aggregates that can be destroyed by a cucurbituril molecule capable of forming an equimolar inclusion complex with metallosurfactant. Disassembly of vesicles was probably associated with an increase in the hydrophilic nature of the ferrocene fragment upon complexation with cucurbituril. Interestingly, well-defined architectures can be reversibly switched using in situ electrochemistry. It is expected that this line of research will lead to the development of a promising new class of responsive vesicles for use in modern materials and drug delivery.

Metalloamphiphiles such as macrobicyclic hexamine ligands were investigated as template agents in the synthesis of mesoporous silicas [68]. Three macrocyclic metallosurfactants containing different alkyl chains and metals **9a**–**9c** were used for creation of the mesoporous silicate compositions. These metallosurfactants form micelles [69], and, in particular, the cobalt complex **9c** forms giant worm-like micelles [38], which can promote the formation of the long cylindrical structures required for the formation of mesoporous silica-based materials from the M41S family. Copper-containing structures of porous silica aggregate at high temperatures during the calcination, while cobalt-containing materials did not aggregate under the same conditions [68]. Apparently, macrocyclic metallosurfactant was more successfully incorporated into the structure of cobalt-containing silica, herewith the metal was distributed throughout, and no signs of aggregation were observed. However, in both cases well-ordered mesoporous materials with a highly developed surface area were obtained.

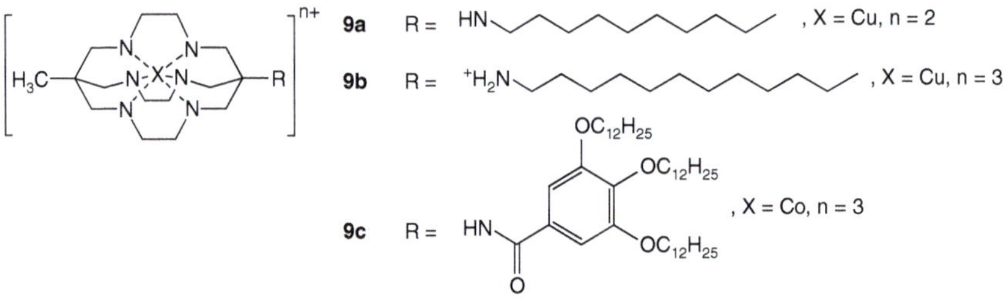

As noted above, the inclusive interaction of macrocycle cavity with alkyl chain of metallosurfactant most often leads to the destruction of the hydrophobic core of metallosurfactant-based aggregates. However, the addition of a macrocycle to a metallosurfactant can initiate joint aggregation, which promotes the spontaneous formation of nanoparticles. In ref. [70], it was demonstrated for the first time that electrostatic interaction between the sulfonate calix[4]resorcinarene and the complex of lanthanum nitrate with two D-16 chains leads to the formation of mixed aggregates that can act as cisplatin nanocontainers (Scheme 12.4). The encapsulation of cisplatin into these aggregates enhances the cytotoxic activity against cancer cells, which is an important step in the development of anti-cancer drugs with minimal side effects.

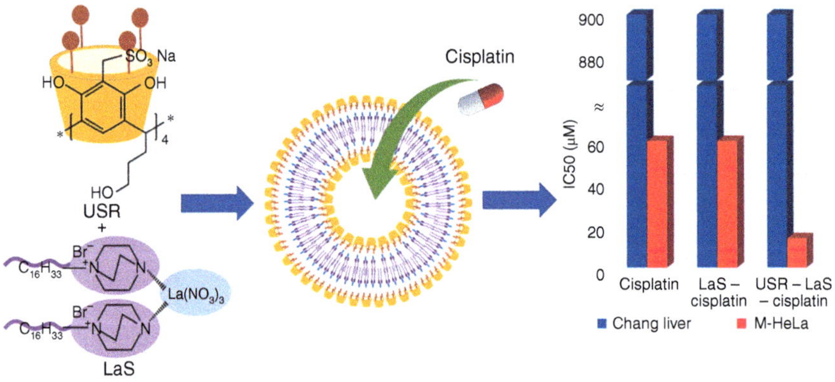

Scheme 12.4 Chemical formulas of sulfonate calix[4]resorcinarene and complex of lanthanum nitrate with two D-16 chains and construction of supramolecular nanoparticles based on them.

In summary, it can be concluded that macrocyclic ligands in combination with metalloamphiphiles have garnered the attention of researchers as new supramolecular metallotectons formed due to associative forces arranging into an organized structure with specific architectural or functional features. Both alkylated derivatives of metallomacrocycles and superamphiphilic complexes (macrocycle–metallosurfactant) can be used as building blocks for creation of these metallotectons. The key advantage of self-assembled functionalized nanoparticles is the non-covalent association of many binding sites and functional groups, which reduces synthetic efforts in the preparation of useful nano objects. These discoveries enrich the library of new macrocyclic host molecules and provide a new platform for the development of potential functional nanostructured materials.

12.4 Nanocomposites Based on Amphiphilic Resorcinarenes and Metal Nanoparticles

Resorcinarenes are a well known class of macrocyclic compounds obtained by condensation of resorcinol with aldehydes [71]. Resorcinarenes are analogues of calixarenes and have found application in the separation technique and catalysis [72–74]. Resorcinarenes tend to self-assemble into a variety of three-dimensional structures both in solutions and in solid state [75]. Since the 2000s, resorcinarenes were intensively studied as a dispersant of metal nanoparticles (MNPs) in water and organic solvents. The nanocomposites based on resorcinarene and MNPs demonstrate high structure resistance and versatile stability [76, 77]. They were subsequently applied to create magnetic nanomaterials, structured optoelectronic films, sensors, and catalytic systems [78].

Hydrophobic resorcinarenes with long alkyl chains on a low rim were applied for stabilizing MNPs in organic media and for creation of nanomaterials with special properties [79]. For example, nanodiamonds about 16 nm in size were selectively isolated from a mixture of diamonds with an amino-substituted resorcinarene cavitand and were applied in the construction of nanofilms with exceptional stability and homogeneously [80]. Gold nanorods dispersed into the organic media with benzylthiol- and dithiocarbamate-modified resorcinarenes demonstrate a solvatochromic effect with a refractive index sensitivity of more than 300 nm/refractive index unit [81]. The article [82] presents ferromagnetic CoNPs spontaneously assembled into nanoscale "bracelets" upon dispersing phenylphosphonate resorcinarene into an organic media.

Amphiphilic resorcinarene are usually utilized to obtain nanocomposites with MNPs in aqueous media. The size, shape, and uniformity of nanocomposites depend on the propensity of resorcinarenes to self-assemble forming different resorcinarene-based architectures in aqueous media [83–85]. Several works have been devoted to the creation of nanocomposites based on resorcinarene with aminoethyl-acetamide groups, which significantly improves the stability of inorganic nanoparticles in water [86]. In the article [87], the effect of the concentration of the aminoethyl-acetamide resorcinarene on stability and catalytic activity of AuNPs is discussed. It was shown that an increase in the concentration of resorcinarene leads to a decrease in the size of AuNPs and an increase in catalytic activity. In other works [88, 89] AuNPs and PtNPs stabilized with the dimethylaminoethyl-acetamide resorcinarene were used for targeted delivery of the anticancer drug methotrexate and as catalysts for photodegradation of toxic dyes. The next article [90] describes the creation of microtubes decorated with different MNPs such as AgNPs, AuNPs, PdNPs, and PtNPs. For the preparation, aminoethyl-acetamide resorcinarenes with long alkyl tails at the lower rim ($C_{11}H_{23}$ and $C_{15}H_{31}$) were used. Unlike resorcinarene with nonyl tails, these resorcinarenes aggregate in aqueous media into microtubes. Microtubes were decorated with MNPs both during the self-assembly of resorcinarene and after their formation. Graphene oxide modified with aminoethyl acetamide resorcinarene with pentyl groups was used to deposit platinum and AuNPs with the formation of uniform and ordered films [91]. The resulting composites were used as cathodic catalysts for oxygen reduction and methanol oxidation, where they showed high catalytic activity.

Supramolecular and polymer nanocontainers based on carboxylate derivatives of resorcinarene (CRA) were used to create catalytically active nanocomposites with AgNPs. Nanocomposites were prepared by chemical reduction of silver nitrate in the presence of stabilizing agents: amphiphilic resorcinarenes with decyl and decenyl tails (CRA-C10) and polymer nanocontainers that were obtained by the reaction of carboxylate derivative of resorcinarene-propanol with phenylboronic acid (p(CRA-B), Scheme 12.5) [92, 93]. The use of amphiphilic resorcinarene CRA-C10 leads to the formation of silver nanocrystals 4–6 nm in size, which are organized into nanoparticles of 15–20 nm size. In the presence of polymer nanocontainers p(CRA-B), the nanocrystals of 15–30 nm in size, formed upon reduction,

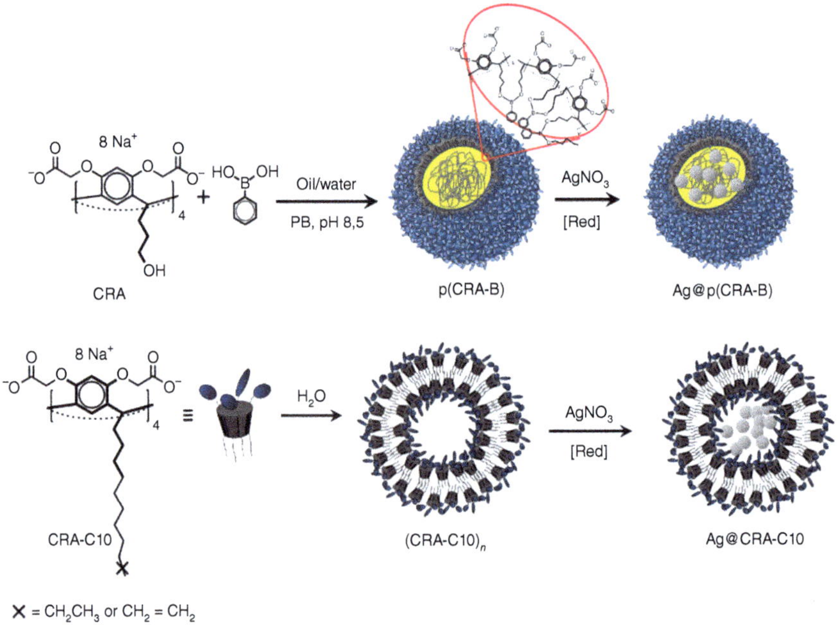

Scheme 12.5 Synthesis of Ag@p(CRA-B) and Ag@CRA-C10.

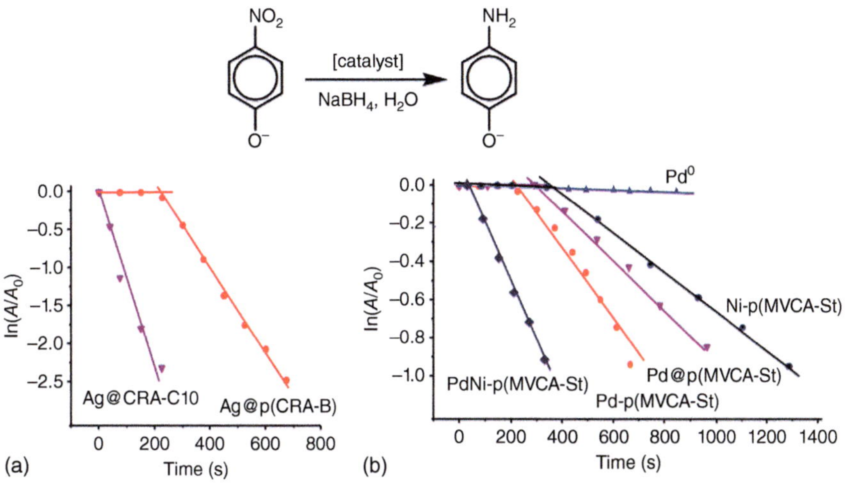

Figure 12.1 Plots of ln(A/A_0) versus time for the catalytic reduction of *p*-nitrophenol with $NaBH_4$ by (a) Ag@CRA-C10 and Ag@p(CRA-B), *C*(*p*-nitrophenol) = 0.1 mM, *C*($NaBH_4$) = 5 mM, *C*(Ag) ≈ 14 μM; and (b) composites of Pd, Ni with p(MVCA-St), *C*(*p*-nitrophenol) = 0.1 mM, *C*($NaBH_4$) = 5 mM, *C*(Metal) ≈ 2 μM.

are encapsulated in the p(CRA-B) cavity to form 60–65 nm nanoparticles. In the model reaction of *p*-nitrophenol reduction for the composites based on polymer nanocontainer Ag@p(CRA-B), an induction period of 210 seconds is observed (Figure 12.1a). The nanocomposites with amphiphilic resorcinarenes Ag@CRA-C10 proceed without an induction period and show greater catalytic activity (Table 12.2). The difference in the catalytic activity of the composites is mainly related to the permeability and lability of the stabilizing shell, and the size of the AgNPs. The absence of an induction period is determined by the lability of the amphiphilic shell. The supramolecular stabilizing layer, consisting of decyl and

Table 12.2 Catalytic activity of nanocomposites based on resorcinarene nanocontainers and MNPs in the reduction of *p*-nitrophenol.

Catalyst	Concentration, *C*(M^0), mM	Apparent rate constant, k_{app}, s^{-1}	References
Ag@CRA-C10	0.014	0.0101	[92]
Ag@p(CRA-B)	0.014	0.0054	[92]
Free Pd NPs	0.0045	0.000 09	[94]
Pd@p(MVCA-St)	0.002	0.0014	[95]
Pd-p(MVCA-St)	0.002	0.0029	[94]
Ni-p(MVCA-St)	0.002	0.0011	[96]
PdNi-p(MVCA-St)	0.002	0.0037	[96]

C(*p*-nitrophenol) = 0.1 mM, *C*($NaBH_4$) = 5 mM, H_2O.

decenyl resorcinarene, facilitates the rapid passage of *p*-nitrophenol to the active surface of AgNPs, on which the reaction of *p*-nitrophenol reduction occurs.

A hollow polymer nanocontainer consisting of viologen cavitands interconnected by alkyl chains (p(MVCA-St), Scheme 12.6) [97] was used to create nanocomposites with PdNPs. In contrast to simple viologen and viologen cavitand [98, 99], p(MVCA-St) better stabilized PdNPs due to its multiply charged shell and flexible structure that adapts to the aggregation of MNPs. The structure of nanocomposites and their catalytic activity depends on the method of obtaining PdNPs. The production of nanoparticles by the mediated electrochemical reduction of $[PdCl_4]^{2-}$ ions leads to the formation of PdNPs encapsulated in the nanocontainers' cavity Pd@p(MVCA-St) [95]. Upon chemical reduction with ascorbic acid, nanocomposites are formed with the arrangement of PdNPs on the surface of the nanocontainer Pd-p(MVCA-St) [94]. In all likelihood, the difference in structure is caused by different rates of nanoparticle formation. In the course of electrochemical reduction, the formed palladium grains can penetrate through the walls of the polymer shell and grow to nanoparticles in the cavity of p(MVCA-St). Upon chemical reduction, nanoparticles are formed in solution and aggregate on the surface of p(MVCA-St) into metal clusters.

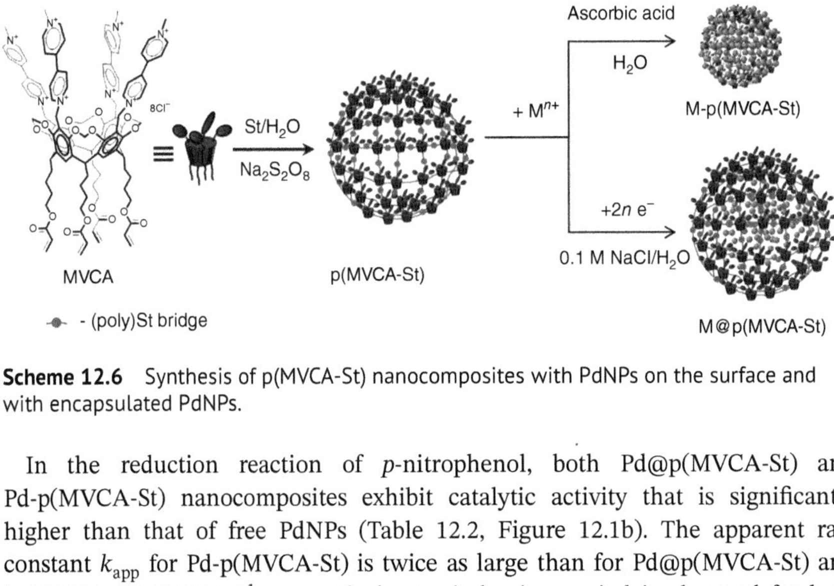

Scheme 12.6 Synthesis of p(MVCA-St) nanocomposites with PdNPs on the surface and with encapsulated PdNPs.

In the reduction reaction of *p*-nitrophenol, both Pd@p(MVCA-St) and Pd-p(MVCA-St) nanocomposites exhibit catalytic activity that is significantly higher than that of free PdNPs (Table 12.2, Figure 12.1b). The apparent rate constant k_{app} for Pd-p(MVCA-St) is twice as large than for Pd@p(MVCA-St) and is 0.0029 and 0.0014 s^{-1}, respectively. An induction period is observed for both composites and is much shorter in the case of Pd-p(MVCA-St). Such a high activity of Pd-p(MVCA-St) is the result of its structure, where PdNPs are packed on the surface of p(MVCA-St), which is the most optimal for catalytic activity. Moreover, MNPs in clusters have a mutual effect on each other, which significantly improves their catalytic properties [100–102].

Ni(II) ions are reduced to amorphous Ni(0), which covers the surface of p(MVCA-St) to form composite Ni-p(MVCA-St) [96]. The reason being that amorphous nickel is a flexible polymeric structure of p(MVCA-St), which prevents Ni(0)

Table 12.3 Suzuki reaction using Pd, Ni nanocomposite with p(MVCA-St).

Catalyst	6-h yield	20-h yield
Pd-p(MVCA-St)	—	72
PdNi-p(MVCA-St)	100	100
Ni-p(MVCA-St)	—	7

C(iodobenzene) = *C*(phenylboronic acid) = 5 mM, $C(K_2CO_3)$ = 15 mM, *V* = 15 ml, H_2O, 25 °C, *C*(M) = 0.05 mM (M = Pd or Ni or PdNi).

from crystallizing [103]. The catalytic activity of Ni-p(MVCA-St) is lower than that of Pd-p(MVCA-St) (Table 12.2). Nickel forms an amorphous film that adsorbs organic molecules well, and for this reason, the reduction reaction begins only five minutes after mixing the reagents. After activation of the nickel surface, the reaction proceeds in the same way as in the presence of a palladium catalyst (Figure 12.1b).

Recent scientific research has shown that bi- and polymetallic composites exhibit better catalytic activity than monometallic composites. The reason is the mutual influence of metals in composites, which leads to a change in their physical, electronic, and chemical properties [104, 105]. The catalytic activity of the bimetallic nanocomposite PdNi-p(MVCA-St) significantly exceeds the activity of monometallic composites (Table 12.2, Figure 12.1b). The reduction reaction of *p*-nitrophenol proceeds without an incubation period and begins immediately after the addition of the catalyst. The reaction takes eight minutes and the apparent rate constant k_{app} is only $0.037\ s^{-1}$.

Currently, the Suzuki cross-coupling reaction is one of the simplest and most versatile methods for producing carbon–carbon bonds. It is widely used in fine organic synthesis to produce biaryls [106]. Mono- and bimetallic nanocomposites Pd-p(MVCA-St), Ni-p(MVCA-St), and PdNi-p(MVCA-St) were studied as catalysts for the reaction of iodobenzene with phenylboronic acid. The results showed that in the presence of nanocomposites, the Suzuki reaction proceeds under mild conditions at room temperature in water (Table 12.3). Bimetallic composites have better catalytic properties than monometallic ones. In the presence of 1 mol% PdNi-p(MVCA-St), the reaction of iodobenzene with phenylboronic acid takes six hours. The lowest activity is exhibited by the composite consisting of monometallic nickel Ni-p(MVCA-St).

12.5 Conclusions

The data on metallosurfactants and their supramolecular systems indicate their improved aggregation characteristics. A variety of morphological forms, high practical potential, and the structure–property relationships established open up the possibility of the most effective use of these compounds for solving technological and biomedicinal problems. Beneficial properties of metallosurfactant-based nanostructures are the combination of features typical for amphiphilic matrix, i.e. the formation of dynamic self-assemblies with controllable characteristics capable of providing the host domains with different active molecules (drugs, food or cosmetic ingredients, pesticides, chemicals) along with advantages of metal-containing derivatives, i.e. redox properties, electrophilic character, charge occurrence, catalytic and luminescence properties, etc. Key benefit of self-assembling platform is the possibility of noncovalent modification of nanocontainers thereby providing them additional functionality. Meanwhile dynamic character of such polyfunctional nanosystems is one of the limitations preventing their wide application *in vivo*. The latter can be improved by their admixing with the covalently bound compounds, e.g. macrocycles or/and polymers.

Macrocycle–metallosurfactant systems are also getting the attention of researchers. Both covalent and noncovalent bonds are used as the driving forces to fabricate these systems and supramolecular self-assemblies based on them. Macrocyclic unit can serve as the donor to coordinate with metal cations supplementing the amphiphilic unit with the supramolecular recognition ability. Non-covalent interactions of metallosurfactants with macrocycles can be used not only for disassembly of metallosurfactant aggregates due to the formation of guest–host complexes between the macrocycle and the metallosurfactant, but also for creating new types of macrocyclic metallosurfactant, avoiding lengthy synthetic procedures using organic solvents. The resulting macrocycle–metallosurfactant compositions have great potential for applications in biomedicine, electroluminescent devices, catalysis, gas adsorption and conversion.

While we mainly focus on biomedical applications of metal-modified nanocontainers, it should be emphasized that they are multifunctional materials that can find applications in a wide variety of modern technologies. One of the most attractive contributions of metal-bearing compositions is related to the fabrication of catalytic systems. Supramolecular catalysts are of special interest in this field due to their biomimetic character, since the host–guest principles responsible for binding the reagents inside the nanoreactors are similar to those of enzyme catalysts. The final part of the review presents the results on the creation of nanocomposite materials based on macrocyclic resorcinarenes and mono- and bimetallic nanoparticles that demonstrate high structure resistance and versatile stability. They were subsequently applied to create magnetic nanomaterials, structured optoelectronic films, sensors, and catalytic systems. Depending on the preparation method, composites with both encapsulated MNPs and MNPs on the surface of polymer nanocontainers have been created. Bimetallic nanocomposites exhibit better catalytic activity than monometallic ones both in the reduction of *p*-nitrophenol and in the Suzuki cross-coupling reaction.

Acknowledgments

The authors thank Russian Science Foundation (project No 19-73-30012) for financial support. A.Y.Z. thanks Kazan Federal University for the allocated subsidies for the state assignment in the sphere of scientific activities (0671-2020-0063).

References

1 Zhang, Y., Fang, F., Li, L. et al. (2020). Self-assembled organic nanomaterials for drug delivery, bioimaging, and cancer therapy. *ACS Biomaterials Science & Engineering* 6 (9): 4816–4833.
2 Chariou, P.L., Ortega-Rivera, O.A., and Steinmetz, N.F. (2020). Nanocarriers for the delivery of medical, veterinary, and agricultural active ingredients. *ACS Nano* 14 (3): 2678–2701.
3 Li, M., Luo Zh., and Zhao, Y. (2018). Self-assembled hybrid nanostructures: versatile multifunctional nanoplatforms for cancer diagnosis and therapy. *Chemistry of Materials* 30 (1): 25–53.
4 Zakharova L. Ya., Pashirova, T.N., Doktorovova, S. et al. (2019). Cationic surfactants: self-assembly, structure-activity correlation and their biological applications. *International Journal of Molecular Sciences* 20 (22): 5534.
5 Kashapov, R., Gaynanova, G., Gabdrakhmanov, D. et al. (2020). Self-assembly of amphiphilic compounds as a versatile tool for construction of nanoscale drug carriers. *International Journal of Molecular Sciences* 21 (18): 6961.
6 Tian, B., Liu, Y., and Liu, J. (2021). Smart stimuli-responsive drug delivery systems based on cyclodextrin: a review. *Carbohydrate Polymers* 251: 116871.
7 Razuvayeva, Y.S., Kashapov, R.R., and Zakharova, L.Y. (2020). Calixarene-based pure and mixed assemblies for biomedical applications. *Supramolecular Chemistry* 32 (3): 178–206.
8 Boyaciyan, D. and von Klitzing, R. (2020). Stimuli-responsive polymer/metal composites: from fundamental research to self-regulating devices. *Current Opinion in Colloid & Interface Science* 44: 193–207.
9 Zare, E.N., Jamaledin, R., Naserzadeh, P. et al. (2020). Metal-based nanostructures/PLGA nanocomposites: antimicrobial activity, cytotoxicity, and their biomedical applications. *ACS Applied Materials & Interfaces* 12 (3): 3279–3300.
10 Kuznetsova, D.A., Gaynanova, G.A., Vasileva, L.A. et al. (2019). Mitochondria-targeted cationic liposomes modified with alkyltriphenylphosphonium bromides loaded with hydrophilic drugs: preparation, cytotoxicity and colocalization assay. *Journal of Materials Chemistry B* 7: 7351–7362.
11 Voloshina, A.D., Gumerova, S.K., Sapunova, A.S. et al. (2020). The structure-activity correlation in the family of dicationic imidazolium surfactants: antimicrobial properties and cytotoxic effect. *Biochimica et Biophysica Acta General Subjects* 1864 (12): 129728.
12 Gabdrakhmanov, D.R., Vasilieva, E.A., Voronin, M.A. et al. (2020). Soft nanocontainers based on hydroxyethylated geminis: role of spacer in self-assembling, solubilization and complexation with oligonucleotide. *Journal of Physical Chemistry C* 124 (3): 2178–2192.
13 Zakharova L. Ya., Kaupova, G.I., Gabdrakhmanov, D.R. et al. (2019). Alkyl triphenylphosphonium surfactants as nucleic acid carriers: complexation efficacy toward DNA decamers, interaction with lipid bilayers and cytotoxicity studies. *Physical Chemistry Chemical Physics* 21: 16706–16717.
14 Pavlov, R.V., Gaynanova, G.A., Kuznetsova, D.A. et al. (2020). Biomedical potentialities of cationic geminis as modulating agents of liposome in drug

delivery across biological barriers and cellular uptake. *International Journal of Pharmaceutics* 587: 119640.

15 Zhiltsova, E.P., Ibatullina, M.R., Mirgorodskaya, A.B. et al. (2020). Polyfunctional nanosystems on amphiphilic and hybrid platform: self-assembly, mesogenic properties and application. *Liquid Crystals and Their Application* 20 (3): 5–25.

16 Mehta, S.K., Kaur, R., and Chaudhary, G.R. (2012). Self aggregation and solution behavior of copper and nickel based surfactants. *Colloids and Surfaces, A: Physicochemical and Engineering Aspects* 403: 103–109.

17 Do Pim, W.D., Ribeiro-Santos, T.A., Jardim, I.S. et al. (2018). Bistable copper(II) metallosurfactant as molecular machine for the preparation of hybrid silica-based porous materials. *Materials & Design* 160: 876–885.

18 Toth, A., Schnedl, S., Painer, D. et al. (2019). Interfacial catalysis in biphasic carboxylic acid esterification with a nickel-based metallosurfactant. *ACS Sustainable Chemistry & Engineering* 7 (22): 18547–18553.

19 Pradeep, Kaur, G., Chaudhary, G.R. et al. (2020). Investigating affordable cobalt based metallosurfactant as an efficient electrocatalyst for hydrogen evolution reaction. *Journal of Colloid and Interface Science* 562: 598–607.

20 Dogra, V., Kaur, G., Jindal, S. et al. (2019). Bactericidal effects of metallosurfactants based cobalt oxide/hydroxide nanoparticles against *Staphylococcus aureus*. *Science of the Total Environment* 681: 350–364.

21 Ambika, S., Manojkumar, Y., Arunachalam, S. et al. (2019). Biomolecular interaction, anti-cancer and anti-angiogenic properties of cobalt(III) Schiff base complexes. *Scientific Reports* 9 (1): 2721.

22 Amiens, C., Chaudret, B., Ciuculescu-Pradines, D. et al. (2013). Organometallic approach for the synthesis of nanostructures. *New Journal of Chemistry* 37: 3374–3401.

23 Daniel, C.I., Albo, J., Santos, E. et al. (2013). A group contribution method for the influence of the temperature in the viscosity of magnetic ionic liquids. *Fluid Phase Equilibria* 360: 29–35.

24 Adawy, A.I. and Badr, E.A. (2014). Synthesis and biological activity of some amide-based cationic surfactant complexes with Co(II) and Cu(II). *IOSR Journal of Applied Chemistry* 7 (9): 9–19.

25 Moroi, Y., Motomura, K., and Matuura, R. (1974). The critical micelle concentration of sodium dodecyl sulfate-bivalent metal dodecyl sulfate mixtures in aqueous solutions. *Journal of Colloid and Interface Science* 46 (1): 111–117.

26 Bhattacharya, S. and Kumari, N. (2009). Metallomicelles as potent catalysts for the ester hydrolysis reactions in water. *Coordination Chemistry Reviews* 253: 2133–2149.

27 Zhiltsova, E.P., Pashirova, T.N., Ibatullina, M.R. et al. (2018). New surfactant-copper(II) complex based on 1,4-diazabicyclo[2.2.2]octane amphiphile. Crystal structure determination, self-assembly and functional activity. *Physical Chemistry Chemical Physics* 20 (18): 12688–12699.

28 Parera, E., Marín-García, M., Pons, R. et al. (2016). Supramolecular arrangement of molybdenum carbonyl metallosurfactants with CO-releasing properties. *Organometallics* 35: 484–493.

29 Garg, P., Kaur, G., Chaudhary, G.R. et al. (2018). Investigating the structural integrity of bovine serum albumin in presence of newly synthesized metallosurfactants. *Colloids and Surfaces B: Biointerfaces* 164: 116–124.

30 Pereira, R.F.P., Valente, A.J.M., Burrows, H.D. et al. (2013). Structural characterization of solid trivalent metal dodecyl sulfates: from aqueous solution to lamellar superstructures. *RSC Advances* 3: 1420–1433.

31 Firouzabadi, H., Iranpoor, N., and Nowrouzi, F. (2005). The facile and efficient Michael addition of indoles and pyrrole to α,β-unsaturated electron-deficient compounds catalyzed by aluminium dodecyl sulfate trihydrate $[Al(DS)_3]$ $3H_2O$ in water. *Chemical Communications* 6: 789–791.

32 Zhiltsova, E.P., Ibatullina, M.R., Lukashenko, S.S. et al. (2017). Complexes of 1-hexadecyl-4-aza-1-azoniabicyclo[2.2.2]octane bromide with transition metal nitrates. Micelle-forming, solubilizing, and adsorption properties. *Colloid Journal* 79 (5): 621–629.

33 Zhil'tsova, E.P., Ibatullina, M.R., Lukashenko, S.S. et al. (2017). Complexes of 1-alkyl-4-aza-1-azoniabicyclo[2.2.2]octane bromides with lanthanum nitrate. Micelle-forming and adsorption properties. *Russian Journal of General Chemistry* 87 (11): 2620–2626.

34 Liang, Z., Wang, C., and Huang, J. (2003). The research on the vesicle formation and transformation in novel gemini surfactant systems. *Colloids and Surfaces, A: Physicochemical and Engineering Aspects* 224: 213–220.

35 Kaur, B., Chaudhary, G.R., and Kaur, G. (2019). Cholesterol-induced physicochemical changes in dodecylamine-based metallosomes: drug entrapping ability and interactions with biological molecules. *Journal of Materials Chemistry B: Materials for Biology and Medicine* 7: 3679–3691.

36 Paul, A., Fallis, I.A., Stokes, E.C. et al. (2018). Structural evolution in metallomicroemulsions – the effect of increasing alcohol hydrophobicity. *Dalton Transactions* 47: 14211–14217.

37 Gonawala, S., Leopoldino, V.R., Kpogo, K. et al. (2016). Langmuir–Blodgett films of salophen-based metallosurfactants as surface pretreatment coatings for corrosion mitigation. *Chemical Communications* 52: 11155–11158.

38 Koutsantonis, G.A., Nealon, G.L., Buckley, C.E. et al. (2007). Wormlike micelles from a cage amine metallosurfactant. *Langmuir* 23 (24): 11986–11990.

39 Owen, T. and Butler, A. (2011). Metallosurfactants of bioinorganic interest: coordination induced self-assembly. *Coordination Chemistry Reviews* 225 (7, 8): 678–687.

40 Griffiths, P.C., Fallis, I.A., Chuenpratoom, T. et al. (2006). Metallosurfactants: interfaces and micelles. *Advances in Colloid and Interface Science* 122 (1–3): 107–117.

41 Ibatullina, M.R., Zhil'tsova, E.P., Lukashenko, S.S. et al. (2018). Krafft temperature of 1-alkyl-4-aza-1-azoniabicyclo[2.2.2]octane bromide complexes with transition metal salts. *Russian Journal of Physical Chemistry A* 92 (4): 797–801.

42 Mukherjee, S., Chakraborty, M., Panda, A.K. et al. (2011). Physicochemistry of bis-alkyltrimethylammonium dichromate, tungstate and molybdate amphiphiles: synthesis, characterization, behaviors at the air–water interface and self-aggregation in aqueous medium. *Colloids and Surfaces, A: Physicochemical and Engineering Aspects* 388: 1–11.

43 Zhiltsova, E.P., Ibatullina, M.R., Lukashenko, S.S. et al. (2018). Spectrophotometric study of quercetin in metallomicellar solutions of 1-hexadecyl-4-aza-1-azoniabicyclo[2.2.2]octane bromide complex with copper dibromide. *Journal of Molecular Liquids* 249: 716–722.

44 Zhil'tsova, E.P., Ibatullina, M.R., Lukashenko, S.S. et al. (2018). Metallomicellar complex of 1-hexadecyl-4-aza-1-azoniabicyclo[2.2.2]octane bromide with copper dibromide for solubilization of nitrofurantoin. *Russian Journal of Organic Chemistry* 54 (3): 431–435.

45 Ibatullina, M.R., Zhil'tsova, E.P., Lukashenko, S.S. et al. (2020). Supramolecular systems of metal complexes of 1-cetyl-4-aza-1-azoniabicyclo[2,2,2]octane bromide for increasing griseofulvin solubility. *Colloid Journal* 82 (1): 8–15.

46 Wani, W.A., Prashar, S., Shreaz, S. et al. (2016). Nanostructured materials functionalized with metal complexes: in search of alternatives for administering anticancer metallodrugs. *Coordination Chemistry Reviews* 312: 67–98.

47 Zhiltsova, E.P., Ibatullina, M.R., Lukashenko, S.S. et al. (2016). Complex of 1-hexadecyl-4-aza-1-azoniabicyclo[2.2.2]octane bromide with copper dibromide: structure, aggregation, and biological activity. *Russian Chemical Bulletin, International Edition* 65 (5): 1365–1371.

48 Ibatullina, M.R., Zhil'tsova, E.P., Lukashenko, S.S. et al. (2018). Metallomicellar systems based on the complexes of 1-hexadecyl-4-aza-1-azoniabicyclo [2.2.2]octane bromide with transition metal nitrates. *Russian Journal of General Chemistry* 88 (11): 2359–2367.

49 Bhar, R., Kaur, G., and Mehta, S.K. (2017). Exploring interactions of copper hybrid surfactants with calf thymus-DNA. *Journal of Molecular Liquids* 241: 715–721.

50 Suganthi, D.R. and Kumaraguru, N. (2019). Interaction between single and double chain ruthenium(II) metallosurfactant complexes and CT-DNA and antitumor study. *Adalya Journal* 8 (12): 142–160.

51 Veeralakshmi, S., Nehru, S., Sabapathi, G. et al. (2015). Single and double chain surfactant–cobalt(III) complexes: the impact of hydrophobicity on the interaction with calf thymus DNA, and their biological activities. *RSC Advances* 5: 31746–31758.

52 Mitchell, N., Kalber, T.L., Cooper, M.S. et al. (2013). Incorporation of paramagnetic, fluorescent and PET/SPECT contrast agents into liposomes for multimodal imaging. *Biomaterials* 34 (4): 1179–1192.

53 Tashiro, S., Kubota, R., Kawagoe, M. et al. (2013). Palladium- or proton-induced submicro spherical aggregation of macrocyclic amphiphiles in aqueous solution. *Dalton Transactions* 42: 15915–15918.

54 Li, Z.-Y., Xu, L., Wang, C.-H. et al. (2013). Novel platinum-acetylide metallocycles constructed via a stepwise fragment coupling approach and their aggregation behavior. *Chemical Communications* 49: 6194–6196.

55 Lebrón, J.A., Ostos, F.J., Moyá, M.L. et al. (2015). Cooperative interaction between metallosurfactants, derived from the $[Ru(2,2'\text{-bpy})_3]^{2+}$ complex, and DNA. *Colloids and Surfaces B: Biointerfaces* 135: 817–824.

56 Lopez-Lopez, M., Montilla, F., Olivares, M. et al. (2013). Interaction between monomers of two surfactants derived from the $[Ru(2,2'\text{-bpy})3]^{2+}$ complex and α, β and γ-cyclodextrins: formation of [2]- and [3]-pseudorotaxanes. *Dalton Transactions* 42: 6171–6181.

57 Liu, N., Wang, B., Liu, W. et al. (2013). Reversible luminescence switching accompanied by assembly–disassembly of metallosupramolecular amphiphiles based on a platinum(II) complex. *Journal of Materials Chemistry C: Materials for Optical and Electronic Devices* 1: 1130–1136.

58 Iza, N., Guerrero-Martínez, A., Tardajos, G. et al. (2015). Using inclusion complexes with cyclodextrins to explore the aggregation behavior of a ruthenium metallosurfactant. *Langmuir* 31 (9): 2677–2688.

59 Gruber, B., Kataev, E., Aschenbrenner, J. et al. (2011). Vesicles and micelles from amphiphilic zinc(II)-cyclen complexes as highly potent promoters of hydrolytic DNA cleavage. *Journal of the American Chemical Society* 133 (51): 20704–20707.

60 Kimura, E., Hashimoto, H., and Koike, T. (1996). Hydrolysis of lipophilic esters catalyzed by a zinc(II) complex of a long alkyl-pendant macrocyclic tetraamine in micellar solution. *Journal of the American Chemical Society* 118 (45): 10963–10970.

61 Perera-Bobusch, C., Hormann, J., Weise, C. et al. (2016). Significantly enhanced proteolytic activity of cyclen complexes by monoalkylation. *Dalton Transactions* 45 (26): 10500–10504.

62 Li, F.-z., Feng, F., Yu, L. et al. (2014). The metallomicelle of lanthanide metal (Ce, La) aza-macrocyclic complexes with a carboxyl branch: the catalytic activity and mechanism in the hydrolysis of a phosphate diester. *Journal of Solution Chemistry* 43 (8): 1331–1343.

63 Gruber, B., Stadlbauer, S., Späth, A. et al. (2010). Modular chemosensors from self-assembled vesicle membranes with amphiphilic binding sites and reporter dyes. *Angewandte Chemie International Edition* 49 (39): 7125–7128.

64 Zhamoitina, A.I., Sauerwein, Y., König, B. et al. (2014). A binary catalytic system based on mixed monolayers of a phospholipid and amphiphilic bis(Zn^{2+}-cyclen). *Colloid Journal* 76 (2): 153–160.

65 Roldán-Carmona, C., González-Delgado, A.M., Guerrero-Martínez, A. et al. (2011). Molecular organization and effective energy transfer in iridium metallosurfactant-porphyrin assemblies embedded in Langmuir–Schaefer films. *Physical Chemistry Chemical Physics* 13: 2834–2841.

66 Allampally, N.K., Bredol, M., Strassert, C.A. et al. (2014). Highly phosphorescent supramolecular hydrogels based on platinum(II) emitters. *Chemistry A European Journal* 20 (51): 16863–16868.

67 Xing, L.-B., Yu, S., Wang, X.-J. et al. (2012). Reversible multistimuli-responsive vesicles formed by an amphiphilic cationic platinum terpyridine complex with a ferrocene unit in water. *Chemical Communications* 48 (88): 10886–10888.
68 Hondow, N., Harowfield, J., Koutsantonis, G. et al. (2012). Metallosurfactants in the preparation of mesoporous silicas. *Microporous and Mesoporous Materials* 151: 264–270.
69 Dittrich, B., Harrowfield, J.M., Koutsantonis, G.A. et al. (2010). Long tailed cage amines: synthesis, metal complexation, and structure. *Dalton Transactions* 39: 3433–3448.
70 Kashapov, R., Razuvayeva, Y., Ziganshina, A. et al. (2020). Supra-amphiphilic systems based on metallosurfactant and calix[4]resorcinol: self-assembly and drug delivery potential. *Inorganic Chemistry* 59 (24): 18276–18286.
71 Sliwa, W. and Kozlowski, C. (2009). *Calixarenes and Resorcinarenes: Synthesis, Properties and Application*. Weinheim: Wiley-VCH.
72 Li, N., Harrison, R.G., and Lamb, J.D. (2014). Application of resorcinarene derivatives in chemical separations. *Journal of Inclusion Phenomena and Macrocyclic Chemistry* 78: 39–60.
73 Zhang, Q. and Tiefenbacher, K. (2013). Hexameric resorcinarene capsule is a brønsted acid: investigation and application to synthesis and catalysis. *Journal of the American Chemical Society* 135 (43): 16213–16219.
74 Zhang, Q. and Tiefenbacher, K. (2015). Terpene cyclization catalyzed inside a self-assembled cavity. *Nature Chemistry* 7: 197–202.
75 Jain, V.K. and Kanaiya, P.H. (2011). Chemistry of calix[4]resorcinarenes. *Russian Chemical Reviews* 80: 75–102.
76 Wei, A., Kim, B., Pusztay, S.V. et al. (2001). Resorcinarene-encapsulated nanoparticles: building blocks for self-assembled nanostructures. *Journal of Inclusion Phenomena and Macrocyclic Chemistry* 41: 83–86.
77 Wei, A. (2006). Calixarene-encapsulated nanoparticles: self-assembly into functional nanomaterials. *Chemical Communications* 15: 1581–1591.
78 Montes-García, V., Pérez-Juste, J., Pastoriza-Santos, I. et al. (2014). Metal nanoparticles and supramolecular macrocycles: a tale of synergy. *Chemistry A European Journal* 20: 10874–10883.
79 Misra, T.K., Chen, T.S., and Liu, C.Y. (2006). Phase transfer of gold nanoparticles from aqueous to organic solution containing resorcinarene. *Journal of Colloid and Interface Science* 297: 584–588.
80 Sheela, V.P., Xiao, W., Han, S. et al. (2011). Resorcinarene amine stabilized nanodiamond dispersions in organic solvents: applications in diamond film growth. *Journal of Materials Chemistry* 21: 6395–6400.
81 Hansen, M.N., Chang, L.-S., and Wei, A. (2008). Resorcinarene-encapsulated gold nanorods: solvatochromism and magnetic nanoshell formation. *Supramolecular Chemistry* 20: 35–40.
82 Tripp, S.L., Pusztay, S.V., Ribbe, A.E. et al. (2002). Self-assembly of cobalt nanoparticle rings. *Journal of the American Chemical Society* 124 (27): 7914–7915.

83 Shalaeva Ya. V., Morozova Ju. E., Gubaidullin, A.T. et al. (2018). Gold nanoparticles, capped by carboxy-calix[4]resorcinarenes: effect of structure and concentration of macrocycles on the nanoparticles size and aggregation. *Journal of Inclusion Phenomena and Macrocyclic Chemistry* 92: 211–221.

84 Ermakova, A.M., Morozova Ju. E., Shalaeva Ya. V. et al. (2017). The supramolecular approach to the phase transfer of carboxylic calixresorcinarene-capped silver nanoparticles. *Colloids and Surfaces A: Physicochemical and Engineering Aspects* 524: 127–134.

85 Shalaeva Ya. V., Morozova Ju. E., Shumatbaeva, A.M. et al. (2019). Binding of l-tryptophan and bovine serum albumin by novel gold nanoparticles capped with amphiphilic sulfonatomethylated calixresorcinarenes. *Journal of Molecular Liquids* 286: 110879.

86 Shen, M., Sun, Y., Han, Y. et al. (2008). Strong deaggregating effect of a novel polyamino resorcinarene surfactant on gold nanoaggregates under microwave irradiation. *Langmuir* 24 (22): 13161–13167.

87 Yao, Y., Sun, Y., Han, Y. et al. (2010). Preparation of resorcinarene-functionalized gold nanoparticles and their catalytic activities for reduction of aromatic nitro compounds. *Chinese Journal of Chemistry* 28 (5): 705–712.

88 Shalaeva Ya. V., Morozova Ju. E., Syakaev, V.V. et al. (2018). Formation of cooperative amidoamino calixresorcinarene – methotrexate nanosized aggregates in an aqueous solution and on the surface of gold nanoparticles. *Supramolecular Chemistry* 30 (11): 901–910.

89 Shalaeva Ya. V., Morozova Ju. E., Gubaidullin, A.T. et al. (2020). Photocatalytic properties of supramolecular nanoassociates based on gold and platinum nanoparticles, capped by amphiphilic calix[4]resorcinarenes, towards organic dyes. *Colloids and Surfaces A: Physicochemical and Engineering Aspects* 596: 124700.

90 Sun, Y., Yao, Y., Yan, C.-G. et al. (2010). Selective decoration of metal nanoparticles inside or outside of organic microstructures via self-assembly of resorcinarene. *ACS Nano* 4 (4): 2129–2141.

91 Zhou, J., Chen, M., and Diao, G. (2013). Assembling gold and platinum nanoparticles on resorcinarene modified graphene and their electrochemical applications. *Journal of Materials Chemistry A* 1: 2278–2285.

92 Sergeeva T. Yu., Nizameev, I.R., Kholin, K.V. et al. (2020). Polymer and supramolecular nanocontainers based on carboxylate derivatives of resorcinarenes for binding of substrates and design of composites for catalysis. *Russian Chemical Bulletin* 69: 351–359.

93 Sergeeva T. Yu., Samigullina, A.I., Gubaidullin, A.T. et al. (2016). Application of ferrocene-resorcinarene in silver nanoparticle synthesis. *RSC Advances* 6: 87128–87133.

94 Sultanova, E.D., Salnikov, V.V., Mukhitova, R.K. et al. (2015). High catalytic activity of palladium nanoparticle clusters supported on a spherical polymer network. *Chemical Communications* 51: 13317–13320.

95 Yanilkin, V.V., Nastapova, N.V., Sultanova, E.D. et al. (2016). Electrochemical synthesis of nanocomposite of palladium nanoparticles with polymer viologen-containing nanocapsules. *Russian Chemical Bulletin* 65: 125–132.

96 Sultanova, E.D., Samigullina, A.I., Nastapova, N.V. et al. (2017). Highly active Pd–Ni nanocatalysts supported on multicharged polymer matrix. *Catalysis Science & Technology* 7: 5914–5919.

97 Sultanova, E.D., Krasnova, E.G., Kharlamov, S.V. et al. (2015). Thermoresponsive polymer nanoparticles based on viologen cavitands. *ChemPlusChem* 80 (1): 217–222.

98 Yanilkin, V.V., Nasybullina, G.R., Sultanova, E.D. et al. (2014). Methyl viologen and tetraviologen calix[4]resorcinol as mediators of the electrochemical reduction of $[PdCl_4]^{2-}$ with formation of finely dispersed Pd^0. *Russian Chemical Bulletin* 63: 1409–1415.

99 Yanilkin, V.V., Nasybullina, G.R., Ziganshina, A.Y. et al. (2014). Tetraviologen calix[4]resorcine as a mediator of the electrochemical reduction of $[PdCl_4]^{2-}$ for the production of Pd^0 nanoparticles. *Mendeleev Communications* 24 (2): 108–110.

100 Imura, Y., Maezawa, A., Morita, C. et al. (2012). Neuron-shaped gold nanocrystals and two-dimensional dendritic gold nanowires fabricated by use of a long-chain amidoamine derivative. *Langmuir* 28 (42): 14998–15004.

101 Ruan, L., Zhu, E., Chen, Y. et al. (2013). Biomimetic synthesis of an ultrathin platinum nanowire network with a high twin density for enhanced electrocatalytic activity and durability. *Angewandte Chemie International Edition* 52 (48): 12577–12581.

102 Yanilkin, V.V., Nastapova, N.V., Fazleeva, R.R. et al. (2018). Electrochemical synthesis of metal nanoparticles using a polymeric mediator, whose reduced form is adsorbed (deposited) on an electrode. *Russian Chemical Bulletin* 67: 215–229.

103 Zhang, W., Tan, Y., Gao, Y. et al. (2014). Amorphous nickel–boron and nickel–manganese–boron alloy as electrochemical pseudocapacitor materials. *RSC Advances* 4: 27800–27804.

104 Shen, S.Y., Zhao, T.S., Xu, J. et al. (2010). Synthesis of PdNi catalysts for the oxidation of ethanol in alkaline direct ethanol fuel cells. *Journal of Power Sources* 195: 1001–1006.

105 Gu, J., Zhang, Y.-W., and Tao, F. (2012). Shape control of bimetallic nanocatalysts through well-designed colloidal chemistry approaches. *Chemical Society Reviews* 41: 8050–8065.

106 Magano, J. and Dunetz, J.R. (2011). Large-scale applications of transition metal-catalyzed couplings for the synthesis of pharmaceuticals. *Chemical Reviews* 111 (3): 2177–2250.

13

Metallosurfactants in Nanoscale Molecular Containers as Sensors

Devika Vashisht[1,2] *and Nicole Pamme*[3]

[1] *Panjab University, Department of Chemistry, Centre for Advanced Studies in Chemistry, Chandigarh 160014, India*
[2] *University of Hull, Department of Chemical Engineering, Hull HU6 7RX, UK*
[3] *University of Hull, Department of Chemistry and Biochemistry, Hull HU6 7RX, UK*

13.1 Introduction

The study of amphiphilic molecules began after the discovery of soap [1]. Advancements in the solubilization of hydrophobic materials began the surfactant chemistry in the twentieth century [2]. The tendency of amphiphilic aggregates to entrap lipophilic entities in their hydrophobic core is a particularly interesting characteristic of such compounds. Over the past few decades, surfactant-based aggregates have attracted interest by researchers from diverse backgrounds. Generally, surfactants, also called as "surface-active agents," comprise a polar head group and a non-polar tail (Figure 13.1). The polar group is hydrophilic and the non-polar long chain organic portion is hydrophobic. Non-ionic, anionic, cationic, and zwitterionic form four different classes of surfactants.

The alkyl chain length of the hydrophobic part of any surfactant contributes strongly to its physical properties. At a particular concentration, known as the critical aggregation concentration (CAC), or critical micelle concentration (CMC) for micelles, the surfactant molecules aggregate to form assemblies such as micelles, bilayers, and vesicles. Surfactants can be dissolved in both organic and aqueous media. They tend to adsorb at interfaces and self-assemble spontaneously. The arrangement of the assemblies also depends on the solvent polarity, hydrophobic chain length, and surfactant properties (Figure 13.2). These are not only crucial to the detergent industry, but also find applications in chemical processes, such as catalysis and as self-assembled fluorescence detectors. Incorporation of transition metal ions into surfactant assemblies, leading to formation of metallosurfactants, can impart interesting and advanced functionalities for diverse applications.

13.1.1 Metallosurfactants

Metallosurfactants are surfactants in which the hydrophilic head usually binds to a metal ion, generally, a d- or f-block metal ion, although in some cases the metal

Metallosurfactants: From Fundamentals to Catalytic and Biomedical Applications, First Edition.
Edited by Surinder K. Mehta and Ravneet Kaur.

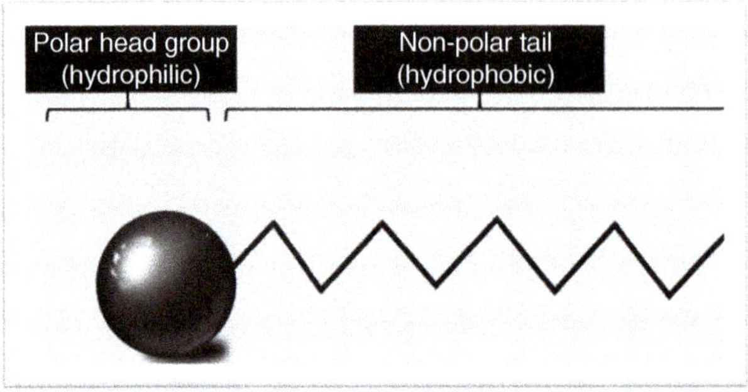

Figure 13.1 Basic components of a surfactant.

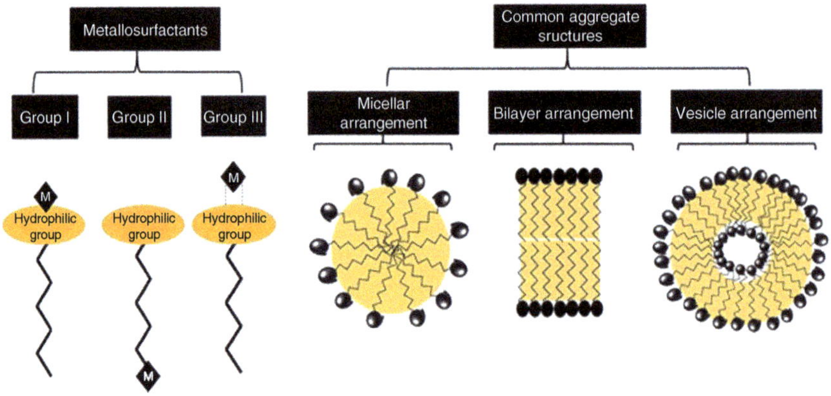

Figure 13.2 Micellar, bilayer, and vesicular arrangements.

ion may be present in the hydrophobic tail part or as a counter ion. This often leads to intriguing applications not possible with the conventional surfactant systems and individual metal ions. The incorporation of metal ions may impart magnetic characteristics or catalytic features [3]. Metallosurfactants can serve as a template for mesoporous materials [4], and have applications in catalysis [5], imaging [6], as a nanoscale carrier for fluorescence sensing [7] and in optoelectronics [8].

Metallosurfactants bring together the adaptability of transition metals with organic skeletons to obtain various supramolecular assemblies. The position of metal ion within the metallosurfactant classifies them into three categories, i.e. group I, group II, and group III metallosurfactants (Figure 13.2). In group I, the metal ions are incorporated into the hydrophilic head group, while in group II, they are positioned in the hydrophobic segment of the surfactant [2]. The metal ions in group III act as counter ions and are thus weakly bound to the surfactant. The aggregation of the metallosurfactants can create a range of assemblies such as micelles, vesicles, and hybrid structures with different shapes and sizes (Figure 13.2). The type of aggregation of the supramolecular assemblies depends on properties such as pH, solvent, and/or concentration of different components [3].

13.2 Metallosurfactants as Nanosized Containers for Self-assembled Molecular Devices

Metallosurfactant aggregates may range in size from micrometer assemblies to nanometer structures. Here, we will focus on metallosurfactant aggregates, which have sizes in the nanometric range. Each aggregate can be considered as a container. Thus, the metallosurfactant assemblies can be treated as nanosized molecular containers capable of entrapping and concentrating lipophilic entities within a small volume at relatively high concentration. Metallosurfactant aggregates are also referred to self-assembled molecular devices. As components with different functionalities are brought together, the nanosized container is formed via self-assembly in the absence of any program-modulated procedure [2]. This chapter will focus on the applicability of metalloaggregates, such as metallomicelles, metallovesicles, and metallosomes, in the field of chemical sensing.

Chemosensors incorporate a material or chemical substance that can selectively bind a specific analyte and a transducer that signals its presence [9]. Among the various signaling procedures, colorimetric and fluorometric detection have advantages such as easy operation, low cost, high resolution, and sensitivity. Metalloaggregates are among the materials that can be employed for chemosensing. The small size of the metalloaggregates enables sensing at small scales with relatively little physical interference. To exemplify, molecular systems with such dimensions are well suited for intracellular monitoring.

Fluorescence-sensing applications are particularly prominent, due to high sensitivity and relative ease of visual readout or instrumental readout [10, 11]. The basic sensing mechanism is composed of three units: (i) a recognition unit, which binds with the analyte, (ii) a fluorescent moiety for signal generation, and, (iii) a transduction unit, which converts the recognition unit into a signal [12]. Fluorescence sensors typically involve compounds with delocalized electrons in conjugated double bonds such as polyaromatic compounds. Such polyaromatic compounds have interaction site that can bind a specific analyte. The challenge with using such compounds is their limited selectivity and tunability, which limits their applications. Polyaromatic compounds generally have interaction sites that can bind multiple analytes, which limits their application for specific tagging. Using the concepts of supramolecular chemistry, groups of more sophisticated and sensitive fluorescence sensors were developed and explored. This includes metallocyclic self-assemblies that produce a modification in emission signal upon interaction with metal cations or inorganic anions in metallocycle voids. However, multi-step synthesis procedures are required, and the non-trivial architecture of the metallocycle supramolecular structures can be a challenge.

Thus, protocols based on self-assembly offer advantages. Self-assembled metalloaggregates can serve as a template that drive the binding between a fluorescent dye and a ligand. This leads to complexation of the metal ion resulting in quenching of the fluorescence emission of the dye. Based on this mechanism, numerous self-assembled chemosensors have been reported.

13.3 Surfactant Aggregates in Chemical Detection

Before proceeding to the application of metallosurfactants/metalloaggregates as chemosensor, we will take a look at the advantages of employing surfactant aggregates in chemical detection.

One attractive feature of self-assembling systems is that covalent linking between two subunits is not required. Instead, the self-assembling approach remarkably simplifies the chemosensor design, by reducing the challenges associated with conventional covalent linking systems. The blend of chemosensors and self-aggregating molecules can also give rise to new properties usually related to the conventional properties detected with aggregates, including the concentration of ionic species and the variety of microenvironment, which affect emissive features of the probe [7].

The localization of the species within aggregates imparts sensitivity and better photophysical features to self-assembled systems. Hughes et al. [13] reported micelle-solubilized fluorophores for identification and detection of explosives such as trinitrotoluene (TNT), Royal demolition explosive (RDX), high melting explosive (HMX), and trinitrophenylmethylnitramine (tetryl). The micellar solution was composed of Tween-80 with solubilized pyrene as fluorophore. The difference in tendency of these aromatic explosive compounds to quench the fluorescence signal of pyrene was exploited. With the pyrene dye being localized inside the micelles and the explosives molecules also entering the micelles, the quenching effect was enhanced compared to free pyrene dye. Interference through quenching by oxygen was also reduced. The micellar-based sensor array showed sensitivity of 19 μM.

A similar report on the use of a micellar sodium dodecyl sulfate (SDS) system for investigation of Cu^{2+} quenching a fluorescent dye, 3-acetyl-4-oxo-6,7-dihydro-12*H*-indolo-[2,3-a]-quinolizine (AODIQ), was published by Mallick et al. [14]. The interaction of Cu^{2+} with AODIQ generally leads to quenching. However, localizing the dye molecules inside the SDS micelles and letting the metallic cations attach to the anionic surface of the micelle significantly increased the quenching efficiency due to the close proximity of the fluorophore and metal ions in the micellar assemblies.

Metal ion detection using aggregates has also been reported by Sasaki et al. [15]. Their detection principle is based on an excimer–monomer equilibrium. This involved vesicular aggregates from distearoylphosphatidylcholine (DSPC) and lipids labeled with a fluorescent functional group (pyrene) and a metal-chelating functional group (iminodiacetic acid [IDA]) and derivatives thereof (Figure 13.3). The metallosurfactant formation took place via coordination with oxygen atoms. The emission spectrum of these lipids displays a sharp monomer emission peak and a broad emission band attributed to the pyrene excimer. The sensing of four divalent metal ions (Cu^{2+}, Co^{2+}, Mn^{2+}, Ca^{2+}) with metal-chelating **lipid 1a** was investigated. Upon addition of a divalent metal ion, the excimer emission of pyrene may convert into monomer emission. The change in ratio of signal intensities of excimer and monomer (E/M) after the addition of a divalent metal ion formed the detection approach.

The authors observed the excimer intensity to decrease with increasing Mn^{2+} concentration. The other divalent metal ions tested displayed a similar behavior but at

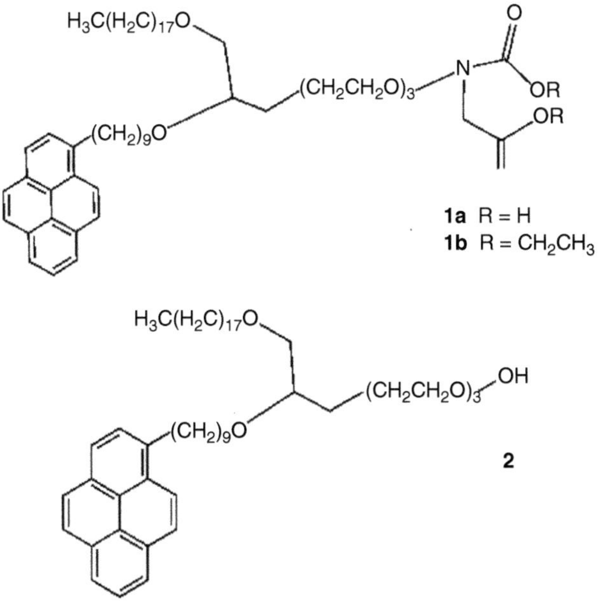

1a R = H
1b R = CH_2CH_3

2

Figure 13.3 Chemical structure of lipids **1a**, **1b**, and **2**.

different metal concentrations. The authors also studied addition of metals to the non-metal-chelating pyrene lipids **1b** and **2** dispersed in DSPC vesicles and concluded that the observed E/M changes happened due to metal chelation by IDA headgroup, rather than metal-induced perturbations or fluorescence quenching. It was proposed that metal ion binding with the IDA head group causes dispersal of lipid **1a** in DSPC matrix resulting in pyrene collision rate reduction and hence the ratio E/M. The DSPC vesicular system was found to be most sensitive for Cu^{2+}.

Roy et al. [16] reported a similar example, whereby four pyrene-containing lipids (Figure 13.4) aggregated in liposomes were fabricated. The pyrene monomer fluorescence signal got quenched on addition of cupric ions. The authors hypothesized the quenching effect was due to cupric ions rather than lipid redistribution in the liposomes. The fluorescence signal did not alter when three other transition metal ions (Ni^{2+}, Hg^{2+}, Zn^{2+}) were utilized, which is indicative of the selectivity of the system.

13.4 Self-Assembled Moieties as Fluorescent Sensors

As illustrated by the examples above, the blending of a fluorescence quenching sensing system with surfactant-based aggregates results in promising features and behaviors for chemosensors.

The major advantage of the self-assembling technique is the ability to modify or change the components with molecules that are readily, even commercially, available to scan and optimize parameters.

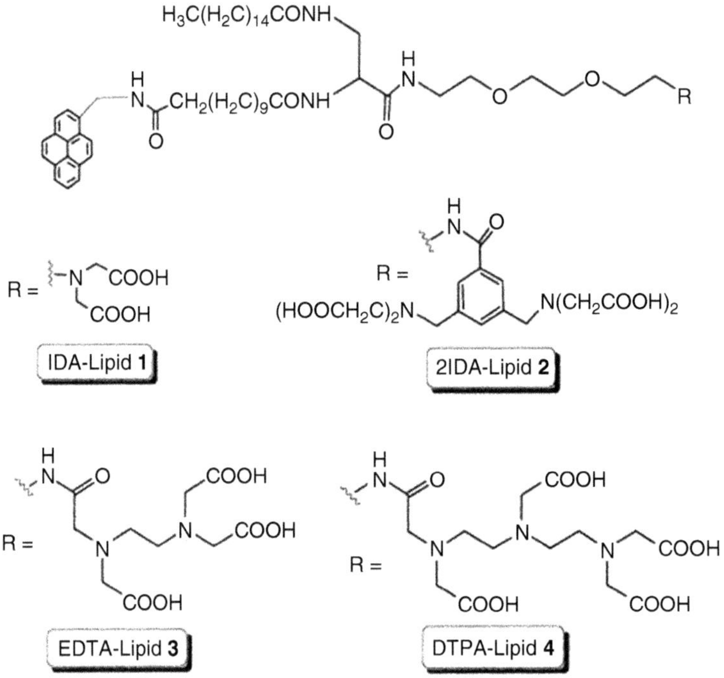

Figure 13.4 Structure of the lipids **1**–**4**. Source: Roy et al. [16].

Grandini et al. [17] proposed a route involving the self-assembly of a fluorescent dye and a receptor inside surfactant-based micellar aggregates. No direct interaction between the receptor and dye took place until they were confined in close proximity inside the small volume of the hydrophobic interior of the aggregate.

A transduction mechanism in the form of fluorescence signal readout is required to signal the binding event. This can be challenging if there is no covalent bond between the receptor and the fluorescent dye. The authors here investigated a system composed of a dye molecule, 8-anilinonaphthalensulfonic acid (ANS), with decylglycylglycine (GlyGly). These were localized in micellar aggregates formed from cetyl trimethylammonium bromide (CTAB). The system enabled the selective binding of Cu^{2+} ions in slightly acidic or neutral pH conditions due to amide nitrogen deprotonation.

This quenched the fluorescence signal of GlyGly–ANS construct. The fluorescence properties of the ANS dye are sensitive to the polarity of the chemical environment; ANS does not fluorescence in water, but does become fluorescent in the less polar micellar environment. The authors suggested that this effect reduced background fluorescence. Varying the ratio of sensing molecules to surfactants, the best readout was achieved at 1 : 2. By decreasing the surfactant concentration, the detection range could be modified. The self-assembly process is simple involving little or no structural modifications of the chemosensor. Rapid screening of experimental conditions can take place to optimize the sensing protocol. Due to the simplicity of this procedure, 16 combinations could be tested with four different surfactants and dyes.

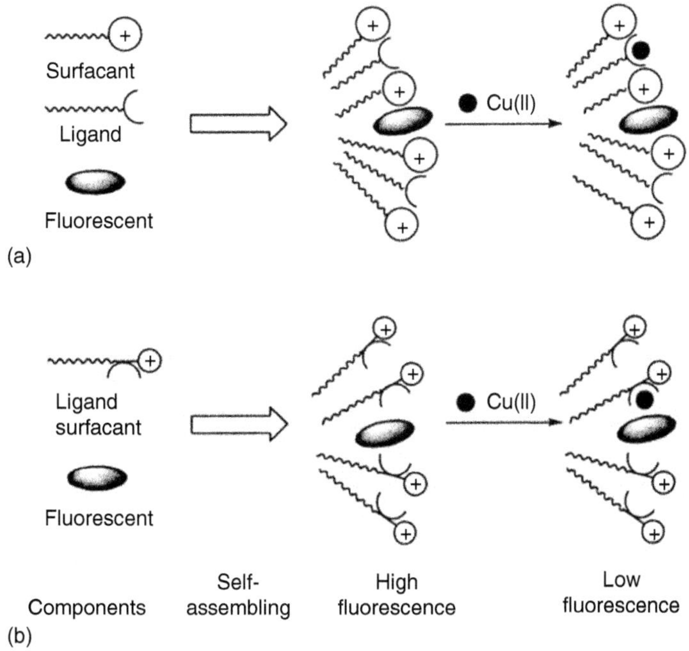

Figure 13.5 Schematic representation of the self-assembled chemosensor. (a) Three-component system; (b) two-component system. Source: Berton et al. [18].

In a further study, Berton et al. [18] used GlyGlu and GlyLys dipeptides. Both the free and copper-complexed form of the dipeptides were amphiphilic and formed homo-aggregates without the addition of any surfactant moieties allowing transition from a three-component system to a two-component system (Figure 13.5a,b). The simple modification of the dye co-micellized within the systems enabled the authors to study, investigate, and optimize interactions of the latter. The results reported surfactant-based self-assembling methodology to be efficient enough to synthesize sensitive sensors.

13.5 Metallosurfactants and Detection Protocol

Moving on to the metallosurfactants, chemosensing systems were synthesized using either the metallosurfactants or metalloaggregates as reported by Tao et al. [19]. They reported a sensor employing quinoline molecules as both structure-directing agents and ligands. In a one-step sol–gel synthesis process, a mesoporous silica material was produced with an amphiphilic quinoline derivative and tetraethylorthisilicate (TEOS) arranging into nanochannels filled by the amphiphilic derivative. Upon addition of Al^{3+} to nanochannels, a fluorescent metallomicelle–silica nanocomposite material was obtained (Figure 13.6). The fluorescence signal of hybrid $QC_{12}Et_3Br$-Al^{3+}/silica complex could be quenched by pyrophosphate and phosphate. $QC_{12}Et_3Br$-Al^{3+} is a quinoline-based amphiphilic ligand featuring a

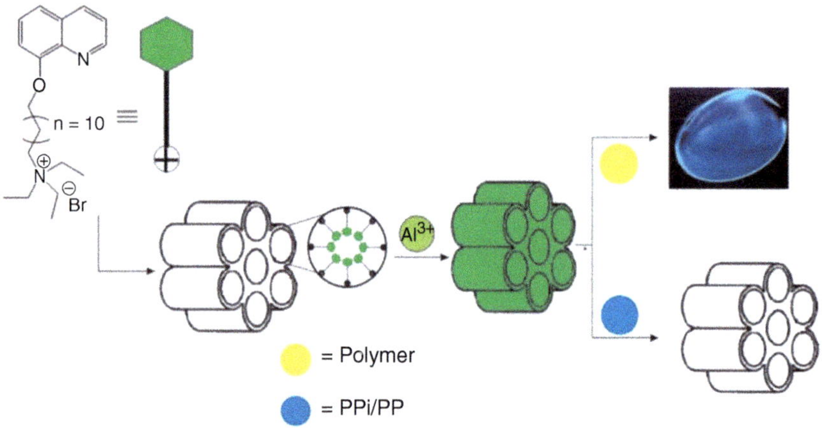

Figure 13.6 Schematic illustration of synthetic process of hybrid $QC_{12}Et_3Br$-Al^{3+}/silica complex, and its application in fluorescent display and ions detection: PPi (Pyrophosphate)/ PP (Phosphate). Source: Tao et al. [19], Scheme 01 / with permission from Royal Society of Chemistry.

12-carbon chain with positively charged head and ligand tail with Al^{3+}. The limit of detection was determined as 10.23 μM with linear range from 57.50 to 1696.00 μM. When mixing the fluorescent powder with methyl methacrylate (MMA), the complex could be polymerized into a film, rod, or other shapes, which could be used for incorporating this material into a display or sensor device.

Guerrero-Martínez [20] along with coworkers presented an interesting example of inclusion of two fluorescent dyes in surfactant self-assembly. The formation of metalloaggregates was carried out by attaching long-chain hydrocarbons to form electron–donor iridium and electron–acceptor ruthenium complexes, both containing dialklybipyridine as hydrophobic part of the aggregate (Figure 13.7). The authors compared the absorption and emission properties of the prepared metallosurfactants with their analogues $[Ru(bpy)_3]Cl_2$ and $[Ir(ppyFF)_2(bpy)]Cl$, where bpy is 2,2′-bipyridine and ppy is 2-phenylpyridine. The dilution-dependent switching of the electron transfer process in the co-micellar aggregate formed the basis of emission signal, which could be tuned by varying the ratio of the inert and functional surfactants.

The pH-sensing ability of the metalloaggregates has also been explored by various groups. Worlinskya et al. [21] tuned the pH-sensing range of [*meso*-tetraaryl-porpholactonato]M(II) complexes by investigating the effect of different metal ions (M = Zn(II), Ni(II), Cu(II), Pd(II), Ag(II), Pt(II)), the aryl group substitution (Ar = C_6H_5, C_6F_5), and introduction of a nitro group. Both β-nitration and pentafluorophenyl had additive effects, and the optical response was found to be stronger for the pentafluorophenyl-substituted derivative. The basic sensing mechanism begins with the nucleophilic attack of hydroxide on lactonic carbonyl. The interactions are stabilized by intermolecular interactions between oxygen and the pentafluorophenyl group. Zinc, silver, and copper complexes constituted more price-competitive options in comparison to platinum.

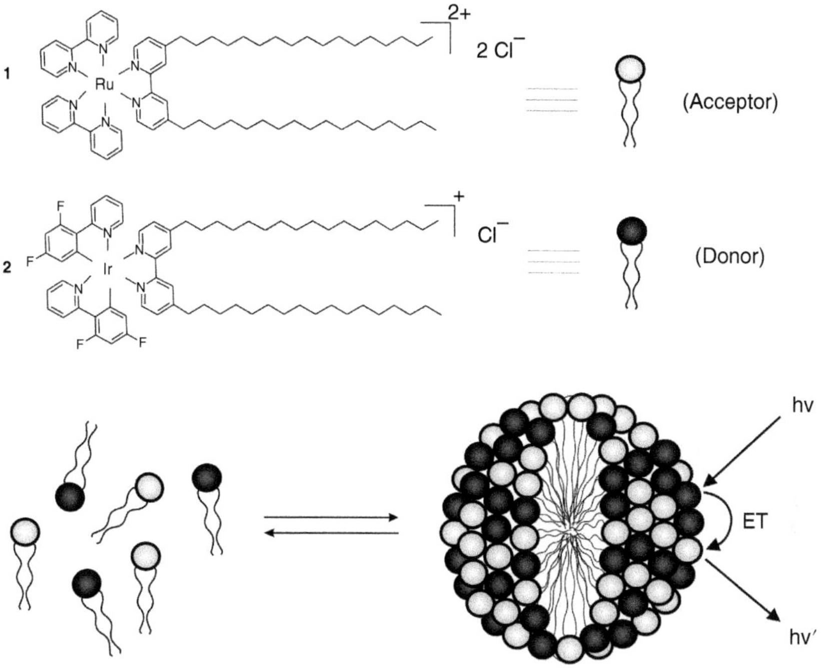

Figure 13.7 Schematic representation of complexes 1 and 2 and their self-aggregation into mixed aggregates [20]. ET: electron transfer. Source: Guerrero-Martínez et al. [20].

Lei et al. [22] fabricated Ru(II) complexed mesostructured silica with the aid of CTAB surfactant template for optical sensing of oxygen. For validation of the method, the authors also checked the optical sensing capability of $[Ru(bpy)_2 phen]Cl_2$ loaded into the matrix. The covalently grafted samples showed minimal leaching. The effect of aging time on the sensitivity and stability was observed to be so small that it could be neglected. Figure 13.8 illustrates the time-resolved decay curves transition from double to single exponential, resulting into linear Stern–Volmer plots.

In a similar report by Zhang et al. [23], oxygen-detecting materials consisting of $[Ru(phen)_3]^{2+}$ decorated to the functionalized MSU-3 fabricated by tetraethoxysilane and functionalized Ru(II) complex $[Ru(Phen)_2Phen\text{-}Si]^{2+}$ (Phen = 1,10-phenanthroline, Phen-Si = 2-[4′-{3-(triethoxysilyl)propyl}phenyl]imidazo[4,5-f]-1, 10-phenanthroline) using Pluronic P123 surfactant as template were synthesized. The results were compared with a matrix incorporated with just $[Ru(phen)_3]^{2+}$. The luminescent signal quenching of the Ru(II) complex by oxygen and the Stern–Volmer analysis was discussed by the authors. The deviation from the linear response could be attributed to different microenvironments in the matrix. Hence, by employing grafting strategy, one can aim to prepare organometallic complex-doped devices.

Metallosurfactants also find applications as imaging agents and even bimodal imaging agents. For example, Xu et al. explored fluorescent nanoparticles with

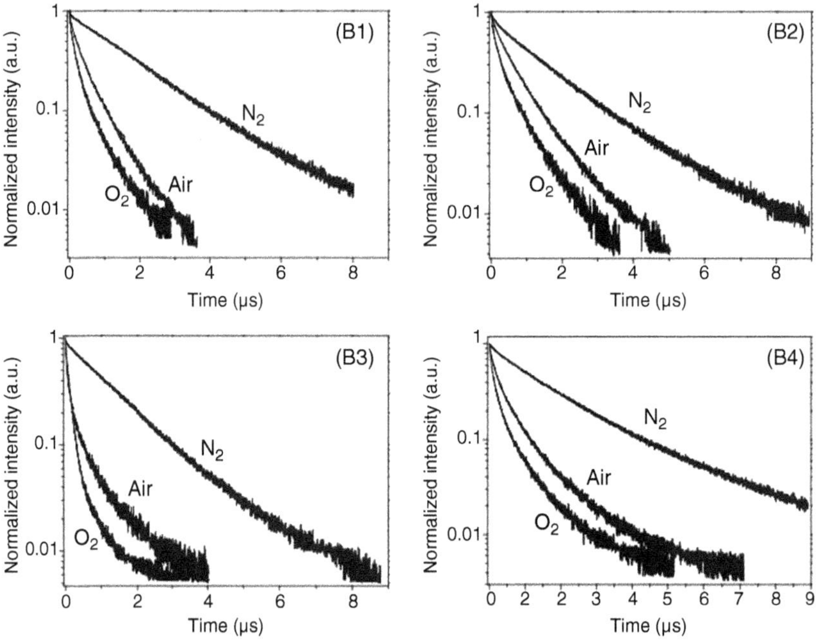

Figure 13.8 Typical time-resolved intensity decay curves of four-month-old Ru(II)-doped bulk samples. B1: a covalently grafted mesostructured bulk xerogel; B2: a covalently grafted amorphous bulk xerogel; B3: a physically incorporated mesostructured bulk xerogel; and B4: a physically incorporated amorphous bulk xerogel. Source: Lei et al. [22].

Gd(III) metallosurfactant and cross-linked organosilica as bimodal imaging agents [24]. The co-assembly of the amphiphilic Gd(III) complex with Pluronic F127 entrapped with conjugated polymeric nanoparticles (CPNs) in the core organosilica displayed a very strong fluorescence intensity using optical measurements (Figure 13.9). A good biocompatibility was suggested by the low cytotoxicity parameters.

In another report by Adams et al. [25], the emissive properties of monolayered RubpySS, IrbpySS, and OsbpySS in the presence of fluorinated surfactants and their capacity for adsorption of bovine serum albumin (BSA) and fetal bovine serum (FBS) were explored. It was anticipated that fluorinated surfactants prevent agglomeration of metal complexes-coated gold nanostructures (Figure 13.10). The lifetime of ruthenium tris-bipyridyl complex coated to gold surface increased significantly.

Dominguez-Gutierrez et al. [26] studied the formation of metallosurfactant assemblies with time-resolved spectroscopy. The authors employed luminescent ruthenium complexes-based metallosurfactant assemblies using $[Ru(bpy)_3]^{2+}$ unit as a head group. The morphology of the aggregates changed with monomer and choice of solvent. A range of complexes were fabricated, and the photophysical behavior of metallosurfactants both in the absence and presence of aggregates was studied in hexane and ethanol solutions. The higher degree of scatter for time profiles in hexane than ethanol indicated micelles or vesicles formation.

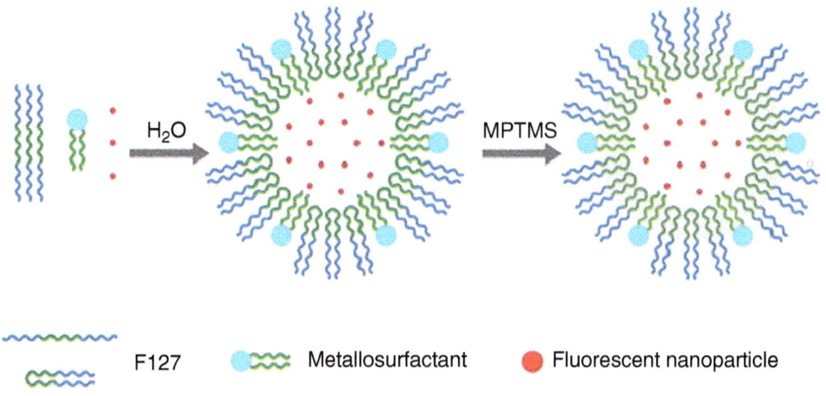

Figure 13.9 Schematic illustration of the fabrication of the bio-modal nanoparticular contrast agents [24]. MPTMS: (3-mercaptopropyl)trimethoxysilane. Source: Xu et al. [24].

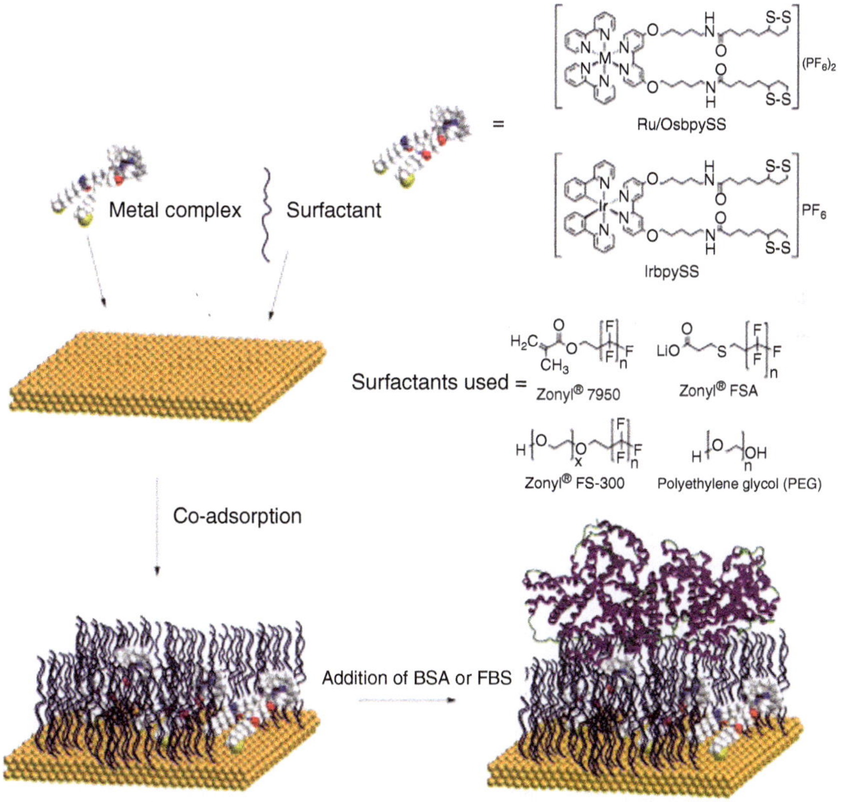

Figure 13.10 Schematic diagram of attachment of complexes and surfactants to gold surface and subsequent addition of BSA or FBS. Source: Adams et al. [25].

Similarly, a review article published by Owen and Butler [27] elaborates on coordination of various transition metal ions to biologically derived amphiphiles relevant to bio-inorganic chemistry.

There are many reports published exploring the redox and photophysical properties of metal complexes featuring self-aggregation tendencies through intermolecular interactions. Probing the environment of the metal center can lead to luminescent metallosurfactants that can potentially be applied in bio-imaging, thin-film optoelectronics, and mesoporous materials. The variability of luminescent properties of amphiphilic complexes has not been investigated much. A recent article by Mauro et al. [28] reported the synthesis of luminescent iridium metallosurfactant complex molecules soluble in an aqueous medium. A cyclometallated phenyl pyridine Ir(III) picolinate phosphorescent core with long alkyl chain terminated at sulfate group was investigated (Figure 13.11). Aggregation behavior was studied by varying the counterion, and results were compared with the parental non-amphiphilic compounds. Sodium derivatives were found to exhibit a strong aggregation tendency. Such compounds show potential applications in the fields like bio-imaging and tagging. Hence, the authors demonstrated the fabrication of a new class of metallosurfactants following the reverse strategy bearing a neutral and phosphorescent core, which possessed solubility in both aqueous and organic media. The counterion was observed to govern the nature of aggregates, and hence different aggregation characteristics were observed in case of tetrabutylammonium and sodium. High zeta potential values confirmed a highly negative surface near the aqueous phase. A significant blue shift in the emission spectra and prolonged excited state lifetimes indicated that the fabricated aggregates were not affected much by the presence dioxygen and water in comparison to the monomeric compounds.

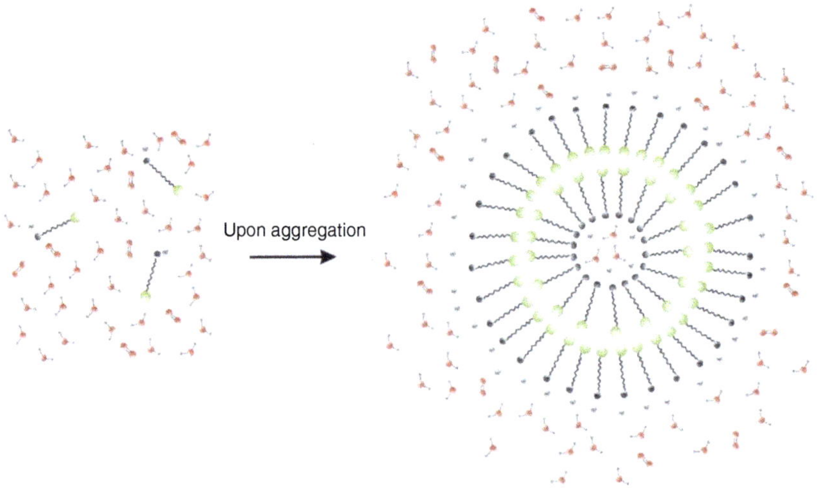

Figure 13.11 Schematic representation of the formation of the iridium aggregates. The luminescent iridium moieties are segregated inside the vesicles, and the charged and highly hydrophilic sulfate groups are facing the water molecules. Source: Mauro et al. [28].

Garcia et al. [29] studied the effect of solvent polarity to probe an iron(II) complex with diamine ligands. The authors synthesized and investigated the solvatochromic behavior of *cis*-bis(4,4′-dinonyl-2,2′-bipyridine)dicyanoiron(II) by using $(NH_4)_2Fe(SO_4)_2{\cdot}5H_2O$ as a precursor for Fe. It was observed that solvatochromic characteristics were related to the complex *cis*-$[Fe(bpy)_2(CN)_2]$, which is insoluble in low-polarity solvents. The attachment of two nonyl groups introduced solvent polarity probing features in the lipophilic systems. Hence, the authors inferred that the complex constituted a new metallosurfactant, which forms vesicles in water-dominant media. The solvatochromic properties make aggregation self-indicated. Thus, the complex exhibits promising collaborative effect of redox reactivity and solvation effect.

The emission behavior of Ru–surfactant complexes was also studied by Nehru et al. [30]. The authors synthesized a series of water-soluble Ru-complexes bearing different head groups and chain sizes. An in-depth analysis of thermodynamic characteristics, CAC, and hydrophobic features was carried out. With the increase in the length of chain and size of aromatic group, a decrease in the CAC values was observed. However, the hydrophobic character and the size distribution increased. The spherical self-assembled ruthenium–surfactant vesicles were monodispersed in nature and featured a green emission with high stability in an aqueous medium. Due to the fluorescent properties of synthesized metallovesicles, they could be employed for biomedical and material applications.

Draeger et al. [31] illustrated the formation of fluorescent emissive metallomicelles derived from tris-bipyridyl ruthenium hexaphosphate with a back-to-back ruthenium complex bilayer featuring a headgroup with the alkyl portion pointing inward and outward. Two 2,2′-(bipyridyl)ruthenium(II) units were combined with amphiphilic dipyridyl and Ru(II) hexafluorophosphate. The luminescent properties were explored by forming a suspension of the complexes in water and exciting the molecules at 640 nm. The granular micro crystallites imparted strong emission to the aggregates. The solution turned turbid in a matter of 15–20 minutes and remained stable for weeks. Redox-active compounds such as phenyl-*p*-benzoquinone, tris(acetylacetonato)manganese(III), manganese(III) *meso*-*p*-tetracarboxylatoporphyrinate (TCPP), and naphthoquinonesulfonate (NQS^-) were found to quench the fluorescence signal of the colloidal solution. Among them, maximum quenching of about 50% could be seen in case of NQS^-. No changes were seen when solid micellar spheres were deposited onto solid gold, mica, and graphite layer when investigated with atomic force microscopy on day 1 and 3 weeks later. Negatively charged quenchers were found to be stronger than positive or neutral moieties. Similarly, Garg et al. [32] synthesized mono and dual alkyl chain chromium-based metallosurfactants and investigated their interactions with xanthene dyes with the aid of fluorescence and UV/vis spectroscopy. Analysis of the photophysical properties of xanthene dyes in monomeric and aggregated form of the metallosurfactants was carried out. Both of the dyes investigated were found to bind through one site with these metallosurfactants. The incorporation of chromium ions in the surfactants resulted in the formation of a metallosurfactant with high thermal stability and affinity for xanthene dyes. This is assumed to be an effective pathway for detection of drugs in biological applications.

Hallett et al. [33] investigated the cellular uptake of fluorescent rhenium complexes bearing thiazole/imidazole group as a function of their lipophilicity. Due to their large emission, these may have applications in cellular imaging. Fluorescence cellular imaging was conducted on a fish parasite (*Spironucleus vortens* [*S. vortens*]) and yeast (*Schizosaccharomyces pombe*). Cellular uptake, toxicity, and localization were influenced by the lipophilicity of the ligands. An increase in lipophilicity of the alkyl chains imparted greater toxicity. Complexes of medium lipophilicity were found to exhibit reasonable uptake at no toxicity. Complexes with appropriate ligands could serve as possible radio-imaging nuclei by sustaining the physical properties for imaging using confocal fluorescence microscopy.

13.6 Conclusions

This chapter has highlighted how metallosurfactant aggregates can serve as nanocontainers, stabilized by electrostatic, ionic, lipophilic, or hydrophobic interactions. By concentrating fluorescent molecules into the small volumes of the container, and exploiting phenomena such as fluorescence quenching, metalloaggregates have applications in the fields of sensing, labeling, bio-imaging, and magnetic resonance imaging. In comparison to the complicated synthesis protocols and "difficult to tune" selectivity associated with conventional non-micellar sensors, the self-assembled sensors are more efficient and "easy to synthesize." A relatively limited number of reports exist on applications of metallosurfactants, leaving plenty of room for researchers to explore and investigate, for drug delivery, resonance imaging, solvochromatics, drug design, and many more.

Acknowledgments

The authors would like to acknowledge Commonwealth Scholarship Commission, UK, for a Commonwealth Split-site Scholarship 2019.

References

1 Hunt, J.A. (1999). A short history of soap. *The Pharmaceutical Journal* 263: 985–989.
2 Pallavicni, P., Diaz-Fernandez, Y.A., and Pasotti, L. (2009). Micelles as nanosized containers for the self-assembly of multicomponent fluorescent sensors. *Coordination Chemistry Reviews* 253: 2226–22240.
3 Griffiths, P.C., Fallis, I.A., Chuenpratoom, T., and Watanesk, R. (2006). Metallosurfactants: interfaces and micelles. *Advances in Colloid and Interface Science* 122: 107–117.

4 Pim, W.D., Ribeiro-Santos, T.A., Jardim, I.S. et al. (2018). Bistable copper(II) metallosurfactant as molecular machine for the preparation of hybrid silica-based porous materials. *Materials & Design* 160: 876–885.

5 Hafiz, A.A. (2005). Metallosurfactants of Cu(II) and Fe(III) complexes as catalysts for the destruction of paraoxon. *Journal of Surfactants and Detergents* 8: 359–363.

6 Zhu, Q., Pan, F., Tian, Y. et al. (2016). Facile synthesis of Gd(III) metallosurfactant-functionalized carbon nanodots with high relaxivity as bimodal imaging probes. *RSC Advances* 6: 29441–29447.

7 Mancin, F., Scrimin, P., Tecilla, P., and Tonellato, U. (2009). Amphiphilic metalloaggregates: catalysis, transport, and sensing. *Coordination Chemistry Reviews* 253: 2150–2165.

8 Jayathilake, H.D., Driscoll, J.A., Bordenyuk, A.N. et al. (2009). Molecular order in Langmuir–Blodgett monolayers of metal–ligand surfactants probed by sum frequency generation. *Langmuir* 25: 6880–6886.

9 Czarnik, A.D. (1993). Fluorescent chemosensors for ion and molecule recognition. *ACS Symposium Series* 538: 1–9.

10 Duuren, B.L. (1963). Effects of the environment on the fluorescence of aromatic compounds in solution. *Chemical Reviews* 63: 325–354.

11 de Silva, A.P. and de Silva, S.A. (1986). Fluorescent signalling crown ethers; 'switching on' of fluorescence by alkali metal ion recognition and binding in situ. *Journal of the Chemical Society, Chemical Communications* 23: 1709–1710.

12 Mancin, F., Rampazzo, E., Tecilla, P., and Tonellato, U. (2006). Self-assembled fluorescent chemosensors. *Chemistry A European Journal* 12: 1844–1854.

13 Hughes, A.D., Glenn, I.C., Patrick, A.D. et al. (2008). A pattern recognition-based fluorescence quenching assay for the detection and identification of nitrated explosive analytes. *Chemistry A European Journal* 14: 1822–1827.

14 Mallick, A., Mandal, M.C., Haldar, B. et al. (2006). Surfactant-induced modulation of fluorescence activity: a simple way to maximize the sensor efficiency. *Journal of the American Chemical Society* 128: 3126–3127.

15 Sasaki, D.Y., Shnek, D.R., Pack, D.W., and Arnold, F.H. (1995). Metal-induced dispersion of lipid aggregates: a simple, selective, and sensitive fluorescent metal ion sensor. *Angewandte Chemie International Edition in English* 34: 905–907.

16 Roy, B.C., Chandra, B., Hromas, D., and Mallik, S. (2003). Synthesis of new, pyrene-containing, metal-chelating lipids and sensing of cupric ions. *Organic Letters* 5: 11–14.

17 Grandini, P., Mancin, F., Tecilla, P. et al. (1999). Exploiting the self-assembly strategy for the design of selective Cu(II) ions chemosensors. *Angewandte Chemie International Edition* 38: 3061–3064.

18 Berton, M., Mancin, F., Stocchero, G. et al. (2001). Self-assembling in surfactant aggregates: an alternative way to the realization of fluorescence chemosensors for Cu(II) ions. *Langmuir* 17: 7521–7528.

19 Tao, S., Fan, P., Wang, Y. et al. (2014). Multifunctional fluorescent material based on metallomicelles trapped in silica nanochannels. *Journal of Materials Chemistry C* 2: 1962–1965.

20 Guerrero-Martínez, A., Vida, Y., Domínguez-Gutiérrez, D. et al. (2008). Tuning emission properties of iridium and ruthenium metallosurfactants in micellar systems. *Inorganic Chemistry* 47: 9131–9133.

21 Worlinskya, J.L., Zaratea, G., Zellerb, M. et al. (2013). Tuning the dynamic high pH sensing range of [*meso*-tetraarylporpholactonato]M(II) complexes by variation of the central metal ion, the aryl substituents, and introduction of a β-nitro group. *Journal of Porphyrins and Phthalocyanines* 17: 836–849.

22 Lei, B., Li, B., Zhang, H. et al. (2006). Mesostructured silica chemically doped with Ru(II) as a superior optical oxygen sensor. *Advanced Functional Materials* 16: 1883–1891.

23 Zhang, H., Lei, B., Mai, W., and Liu, Y. (2011). Oxygen-sensing materials based on ruthenium(II) complex covalently assembled mesoporous MSU-3 silica. *Sensors and Actuators B: Chemical* 160: 677–683.

24 Xu, K., Xu, N., Zhu, Y. et al. (2020). Co-assembly of Gd(III)-based metallosurfactant and conjugated polymer nanoparticles in organosilica cross-linked block copolymer micelles for highly efficient MRI and fluorescent bimodal imaging. *Particle & Particle Systems Characterization* 37: 2000044. (1)–(8).

25 Adams, S.J., Carrod, A.J., Rochford, L.A. et al. (2018). Surfactant-enhanced luminescence lifetime for biomolecular detection on luminescent gold surfaces decorated with transition metal complexes. *Chemistry Select* 3: 3251–3257.

26 Dominguez-Gutierrez, D., Paoli, G.D., Guerrero-Martinez, A. et al. (2008). Inverted aggregates of luminescent ruthenium metallosurfactants. *Journal of Materials Chemistry* 24: 2762–2768.

27 Owen, T. and Butler, A. (2011). Metallosurfactants of bioinorganic interest: coordination-induced self-assembly. *Coordination Chemistry Reviews* 255: 678–687.

28 Mauro, M., De Paoli, G.D., Otter, M. et al. (2001). Aggregation induced colour change for phosphorescent iridium(III) complex-based anionic surfactants. *Dalton Transactions* 40: 12106–12116.

29 Garcia, P., Marques, J., Pereira, E. et al. (2001). A novel self-indicative vesicle based on an iron(II) complex. *Chemical Communications*: 1298–1299.

30 Nehru, S., Veeralakshmi, S., and Arunachalam, S. (2017). Synthesis, characterization and self-assembly behaviour of emissive-ruthenium complexes. *New Journal of Chemistry* 41: 13830–13837.

31 Draeger, C., Bottcher, C., Messerschmidr, C. et al. (2000). Isolable and fluorescent mesoscopic micelles made of an amphiphilic derivative of tris-bipyridyl ruthenium hexafluorophosphate. *Langmuir* 16: 2068–2077.

32 Garg, P., Kaur, G., and Chaudhary, G.R. (2018). Chromium-based metallosurfactants: synthesis, physiochemical characterization and probing of their interactions with xanthene dyes. *New Journal of Chemistry* 42: 1141–1150.

33 Hallett, A.J., Placet, E., Prieux, R. et al. (2018). Exloring the cellular uptake and localization of phosphorescent rhenium fac-tricarbonyl metallosurfactants as a function of lipophilicity. *Dalton Transactions* 47: 14241–14253.

Index

Metallosurfactants: From Fundamentals to Catalytic and Biomedical Applications, First Edition.
Edited by Surinder K. Mehta and Ravneet Kaur.

t